Microbes and Microbiomes for Clean and Green Environment

(Volume 2)

Microbes as Agents of Change for Sustainable Development

Edited by

Govindaraj Kamalam Dinesh

&

Shiv Prasad
Division of Environment Science
ICAR-Indian Agricultural Research Institute
New Delhi-110012, India

Ramesh Poornima

&

Sangilidurai Karthika
Department of Environmental Sciences
Tamil Nadu Agricultural University, Coimbatore
India

Murugaiyan Sinduja
National Agro Foundation, Taramani, Chennai
Tamil Nadu, India

Velusamy Sathya
Tamil Nadu Pollution Control Board, Chennai
Tamil Nadu, India

Microbes and Microbiomes for Clean and Green Environment

(Volume 2)

Microbes as Agents of Change for Sustainable Development

Editors: Govindaraj Kamalam Dinesh, Shiv Prasad, Ramesh Poornima, Sangilidurai Karthika, Murugaiyan Sinduja & Velusamy Sathya

ISBN (Online): 978-981-5322-34-7

ISBN (Print): 978-981-5322-35-4

ISBN (Paperback): 978-981-5322-36-1

need for a court order if at any point you breach any terms of this License Agreement. In no event will any delay or failure by Bentham Science Publishers in enforcing your compliance with this License Agreement constitute a waiver of any of its rights.

3. You acknowledge that you have read this License Agreement, and agree to be bound by its terms and conditions. To the extent that any other terms and conditions presented on any website of Bentham Science Publishers conflict with, or are inconsistent with, the terms and conditions set out in this License Agreement, you acknowledge that the terms and conditions set out in this License Agreement shall prevail.

Bentham Science Publishers Pte. Ltd.
80 Robinson Road #02-00
Singapore 068898
Singapore
Email: subscriptions@benthamscience.net

BENTHAM SCIENCE

CONTENTS

    Joseph Ezra John, Boopathi Gopalakrishnan, Senthamizh Selvi, Murugaiyan Sinduja,
    Chidamparam Poornachandhra, Ravi Raveena and E. Akila

PREFACE

The book series "**Microbes and Microbiomes for a Clean and Green Environment**" explores the current state of polluted and degraded ecosystems, emphasizing the vital roles that microbes and microbiomes play in managing natural resources and restoring ecosystems. It also highlights the significance of these microorganisms in generating renewable energy, reducing greenhouse gas emissions, mitigating climate change impacts, sustaining marine, mangroves, and wetlands ecosystems, promoting sustainable industrial practices, and contributing to socio-economic development and the security of human and animal health.

We are entering a unique era characterized by various environmental challenges contributing to adverse effects on the ecosystem of the planet and impacting human health and quality of life. At the start of the twenty-first century, a significant concern within the ecological context is the degradation of ecosystems and environmental imbalances, primarily caused by increasing human activities. To tackle this pressing issue, restoration ecology has emerged, which focuses on the potential of microbes and microbiomes as key solutions for creating a clean and green environment. The practical applications of this science offer cost-effective and viable options.

One of the primary objectives of the UN Convention on Biological Diversity from 2011 to 2020 was to restore at least 15% of the world's damaged ecosystems. In 2011, world leaders launched the "Bonn Challenge," committing to rehabilitating 150 million hectares of deforested and degraded land. Furthermore, in 2015, the UN formalized these global commitments by endorsing the 2030 Sustainable Development Goals, which emphasizes ecological restoration's importance.

Microorganisms are remarkably diverse and essential for sustaining ecosystems on a global scale. They provide critical services that enhance productivity and maintain a stable environment for human life.

Volume 2, "**Microbes as Agents of Change for Sustainable Development,**" explores the pivotal role of microbes and microbiomes in restoring degraded ecosystems and advancing sustainable practices.

In Chapter 1, Sinduja *et al.*, emphasize the essential role of microorganisms in sustaining Earth's biogeochemical cycles. They explore the complexity of microbial communities and their contributions to ecosystem processes. Additionally, the chapter discusses strategies for effectively managing natural resources and highlights the impact of beneficial soil microbes on nutrient cycling.

Bioleaching is increasingly used for metal extraction and effective bioremediation of polluted sites. This technique is cost-effective and environmentally friendly, helping to restore damaged ecosystems to their original state. In Chapter 2, Poornima *et al.*, provide a comprehensive overview of bioleaching, covering its various types, the microbes involved, the pathways of bioleaching, and the role of these microbes in the bioremediation of polluted habitats.

In Chapter 3, Sajish *et al.*, discuss the fundamental principles of microbial fuel cells (MFCs), types of bioreactors, factors that influence the development of MFCs' performance, and the crucial role microbes play in catalyzing these systems. They also explore various approaches to enhancing the overall efficiency of MFCs for practical applications, including genetic engineering, biofilm engineering, and electrode engineering.

Recent concerns about energy crises, rising pollution, and unpredictable climate change have made bioenergy an essential alternative to fossil fuels. In this context, Chapter 4, authored by Oyelade *et al*., comprehensively reviews how sustainable bioenergy production through microbes utilizing various biomass feedstocks generates clean and green energy and how, ultimately, it can help mitigate environmental issues, restore ecosystems, and achieve energy security.

Microbes play a significant role as either generators or consumers of greenhouse gases such as carbon dioxide (CO_2), methane (CH_4), and nitrous oxide (N_2O) through various processes. Sethupathi *et al*., in Chapter 5, discuss the role of microbes and the microbiome in the emission of significant greenhouse gases like CO_2, CH_4, N_2O, and NH_3. This chapter also discusses the potential of the microbiome in mitigating these greenhouse gases.

Climate change is now a reality, largely due to the release of carbon dioxide from soil into the atmosphere. In Chapter 6, Al-Jawhari *et al*., examine the balance of soil CO_2, the environmental impacts of climate change, and the significance of the soil carbon cycle, especially the roles played by microbial decomposers. The chapter also addresses the interconnections between the carbon cycle in the soil, the ocean, and ecosystem restoration within the context of climate change.

Microorganisms are pervasive and comprise the "unseen majority" of marine environments. Although marine isolates have been the subject of laboratory-based culture methods for more than ten years, we still don't completely understand their ecology. Thus, in Chapter 7, Poornachandhra *et al*., explore marine microbial diversity, its utilization in bioremediation, and its role in ecosystem sustainability.

Mangroves and wetlands are essential ecosystems that offer numerous ecological and economic benefits. Unfortunately, human activities have led to the rapid degradation of these crucial areas. In Chapter 8, Haghani *et al*., analyze the key characteristics of the microbial communities that inhabit mangroves and wetlands. They describe the biochemical transformations performed by these microorganisms and highlight the complexity of their interactions within these ecosystems.

In Chapter 9, Jerome *et al*., highlight the importance of forest microbiomes in ecosystem restoration and sustainability. Forest microorganisms are crucial in how plants interact with their soil environment and are vital for accessing essential soil nutrients. This chapter examines the above-ground and below-ground ecosystems of a forest microbiome, emphasizing the significance of soil microorganisms and their diverse relationships, including parasitism, mutualism, and commensalism.

The sustainable industrial revolution is the way forward to help humankind prolong its existence on Earth. In Chapter 10, John *et al*., enlighten us on the role of the microbiome in a sustainable industrial production system. They discussed the energy sector's current status, microbes' role in organic and amino acid production, and microalgae's role in sustainable agriculture.

The microbiome plays a vital role in human development, immunity, and nutrition, where beneficial bacteria establish themselves as colonizers rather than destructive invaders. In chapter 11, Pradyutha *et al*., introduce microbes' role in human and animal health security.

This chapter also discusses various human and animal diseases and the potential of microbiota, such as probiotics, in disease treatment.

We sincerely thank all the authors for their outstanding contributions. Our gratitude also extends to the entire team at Bentham Science Publishers, especially Mrs. Fariya Zulfiqar (Manager Publications), for her exceptional management of this book throughout all stages of publication. We are confident that this volume in the book series will be widely appreciated by researchers and professionals alike.

Govindaraj Kamalam Dinesh
&
Shiv Prasad
Division of Environment Science
ICAR-Indian Agricultural Research Institute
New Delhi-110012, India

Ramesh Poornima
&
Sangilidurai Karthika
Department of Environmental Sciences
Tamil Nadu Agricultural University, Coimbatore
India

Murugaiyan Sinduja
National Agro Foundation, Taramani, Chennai
Tamil Nadu, India

Velusamy Sathya
Tamil Nadu Pollution Control Board, Chennai
Tamil Nadu, India

List of Contributors

A. Manikandan	Institute of Ecology and Earth Sciences, University of Tartu, Tartu, Estonia
A. Ch. Pradyutha	Department of Microbiology, Raja Bahadur Venkata Rama Reddy Women's College, Narayanguda, Hyderabad, Telangana, India
B.N. Brunda	Division of Microbiology, Indian Agricultural Research Institute, New Delhi, India
Boopathi Gopalakrishnan	School of Atmospheric Stress Management, ICAR-National Institute of Abiotic Stress Management, Maharashtra, India
Chidambaram Poornachandhra	Department of Environmental Sciences, Tamil Nadu Agricultural University, Coimbatore, Tamil Nadu, India
Deepasri Mohan	Division of Environmental Sciences, Sher-e-Kashmir University of Agricultural Sciences & Technology of Jammu, Jammu and Kashmir, India
E. Akila	Department of Agricultural Engineering, Tamil Nadu Agricultural University, Coimbatore, Tamil Nadu, India
Govindaraj Kamalam Dinesh	Division of Environment Science, ICAR-Indian Agricultural Research Institute, New Delhi-110012, India Division of Environmental Sciences, Department of Soil Science and Agricultural Chemistry, SRM College of Agricultural Sciences, SRM Institute of Science and Technology, Baburayanpettai-603201, Chengalpattu, Tamil Nadu, India INTI International University, Persiaran Perdana BBN, Putra Nilai, 71800 Negeri Sembilan, Malaysia
Ganesan Karthikeyan	Department of Environmental Sciences, Tamil Nadu Agricultural University, Coimbatore, Tamil Nadu, India
Helen Mary Rose	Division of Environment Science, ICAR-IARI, New Delhi, India
Ihsan Flayyih Hasan AI-Jawhari	Department of Biology, College of Education for Pure Sciences, University of Thi-Qar, Iraq
Jerome O. Ihuma	Department of Biological Science, Faculty of Science and Technology, Bingham University, Karu Nasarawa State, Nigeria
J. Sampath	Department of Environmental Sciences, Tamil Nadu Agricultural University, Coimbatore, Tamil Nadu, India
Joseph Ezra John	Department of Environmental Sciences, Tamil Nadu Agricultural University, Coimbatore, Tamil Nadu, India
K. Mathiyarasi	Division of Environment Science, Indian Agriculture Research Institute, New Delhi, India
Karthika Ponnusamy	Department of Microbiology, College of Basic Science & Humanities, Chaudhary Charan Singh Haryana Agricultural University, Haryana, India
Kovilpillai Boomiraj	Climate Research Centre, Tamil Nadu Agricultural University, Coimbatore, Tamil Nadu, India
Kamyar Amirhosseini	Department of Soil Science, Faculty of Agriculture, College of Agriculture and Natural Resources, University of Tehran, Tehran, Iran

Murugaiyan Sinduja	National Agro Foundation, Taramani, Chennai, Tamil Nadu-600113, India
Murugesan Kokila	Division of Environment Science, ICAR-IARI, New Delhi, India
Muthusamy Shankar	Division of Plant Genetic Resources, ICAR-Indian Agricultural Research Institute, New Delhi, India
Malgwi T. Doris	Department of Community Medicine, Nnamdi Azikiwe University, Nnewi, Anambra State, Nigeria
Omolora Victoria Oyelade	Department of Physics, Faculty of Science and Technology, Bingham University, Karu Nasarawa State, Nigeria
P.M. Brindhavani	Adhiyamman College of Agriculture and Research, Krishnagiri, Tamil Nadu-635105, India
Periyasamy Dhevagi	Department of Environmental Sciences, Tamil Nadu Agricultural University, Coimbatore, Tamil Nadu, India
R. Kalpana	National Agro Foundation, Research & Development Centre, Anna University Taramani Campus, Taramani, Chennai, Tamil Nadu, India
Ragul Subramaniyan	Plant Variety Examination Research Associate (PVERA), Protection of Plant Varieties & Farmers' Rights Authority, New Delhi, India
Ramesh Poornima	Department of Environmental Sciences, Tamil Nadu Agricultural University, Coimbatore, Tamil Nadu, India
R. Kaveena	Swamy Vivekananda College of Pharmacy, Tiruchengode, India
Ravi Raveena	Department of Environmental Sciences, Tamil Nadu Agricultural University, Coimbatore, Tamil Nadu, India
Velusamy Sathya	Tamil Nadu Pollution Control Board, Chennai, Tamil Nadu, India
Sangilidurai Karthika	Department of Environmental Sciences, Tamil Nadu Agricultural University, Coimbatore, Tamil Nadu, India
Selvaraj Keerthana	Department of Environmental Sciences, Tamil Nadu Agricultural University, Coimbatore, Tamil Nadu, India
Sagia Sajish	Division of Microbiology, Indian Agricultural Research Institute, New Delhi, India
Sethupathi Nedumaran	Division of Environment Science, ICAR-IARI, New Delhi, India
Sudhakaran Mani	Department of Environmental Scienc, JKK Munirajah College of Agricultural Science, Namakkal, India
S. Akila	National Agro-foundation Research & Development Centre, Chennai, India
Senthamizh Selvi	Department of Agricultural Microbiology, Tamil Nadu Agricultural University, Coimbatore, Tamil Nadu, India
S. Chaitanya Kumari	Department of Microbiology, Bhavan's Vivekananda College of Science, Humanities & Commerce, Sainikpuri, Secunderabad, Telangana, India
Thangaraj Gokul Kannan	Department of Environmental Sciences, Tamil Nadu Agricultural University, Coimbatore, Tamil Nadu, India
Tayo I. Famojuro	Department of Pharmacognosy, Faculty of Pharmaceutical Sciences, Bingham University, Karu Nasarawa State, Nigeria
Zahra Haghani	Department of Soil Science, Faculty of Agriculture, College of Agriculture and Natural Resources, University of Tehran, Tehran, Iran

Significance of Microbiome in Natural Resource Management

CHAPTER 1

Role of Microbes and Microbiomes in Natural Resource Management and the Regulation of Biogeochemical Processes and Nutrient Cycling

Murugaiyan Sinduja[1],*, P.M. Brindhavani[2], Govindaraj Kamalam Dinesh[3,4,5], Joseph Ezra John[6], K. Mathiyarasi[7], Velusamy Sathya[8], R. Kalpana[9] and **Ragul Subramaniyan[10]**

[1] *National Agro Foundation, Taramani, Chennai, Tamil Nadu-600113, India*

[2] *Adhiyamman College of Agriculture and Research, Krishnagiri, Tamil Nadu-635105, India*

[3] *Division of Environment Science, ICAR-Indian Agricultural Research Institute, New Delhi-110012, India*

[4] *Division of Environmental Sciences, Department of Soil Science and Agricultural Chemistry, SRM College of Agricultural Sciences, SRM Institute of Science and Technology, Baburayanpettai-603201, Chengalpattu, Tamil Nadu, India*

[5] *INTI International University, Persiaran Perdana BBN, Putra Nilai, 71800 Negeri Sembilan, Malaysia*

[6] *Department of Environmental Sciences, Tamil Nadu Agricultural University, Coimbatore, Tamil Nadu, India*

[7] *Division of Environment Science, Indian Agriculture Research Institute, New Delhi, India*

[8] *Tamil Nadu Pollution Control Board, Chennai, Tamil Nadu, India*

[9] *National Agro Foundation, Research & Development Centre, Anna University Taramani Campus, Taramani, Chennai, Tamil Nadu, India*

[10] *Plant Variety Examination Research Associate (PVERA), Protection of Plant Varieties & Farmers' Rights Authority, New Delhi, India*

Abstract: Life on Earth is possible due to the vital elements and energy transformations referred as biogeochemical cycle. Microorganisms play an essential role in moderating the Earth's biogeochemical cycles; nevertheless, despite our fast-increasing ability to investigate highly complex microbial communities and ecosystem processes, they remain unknown. Microbes are crucial in nutrient cycling and energy transfers between ecosystems and the tropics, but research on their intricate functions is still restricted due to technological inabilities. A better understanding of microbial communities based on ecological principles may improve our ability to predict ecosystem process rates using environmental variables and microbial physiology. We

---

* **Corresponding author Murugaiyan Sinduja:** National Agro Foundation, Taramani, Chennai, Tamil Nadu-600113, India; E-mail: seethasinduja@gmail.com

Govindaraj Kamalam Dinesh, Shiv Prasad, Ramesh Poornima, Sangilidurai Karthika, Murugaiyan Sinduja & Velusamy Sathya (Eds.)

explored the ecological role of microorganisms participating in biogeochemical cycles, hoping to delineate the role of microbes and microbiomes in biogeochemical cycles. Insights into these aspects can help us mitigate the effects of climate change and other future uncertainties by regulating the microbial-dependent biogeochemical cycle.

Keywords: Environment, Biogeochemical cycling, Microorganisms, Climate change, Ecosystems.

INTRODUCTION

In natural resource management, microorganisms play a prominent role in the biogeochemical cycling of nutrients. Microbiomes have demonstrable effects on the chemical makeup of the biosphere and its surrounding atmosphere, and they are deservedly recognized for their capacity to fix carbon and nitrogen into organic matter. Acclimatization typically begins with a higher commitment to obtaining and mobilizing stored resources when some factors become restricted [1]. The biogeochemical cycling of nutrients relies heavily on microbes. They are lauded for their ability to fix carbon and nitrogen into organic matter, and microbial-driven processes have visibly altered the chemical composition of the biosphere and its surrounding atmosphere [2]. Because soil quality is constantly deteriorating, a healthy soil system is now the outcome of physical, chemical, and biological soil quality indicators that are connected in a complicated network. The interests of the community and the needs of farmers are balanced by healthy soils. By preventing toxic compounds from being released into the environment, squelching infections, and preserving environmental sustainability, soil organic matter (SOM) improves soil health and quality [3]. In order to produce food sustainably, it refers to interactions between internal and exterior soil components. Effective soil microorganisms are essential for the establishment of the soil-plan-microbe interaction because they stimulate numerous biological processes and different pools of carbon (C) and macro- and micronutrients. The soil system has an enormous variety of microorganisms [4].

This chapter emphasizes the role of microbes and microbiomes in natural resource management by regulating biogeochemical processes and nutrient cycling. Although global understanding of microbes and microbiome dynamics is quickly rising, research on rhizospheric complexes is restricted despite their relevance in regulating soil-plant systems. Microorganisms in the soil consume organic matter, including dead organisms, and play an essential role in organic matter breakdown and nutrient cycle [5]. The nutrients are released by the breakdown of the organic molecule, allowing plants to absorb nutrients from the soil *via* their roots. Biogeochemical cycles transport nutrients throughout the ecosystem [6]. An ecosystem's biotic (living) and abiotic (non-living) components can exchange chemical elements like carbon or nitrogen in a process known as a biogeochemical cycle [7]. The elements that move through an ecosystem's processes are not wasted; rather, they are recycled or saved in reservoirs (sometimes referred to as "sinks"), where they can be kept for a long time. These biogeochemical cycles transfer substances from one organism to another and from one region of the biosphere to another, including elements, chemical compounds, and other kinds of matter. Ecosystems have a variety of biogeochemical cycles as part of the overall system [8]. A great example of a molecule cycled within an ecosystem is water, which is constantly recycled through the water cycle. Water vapor rises into the atmosphere, cools, and then eventually returns to Earth as rain (or other types of precipitation). Cycling is typical of all significant aspects of life.

Microorganisms are crucial in the biogeochemical cycling of nutrients. Microorganisms are weak despite the elements' immutability and their vast capability for molecular alterations [9]. This paper discusses the effects of elemental limitation on microorganisms with an emphasis on certain genetic model systems and representative bacteria from the ocean ecosystem. Studies on the genome and proteome reveal evolutionary adaptations that enhance growth in response to ongoing or recurrent elemental constraints [10]. Changes in protein amino acid sequences that considerably lower cellular carbon, nitrogen, or sulfur requirements are among them. These modifications range from dramatic (such as eliminating a requirement for a hard-to-find component) to quite modest. Acclimatization typically begins with a stronger commitment to obtaining and mobilizing stored resources when some factors become restrictive. The cell turns to austerity tactics like elemental recycling and sparing if elemental limitation continues. Research in the fields of ecology, biological oceanography, biogeochemistry, molecular genetics, genomics, and microbial physiology has shed new light on these essential cellular features [11]. This chapter also highlights many research studies findings that are devoted to the conservation of natural resources, global food security, and sustainable agriculture [12].

NATURAL RESOURCE MANAGEMENT – NEED OF THE HOUR

Natural resources are the elixir for living organisms, as human life's existence is highly dependent on the ecosystem and the services it provides to humankind. These natural resources include air, water, land, minerals, flora, fauna, *etc* [13]. They provide the fundamental backing to life by providing goods for sustenance and consumption. Natural resource management (NRM) is the efficient and sustained usage of these valuable resources, which otherwise would lead to depletion or reduction in their existence [14]. Increased human population and scientific developments in the recent decade have led to increased interaction between humans and the environment, eventually leading to increased usage of resources. Thus, problems like food crises, scarcity of resources, mainly water, biodiversity loss, deforestation, and pollution have emerged [15]. These adverse effects are irreversible, and as such, they cause severe damage to future generations.

Global biodiversity is seriously threatened by the illegal exploitation of natural resources. Infringements on property rights, such as taking resources from private property or protected areas without permission, illegal land occupation, and violations of resource use laws, such as exceeding set limits, using resources out of season, and using forbidden extraction techniques without the necessary permits or in forbidden areas, are some examples of these illegal activities. Illegal resource use also includes illegal resource harvesting, such as protected species. Our social dependency on natural resource use continues unabated, to the point where natural resource sustainability has taken precedence in policy and executive considerations [16]. Management includes choosing alternate options to reduce the destruction of non-renewable resources, like opting for wind power instead of natural gas, creating watersheds, *etc*. The interdependence of microbiomes in environmental and food systems demonstrates that microbiome innovations have the potential to enhance circularity-based food [12], feed, and biofuel production. Even though there are numerous technological possibilities, preserving natural resources is still crucial if we want future generations to have access to all the resources they need to exist.

STRATEGIES FOR PROPER MANAGEMENT OF PREVAILING NATURAL RESOURCE – SOIL

The relationship between people and natural landscapes is critical to natural resource management. It integrates biodiversity conservation, land use planning, water management, and the long-term viability of various industries, including forestry, agriculture, mining, tourism, and fisheries [17]. The nation's current agrarian crisis is a result of the extraordinary loss of natural resources, the basis for human existence, progress, and prosperity [18]. Some of these resources include land, water, biodiversity and genetic resources, biomass resources, forests, livestock, and fisheries. Despite pressures from the population and the economy, unmindful agricultural intensification, excessive use of marginal lands, unbalanced fertilizer use, loss of organic matter, declining soil health, extensive conversion of prime agricultural lands to non-agricultural uses, inefficient and wasteful irrigation water use, depleting aquifers, salinization of fertile lands and waterlogging, deforestation, biodiversity loss, genetic erosion, and climate change are still prevalent [19].

The conversion of N between organic and inorganic forms by soil microorganisms, mainly bacteria and fungi, enhances plant mineral uptake [20]. The fundamental processes that ensure the productivity and stability of agroecosystems are aided by microbial communities [21]. One advantage of soil conservation techniques like cover crops and minimal tillage is increased soil life, which breaks down organic matter and releases nutrients for plant uptake. The organism that breaks down organic soil impurities, the soil microbe plant complex, may be impacted by various factors through interactions [22]. The kind of soil, level of calcium in the soil, amount of organic carbon, temperature, moisture, oxygen content, electrical conductivity, and pH are all variables that might affect the makeup and effectiveness of soil microbial communities [23]. For healthy soil, especially on organic farms where biological soil activities cannot be replaced by synthetic additives, the biological component of the soil is essential [24].

The term "soil biological community" refers to the collection of living things found in soil, including worms, insects, nematodes, plant roots, mammals, and bacteria. The breakdown of agricultural residues, the support of plant development, and the cycling of nitrogen and carbon are just a few of the crucial jobs that soil microorganisms (bacteria, fungi, and archaea) perform in soils [25]. Pathogenic microorganisms have a negative impact on crop health and yield and, in the worst situations, can completely destroy a crop. Therefore, not all microbial contributions are beneficial. The microbial component of soil is perhaps the most challenging to monitor and regulate, despite the fact that bacteria clearly play a large impact on soil health and crop performance [26]. However, it can be difficult to properly manage the biological aspect of soils, particularly the microbial component [27]. Farmers routinely handle the physical, chemical, and biological characteristics of soil directly (such as pH, nutrient content, and soil structure). Without specialized gear, microbes are too small to be seen or counted, and many of them are challenging to collect or even identify.

Additionally, microbial communities and their agronomic functions are dynamic, complicated, and challenging to interpret for use in the field [28]. For organic agricultural soils that depend on microbes for nutrient provision, organic material breakdown, and biocontrol, microbial management, on the other hand, has the potential to pay for itself. These management techniques include both the addition of known beneficial soil microbes and the suppression of undesirable soil microbes [29]. These methods also differ in price, labor and equipment requirements, scope of application, and quantifiable effectiveness [30]. We also provide widespread crop management techniques that influence soil microbial communities and address other agronomic requirements. Farmers may directly add microbes to their soils for a variety of reasons. Theoretically, these extra microbes can help with nutrient availability (via biofertilizers or bio-stimulants), pest control (via biopesticides), or plant growth stimulation (via hormone-signaling PGPs or bio-stimulants) [31].

Farmers can introduce specific microorganisms that directly benefit a certain crop, boost nutrient availability, or increase the ratio of beneficial microbes in their soils by using microbial additions [32]. The soil microbial communities can be impacted by soil management techniques to achieve other agronomic objectives. It is likely that farmers primarily affect soil microbes using these management techniques. Some examples are tillage, crop rotation, mixed cropping and under-seeding, cover cropping, and organic mulches [33]. It is essential to consider how these practices affect soil microbes, especially when designing a farm system incorporating microbial enhancements or suppression tactics. However, it can be challenging to predict how the soil microbial community reacts overall to these practices [34].

Soil disturbances, the addition of carbon and nutrients to the soil (for instance, through the addition of organic fertilizer, decomposing plant matter, or living roots), and their diversity can all have an impact on the total number of microorganisms in the soil (measured as microbial biomass), as well as their diversity and the functions of the microbial community [35]. Additionally, they might affect the number or operation of various microbial groups. Another factor is the farmland's agricultural history, which may be significant if there are residual effects from earlier soil management or pest control techniques [36]. For instance, if there is an excess of mineral phosphorus in the soil as a result of substantial phosphate inputs, increasing the number of phosphorus-solubilizing microorganisms or adding more phosphorus may increase the quantity of phosphorus available to crops. There are complex and reliant interactions between soil's physical, chemical, and biological characteristics. Soil management practices may, therefore, concurrently improve all of these soil components or result in a mix of benefits and drawbacks. The physical, chemical, and biological features of the soil are influenced by a number of variables, including soil type, climate, crop type, past land use, and soil management.

CONSIDERATIONS FOR MANAGEMENT OF NATURAL RESOURCES

▪ Specific goals in natural resource management can occasionally be fulfilled using quantitatively successful microbial management practices. In contrast to other components of organic agriculture systems, microbial management offers farmers little tools for tracking the immediate impacts of their activities [37]. In fact, specific criteria for labeling a complex microbial community in a farm system as "good" or "bad" continue to perplex academics and agricultural specialists. The following are some essential considerations in light of the possible benefits and problems of microbial control tactics in natural resource management.

- Microbes can dramatically alter yields, causing anything from an increase in crop productivity to a complete crop loss. They play essential roles in the health of plants and soil.
- Regardless of whether current research has optimized microbial contributions to soil systems, soil microbial interventions may have positive, neutral, or detrimental consequences on an agricultural system.
- The makeup and activity of existing soil microbes, as well as other soil features, are likely to influence how managing soil microorganisms affects soil or plant health.
- Growers are encouraged to be more selective when utilizing time- or money-intensive practices, such as commercial inoculants. Low-cost approaches, such as simply airing a greenhouse to reintroduce microorganisms following soil sterilization, may be employed as a common practice by growers.
- Farms are encouraged to use soil-building techniques such as composting, growing cover crops, and minimizing soil disturbance because they positively impact the biological communities that live in the soil.

With growers' and researchers' growing interest in managing soil microbes, we expect to see more microbial products and professional recommendations in the coming years. Many popular grower practices target microbes, such as farmscape or biodynamic farming. There may also be new applications for microbes, such as microbes that promote plant drought tolerance or resistance to heat stress.

BENEFICIAL APPLICATIONS OF MICROBIAL RESOURCES IN NATURAL RESOURCE MANAGEMENT

Microbes have produced significant social and economic benefits. The key topics covered are green chemistry and engineering, environmental bioremediation, renewable energy, natural medicine, and organic food production and processing. It is crucial to develop agricultural microbial resources. New agricultural production technology research and development has advanced significantly in recent years. Its major components are microbiological feed, microbiological fertilizers, microbiological insecticides, and microbiological food. Extreme energy depletion, resource scarcity, and environmental pollution are problems that have arisen due to people's rampant exploitation of natural resources and overreliance on fossil fuels. Environmental pollution is mainly caused by traditional chemical methods of industrial production and discharge. A sustainable civilization should rely less on non-renewable resources and limit the pollution from fossil fuels. Utilizing all available natural resources is crucial, as is switching from the outdated, polluting chemical sector to the cutting-edge economy.

The aesthetic trend toward urbanization and industrialization has impacted natural ecosystems. The primary resources that this revolution will impact are water and land resources. These resources are being depleted and deteriorating due to numerous anthropogenic activities. Since land degradation affects 1 to 6 billion hectares of arable land worldwide, it poses a serious challenge to sustainable agriculture and food security. The main contributors to soil deterioration are soil salinization, organic and inorganic pollutants, soil erosion, waterlogging, and inadequate nutrient supply. The fundamental concern in the world, and particularly in developing nations, is the ecological rehabilitation and management of land resources. There are numerous options to repair marginal and severely degraded soils. These comprise various organic and inorganic substances and hazardous heavy metals that continue to affect the soil properties, plants, and food quality today [12]. The microbial association is a different idea to reduce the cost of environmentally acceptable soil treatment, like halophytic plant growth-promoting bacterium (PGPR), which increases plant hormone production and helps plants better survive salinity.

Similarly, utilizing bacterial consortium to reduce inorganic metal concentrations and decompose soil organic contaminants has enormous ecological and economic advantages. Mycorrhizae, a type of plant-fungal relationship, are thought to play a vital part in improving nutrient and water intake and protecting plants from root infections, which is vital for managing deteriorated soil. The science of natural resource management places a great deal of importance on the screening of objectively specific microorganisms for the management of damaged soil.

Measurable progress toward specified objectives can be made with the aid of microbial management techniques. Any intervention, though, can potentially have complicated and unexpected results. It might be challenging to determine if direct or indirect methods of influencing microbial populations have had the desired effect. Contrary to other components of organic agriculture systems, microbial management leaves farmers with limited instruments to track the immediate effects of their interventions. In fact, definitive standards for categorizing a complex microbial population as "good" or "poor" in a farm system continue to elude academics and agricultural experts. Here are some crucial factors to take into account, given the advantages and difficulties of microbial management strategies:

- Microbes can dramatically alter yields, from a rise in agricultural output to a complete crop loss. They play vital roles in the health of plants and soil.
- Although microbial contributions to soil systems have not yet been optimized by study, soil microbial interventions may have positive, neutral, or adverse effects on a farming system.
- The makeup and activities of the soil's existing microorganisms and other soil properties will probably determine how much managing soil microbes will affect the health of the soil or plants.
- Pathogens and other microorganisms having negative or neutral effects can be introduced into a system alongside beneficial ones. Cover crops, for example, can support helpful microorganisms and others that are not specifically favorable to the target crop. Similarly, increasing microbial diversity may not always increase positive soil microbe services.
- Soil-building activities such as adding compost, growing cover crops, and avoiding soil disturbance generally positively affect soil biological communities and are recommended on farms to improve soil health.

The Influence of Soil Microbes and Microbiomes on Natural Resource Management

Water, land, food, plants, animals, and soils are the natural resources that are most important to people. Managing natural resources may encompass crucial tasks, such as maintaining, protecting, and conserving the ecosystem [38]. The extensive use of chemical fertilizers and pesticides in the current trend makes sustainability in sustaining the ecosystem a challenging challenge. In addition, anthropogenic activities like urbanization and industrialization produce more garbage and endanger ecosystems. These human-made activities contribute directly to the process of land degradation. Degradation of the land can lead to soil exhaustion, salinization, and desertification [39]. Waste was eliminated, and ecosystems were restored using various methods, although the results were mixed.

Beneficial soil microorganisms (BSMs) have been discovered as viable candidates that could aid in environmentally sustainable management. These microorganisms have a variety of mechanisms that can be used commercially to create biotechnology to address the main environmental problems. Plant-associated microbes can be exploited to solve soil salinity, fertility, degradation, and habitat loss issues. Numerous species, including bacteria, fungi, algae, insects, annelids, and other invertebrates, are found in soil and exhibit close relationships with both plants and one another [40]. Microbial entities stand out because, through various methods, they are actively involved in boosting soil fertility, encouraging plant growth, and reducing biotic and abiotic stressors. By absorbing nutrients, BSMs promote plant growth. They also create complex soil matrices and aid in plant defense responses by secreting a variety of metabolites.

BSMs can also be resilient to harmful environmental factors such as salt stress, drought stress, weed infestation, nutrient deficit, and heavy metal contamination. Researchers have recently discovered that soil bacteria have destructive and valuable functions in the soil ecosystem. However, BSMs have attracted significant attention for their abilities to promote plant development and their roles in the breakdown of organic wastes, detoxifying harmful compounds like pesticides, and reducing soil stress [41, 42]. Microbes play an essential role in natural resource management by cleaning up all the dead organic material. Without them, the preservation of soil fertility is not possible.

Beneficial Soil Microbes (BSMs)

The natural physical covering of the Earth's surface represents the interface of three material states: solids (geological and dead biological materials), liquids (water), and gases (air in soil pores). It is regarded as the bedrock of all terrestrial ecosystems. In soil captivity, microorganisms such as bacteria, archaea, and fungi interact with one another and contribute to ecosystem functioning. Their direct connection with the plant's root enables mineral uptake from the soil, organic matter decomposition, nutrient acquisition, plant growth stimulation, and phytopathogen suppression [43]. By limiting the growth of harmful bacteria, BSMs promote soil health.

Plant Growth Promoting Rhizobacteria (PGPR)

PGPR are potential microorganisms that colonize plant roots and stimulate plant development either directly or indirectly [44]. These soil bacteria have the ability to colonize roots and stimulate plant growth. Azoarcus, Azospirillum, Rhizobium, Azotobacter, Arthrobacter, Bacillus, Clostridium, Enterobacter, Gluconoacetobacter, Pseudomonas sp., and Serratia sp. are all PGPR species [45]. Much recent study has been conducted to better understand plant-microbe interactions [46]. The production of phytohormones, the fixation of atmospheric nitrogen (N_2), the synthesis of iron chelators known as siderophores, and the solubilization of inorganic minerals such as phosphorus (P), potassium (K), and zinc (Zn) to make them more available for plant growth are all examples of direct growth promotion [47]. PGPR are also recognized as potential microorganisms capable of protecting plants in normal and stressed environments from various environmental challenges [48, 49]. Initially studied solely to increase crop yield, multiple studies now show that PGPR plays a vital role in the normal functioning of agroecosystems [50]. According to research, they can also be used to restore degraded land, improve soil quality, reduce environmental contaminants in soils, and prevent climate change [51].

Cyanobacteria

Photosynthetic prokaryotes, known as cyanobacteria, are common in nature. They are frequently found in wetlands, streams, lakes, ponds, springs, and rivers. Additionally, cyanobacteria are a crucial part of soils [52]. Due to their Role in N_2 fixation and status as a natural biofertilizer, cyanobacteria have demonstrated their significance in preserving fertility [53]. Sustainable agriculture has made use of symbiotic or free-living cyanobacteria. Effective nitrogen-fixing cyanobacteria were discovered in several agro-ecological locations and used for rice production, including *Nostoc linkia, Anabaena variabilis, Aulosira fertilisima, Calothrix sp., Tolypothrix sp.,* and *Scytonema sp* [54].

ROLE OF BSMS IN ENVIRONMENTAL MANAGEMENT

The entire planet is coping with very difficult environmental issues. The primary sources of environmental pollution include the excessive use of fossil fuels, waste products from numerous human activities, land deterioration, and climate change caused by greenhouse gas emissions. Most issues are human-made and brought on by population growth, industrialization, urbanization, and deforestation [55]. The use of BSMs in resolving environmental issues has now been demonstrated by research, and they have been highlighted as viable tools for achieving the objective of a sustainable environment [56]. For bioremediation, biological agents are used, such as microorganisms (micro-remediation), plants (phytoremediation), or both (rhizoremediation). *In situ* bioremediation, which has been used for a while, includes stimulating the local microbial community to break down pollutants. The production of various natural compounds by plant-associated bacteria, such as endophytes and PGPR, improves the bioremediation of environmental soils [57].

BSMs such as *Pseudomonas putida, Azospirillumli poferum, Enterobacter cloacae*, and *P. fluorescens* have been shown to be capable of cleaning up soil contaminated with polycyclic aromatic hydrocarbons (PAHs), total petroleum hydrocarbons (TPHs), and trichloroethylene (TCE) (Table **1**). The mining industry releases numerous heavy metals into the soil, including zinc, lead, copper, and cadmium, creating a severe threat to environmental degradation [58]. Traditional methods for handling metal-containing wastes, such as heat procedures, physical separation, electrochemical treatments, washing, stabilization/solidification, and burial to clean polluted soils, are prohibitively expensive and have adverse environmental effects [59]. According to studies, organic contaminants can be directly impacted by BSMs. Some plants and bacteria have evolved the unique ability to tolerate heavy metals and are used to clean up metals [59, 60].

Table 1 . BSMs' role in crop growth under polluted soil.

Plants	Microbes	Toxic metals	Results
Pisum sativum	*Rhizobium* sp., *Microbacterium* sp.	Chromium	Increased nitrogen concentration in plants (54%) decreased the toxicity of chromium.
Scripus mucronatus	*Brevundumonas diminuta, alcaligenus faecalis*	Mercury	Increase phytoremediation, decrease toxicity in soil
Helianthus annuus and *Triticum aestivum*	*Bradyrhizobium japonicum* (CB1809)	Arsenic	Plant biomass excess, growth in high arsenic concentrations
Brassica napus	*Bacillus megaterium*	Lead	Excessive plant biomass, growth in high arsenic concentrations
Prosopis juliflora, Lolium mltiforum	*Bacillus, Staphylococcus, Aerococcus*	Chromium, Cadmium, Copper, Lead and Zinc	Improve the efficiency of phytoremediation, plant biomass excess, growth under high arsenic concentrations)

Microorganisms use chemical and physical processes to create structural alterations or complete degradation of the target molecule. BSMs can break down, convert, or accumulate a wide range of chemicals due to their high catabolic diversity. Hydrocarbons (oil), polychlorinated biphenyls (PCBs), polyaromatic hydrocarbons (PAHs), and radionuclides are all examples [40]. BSMs are known to produce peroxidases, dioxygenases, P450 monooxygenases, laccases, phosphatases, dehalogenases, nitrilases, and nitro reductases [61]. Several VAM fungi produce xylanases, mannoses, and other enzyme complexes that may partially degrade potentially toxic compounds [62]. *Providencia stuartii*, a strain of bacteria discovered from agricultural soil, can digest the herbicide chlorpyrifos [63]. DDT is known to be degraded by several PGPFs, including *Trichoderma viride, Fusarium oxysporum*, and *Mucor alternans* (DDT). As model organisms for lignin biodegradation, white-rot fungi, primarily *Phanerochaete chrysosporium* and *Trametes versicolor*, are utilized [64]. In addition to *Pleurotus ostreatus, T. versicolor, Bjerkandera adusta, Lentinula edodes, Irpexlacteus, Agaricus bisporus, Pleurotus pulmonarius,* and *Pleurotus tuber-regium*, a variety of additional white-rot fungi can also break down persistent xenobiotic chemicals [64].

IMPORTANCE OF BIOGEOCHEMICAL PROCESS TO EMBRACE THE NATURAL RESOURCE MANAGEMENT

The biogeochemical process denotes the cycling of elements (C, H, O, N, P, and S) across various ecosystems that govern the Earth's dynamics. Cycles of elements and Biogeochemical cycles are essential to life's subsistence because they convert energy and matter into usable forms that help the ecosystem's function. These elements are found in various reservoirs at varying degrees and come in various chemical forms, both organic and inorganic. Those reservoirs are known as natural resources since they benefit humans in numerous ways through technology, and sustainable use favors life on Earth as we know it [65]. The element transitions are interdependent, and physical phenomena (dissolution, precipitation, volatilization, *etc*.) ensure the conversion of biological components and their movement between the various compartments, namely, biosphere, lithosphere, hydrosphere, and atmosphere [66].

Among the elements, carbon, oxygen, and hydrogen are vital for all living organisms. In simpler terms, all living things assimilate carbon from these reservoirs and release it into the atmosphere through metabolism, which is again transferred into the soil or other reservoirs. Major carbon reservoirs are the Earth's crust, ocean sediments, and certain autotrophs. The oxygen and hydrogen go alongside the carbon cycle, converting elements into matter and matter into energy [67]. The other essential elements of the biogeochemical cycle are nitrogen, phosphorus, and sulfur. Nitrogen is an integral part of living organisms as protein and genetic material. Phosphorus, being immobile, is present in large quantities in rocks and soil. Once nitrogen and phosphorus enter the water bodies, the change in nutrient flux leads to eutrophication, thus altering the other cycles in the ecosystem [68]. Similarly, sulfate is present in the Earth's crust, and molten magma is released into the atmosphere through volcanic eruptions and deep-sea ruptures. While the sulfur cycle slowly replenishes sulfur back into the Earth's crust, its emission through anthropogenic sources has risen to dangerous levels [69].

The carbon cycle has two sub-cycles [1]: The ecological cycle, where carbon is transferred across the trophic level of the ecosystem and returns to the soil [2]. The geological cycle, where a certain fraction of the carbon from the ecological cycle is sent into Earth's crust for a long duration. The changes in the ecological side of the carbon cycle do not pose a significant threat to humankind or its interests. Nevertheless, a small flux in the geological subset could affect the Earth's dynamics by altering the climate and all its repercussions. Carbon is predominantly available as carbon dioxide in the atmosphere at an optimum concentration crucial for life on Earth. The carbon emitted through the utilization of natural resources, like coal, oil, natural gas, mining *etc.*, has increased the atmospheric concentration to levels where the existing rate of carbon replenishment would fail to maintain balance [70]. This advocates for sensible climate actions in the years to come.

Agricultural activities have altered the nitrogen and phosphorus cycle to a larger extent. As mentioned earlier, these elements in excess are responsible for the eutrophication of lakes, leading to an algal bloom that degrades an essential natural resource, *i.e.*, water. The prevention of agricultural runoff with this nutrient will help nature stabilize its state in the long run. Sulfur emitted by geological activities is brought down by rains into water bodies that settle along with other sediments [71]. A slight portion enters the biosphere through assimilation and finds its way to the Earth's crust. However, due to the changes in other cycles, the sulfate cycle changes as per the prevailing conditions. Considering the above information, the thought of natural resource management seems crucial in the forthcoming days.

Humankind has made its way into the current civilization by selecting the right candidate for the job, *e.g.*, domestication of animals for food and agriculture. In this impeding rigor task of managing the biogeochemical cycle for managing natural resources, the prime candidate can only be the microorganism [72]. They have their hands in all the cycles, which indicates their multiplicity and effectiveness (Fig. **1**). Soil and sediments are the most abundant source of microorganisms because they contain minerals, nutrients, gases, plant roots, and decaying organic matter, all of which work together to cycle nutrients and support life [73]. Their spatial organization is critical for biogeochemical cycling. Understanding the micrometer scale interactions between soil particles and their microbial inhabitants is required to understand and tweak biogeochemical processes [74] fully.

Disentangling Microbes and Microbiome's Role in Biogeochemical Process

The role of microorganisms in the biogeochemical cycle is undeniably essential and extensive. Initially, the culturable microbes were isolated from each ecosystem and studied for characteristics that emphasized their activity [67]. The term microbes describes bacteria in general rather than fungi, mycorrhizae, and virus units. The recent advent of genomic technologies has paved the way for a quick and reliable way to ascertain microorganisms in any given state in a unit of time. This encouraged the research into microbial-mediated processes in a biogeochemical cycle that could help us understand their pivotal role. The aftermath of these modern sequencing technologies in genomic studies is notably impressive [75]. The microbiome concept emerged due to the reliability and affordability of identifying microorganisms, even the unculturable ones.

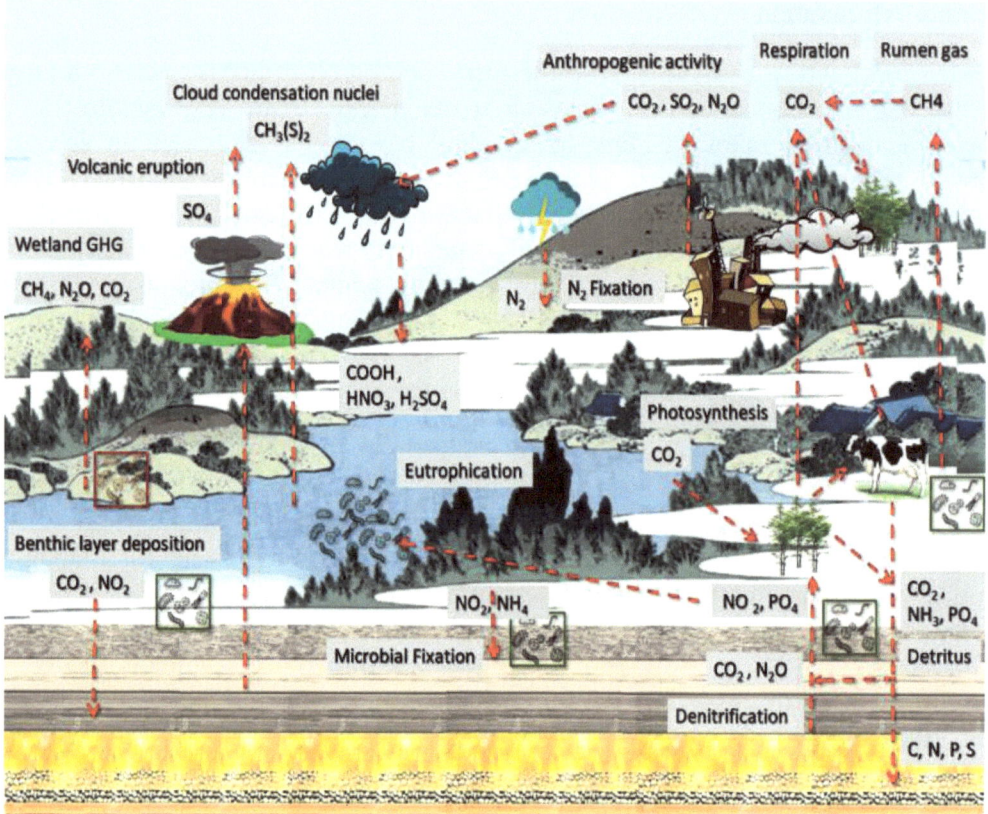

Fig. (1). Biogeochemical cycle depicting the role of microbes.

This narrowed investigation causes dismissal and even ignorance of other organisms' contributions to the balance of the ecosystem. For example, previous descriptions of microbial communities, such as 'commensal' or 'microbiota' in earlier studies, have been offhandedly replaced by 'microbiome'. However, these studies only refer to a part of the microbial diversity, a community that is undoubtedly present and identified through 16s rRNA sequencing [76]. Thus, the prevalence of the word "microbiome" to describe assemblages of Archaea and Bacteria, excluding all others, has been enabled by both technology and the term polysemy [72]. However, technology to define the diversity of eukaryotes or explore a sub-set such as fungi is also available. The literature survey for the term "microbiome" in conjunction with genetic markers "16s", "18s", and "ITS" indicated that only 2% of studies had utilized all the genetic markers, and 97%

applied 16s for their investigation on the microbiome [76]. However, with the exceeding information about the roles of individual entities, we can assume the roles and complexity of the microbiome. The microbiome is at its best in soil aggregates, even throughout the formation process, which stabilizes the microscale architecture and geochemical cycling in the soil matrix [77, 78]. Individual aggregate community-level characterization would be a useful tool for understanding how microbial community structure impacts geochemical interactions and nutrient cycling in soil.

Nevertheless, the role of nitrifying, denitrifying, methanogenic, methylotrophic, calcifying, mineral solubilizing, nitrogen-fixing, photosynthetic, and organics-degrading bacteria in transforming the elements and matter has been well established for many years (Fig. **2**). Similarly, the assistance of fungi and other eukaryotes in the intricate transfer of energy and nutrients is studied. Fungi, with their extensive network of hyphae, had a significant role in the detritus cycle. Interestingly, high viral abundances have long led to speculation that viruses might exert pressure on soil microbial ecology, impacting soil microbes there by the cycle. Irreversible attenuation of virus units to soil components can lead to inactivation [79 - 81], which questions their perpetual role in the interactions. Studies on Arctic permafrost soils identified viral genomes and tracked them across space and time to demonstrate that soil viruses experienced varying stresses throughout the thaw gradient, hindering carbon cycling and the utilization of metabolic genes [82]. Similar effects are likely to shape soil communities where the nutrient transfer is influenced by virus propagation. The lytic enzymes from the viral assemblages break the cells and release the nutrients and matter in the bacterial cells. In addition, the organic matter that escapes the degradation by the microbe enters into the soil's long-term carbon pool in the form of humic and fulvic acids [83]. The situation is directly facilitated by the spike in the viral population in the soil.

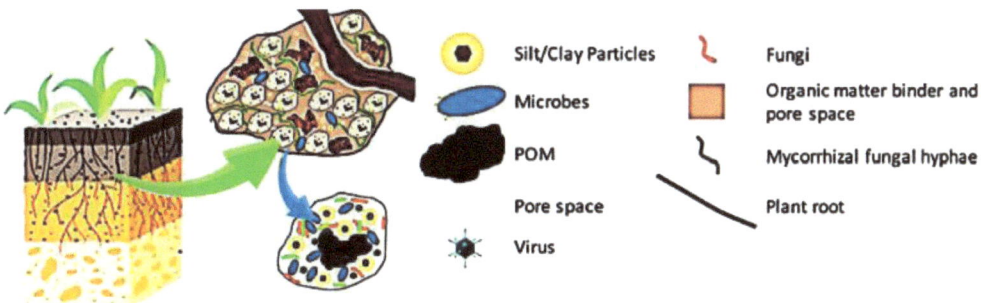

Fig. (2). Location of particulate, humified organic matter and soil microbiome in soil aggregates and their role in soil formation. *Note: Macro-aggregate - 1 mm; Micro-aggregate - 0.1 mm; Organic matter-clay complex - 10 µm.*

Existing technology can characterize soil microbiomes at the aggregate scale, but future work must overcome some challenges to isolate and characterize individuals and communities [84]. Strategies for keeping entire aggregate communities during enrichment and isolation, identifying individual cells, and minimizing biases at all sizes are among the promising approaches. This could help to understand the functional relationships between cells, viruses, soil particles, and their metabolites in biogeochemical interactions. These knowledge gaps necessitate a basic study on the interactions between diverse features of soil microbiomes and the ecological services they sustain.

SPECIAL EMPHASIS ON THE ROLE OF SOIL ENZYMES IN NUTRIENT CYCLING

Soil Enzymes and their Classification

Soil is an important factor that serves nutrients for microorganisms and plays a vital role in geochemical cycles such as carbon, nitrogen, and phosphorous [85, 86]. The enzymes secreted primarily by microorganisms inhabiting soil are called soil enzymes. There are two categories of enzymes based on their secretion: **1.** Constitutive enzymes are always present in the organism regardless of its metabolism—glycolytic enzymes such as hexokinase, phosphofructokinase, and pyruvate kinase come under this category; **2.** Inductive enzymes are activated only when their substrate is in the cell; β galactosidase and nitrate reductase come under this category. Quantitative measurement of soil enzyme activities can help us understand these biological transformations by allowing us to evaluate the activity that exists in soil [87]. Soil enzymes mediate numerous cell processes and catalysts of functions such as organic matter transformation, organic nutrient release for plant growth, nitrogen fixation, detoxification, nitrification, and denitrification [88]. Soil enzyme analysis is a potential indicator of soil biological status or capacity to carry out enzyme-catalyst processes [89].

Nutrient Cycling

Nutrients are essential for the growth of all living forms, and they are obtained from the environment to one organism followed by another; this movement process is known as nutrient cycling, such as carbon, nitrogen, and oxygen. The enzymes are commonly classified into six categories based on their reaction *Viz.*, Oxidoreductase (involves in oxidation and reduction process), transferase (transfer group of atoms from donor to acceptor), hydrolase (cleaves hydrolytic bonds), lysates (cleaves bond other by hydrolysis or oxidation), isomerase (involves isomerization), and ligases (forms bond by ATP cleavage) (Fig. **3**).

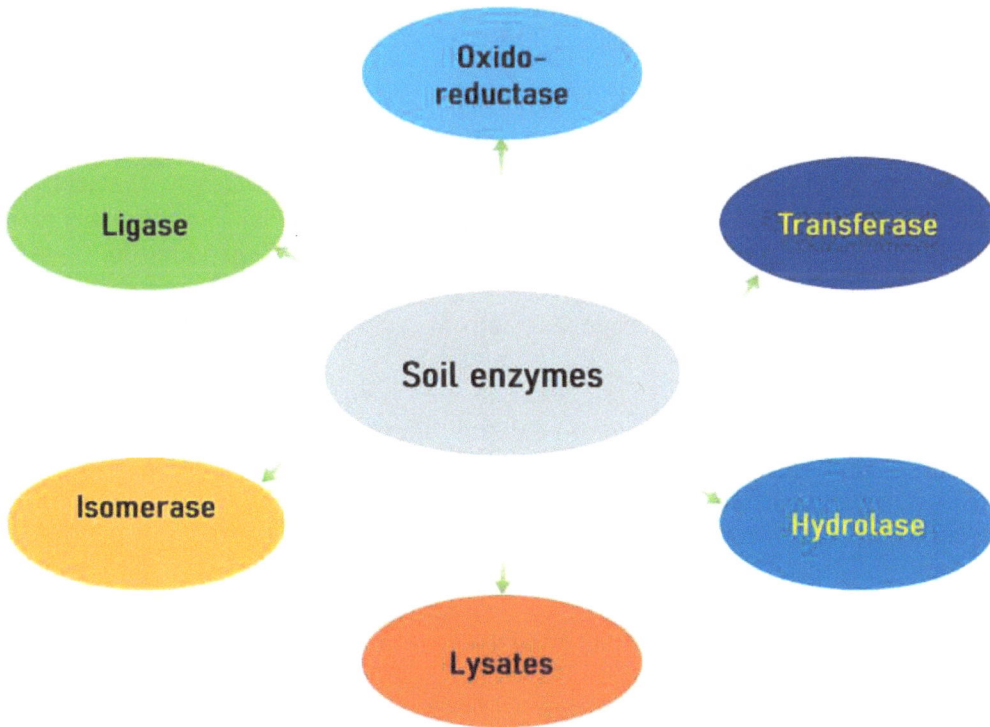

Fig. (3). Role of soil enzymes.

Role of Soil Enzymes in Nutrient Cycling

Soil enzyme activity can be defined by environmental factors such as water, ion concentration, and organic matter content [86]. The essential processes (decomposition of organic matter, transformation, and release of inorganic elements) that occur in soil are majorly employed by enzymes; hence, they play a chief role in the soil and thus act as an indicator to describe the status of soil as healthy or poor. Soil enzyme production can be increased when complex nutrients are abundant and simple nutrients are scarce [90]. Nutrient-releasing enzymes frequently correlate negatively with the concentrations of available nutrients in an ecosystem [91]. Phosphatase, glucosidase, and cellulase production were stimulated by adding organic phosphorous, cellobiose, and cellulose [92], demonstrating that enzyme production can be an inducible response. Microbes manufacture several enzymes constitutively, according to Koroljova-Skorobogatko *et al*., 1998 [93] and Klonowska *et al*., 2002 [94], which may allow them to identify complicated environmental resources. These constitutive enzymes also produce low concentrations of microbially available products, which leads to the induction of additional enzyme synthesis when complex substrates are abundant. The changes in microbial activity due to abiotic or biotic, including anthropogenic activities, change the concentration and type of enzymes in the environment, and the following are the common enzymes: dehydrogenase, urease, phosphatase, and sulfatase.

Dehydrogenase is an oxidoreductase that oxidizes organic materials by transferring protons and electrons from substrates to acceptors within the cell [85], and this enzyme directly signals soil microbial activity [95]. According to Ross (1971), dehydrogenase activity appeared to be more dependent on the soil's metabolic state or the biological activity of the microbial population than on any free enzyme present. Dehydrogenase activity can be measured by triphenyl tetrazolium chloride, making organic matter readily available to microorganisms, which are later converted to formazan. By this measurement, we can study the effect of microorganisms, agronomical practices, and the use of fertilizer and manure on the soil. The vegetation type and organic matter added to the soil also influence urease activity [96]. The changes that occur in soil quality by nutrient management practices can be measured by urease activity. Examining urease enzymes gives knowledge on the management of urea fertilizer or N biofertilizers, such as *Rhizobium* sp., *Azospirillum* sp., *Azotobacter* sp., and *Gluconacetobacter* sp., particularly in tropical and subtropical countries.

Enzyme Activity and Management Practices

Soil biological and biochemical properties have been considered sensitive and early indicators of soil ecological stress or other environmental changes multiple times [97]. Dehydrogenase activity is assumed to reflect the complete range of oxidative activity in soil microflora because it is only found in living cells. As a result, it has the potential to be a trustworthy indication of microbial activity [98]. Crop rotation, mulching, tillage, fertilizer and pesticide use, and soil organic matter concentration can all have an effect on enzyme activity. The use of balanced nutrient and manure treatments raised soil organic matter and MBC status, which in turn increased enzyme activity [99]. The activities of acid phosphatase and glucosidase have frequently been utilized as markers of changes in the quantity and quality of SOC.

Burning and nitrogen fertilizer application in tall grass can potentially affect the activity of soil enzymes. Soil organic content modulates enzyme-producing microbial populations, altering nutrient availability for plant uptake [100]. These microbial populations are also sensitive soil quality indicators [97]. A long-term investigation discovered a large relationship between residue burning and acid phosphatase activity but only weak relationships with several other soil enzymes and a strong relationship between dehydrogenase activity and microbial biomass carbon [88]. Urease activity is highly connected to the organic matter content of the soil and the microbial biomass [101, 102]. Bremner, 1978 studied soil parameters and stated that pH and cation exchange capacity were used to determine the level of enzymatic activity in soils [103, 104].

RESEARCH GAPS, FUTURE PERSPECTIVES, AND CONSTRAINTS

In earlier studies, several attempts were made to predict nutrient cycling based on environmental facts and data, which failed to include the microbial component. As a result, the realities were far from predicted. Beyond models based on environmental and physiological variables, the unique microbial community structure may improve forecasts of ecosystem process rates [105]. While the amount of work generated in recent years regarding microbiomes is enormous, many questions remain unanswered. Soil microbiome institutes ecosystem health, and plant associations significantly impact the ecosystem's transitions [106]. Evolutionary events such as vicariance, species separation, and dispersal *via* various barriers are assumed to be responsible for the distribution of specialized biota. Future research could evaluate the complex interconnected capabilities; nevertheless, it will never be an individual study since the variables that influence soil microbiomes, biogeographical distribution, succession, and nutrient cycling activities must be elucidated. Such multidisciplinary studies have many opportunities to maintain soil health and ecosystem productivity. However, more advances in high-throughput technologies and omics approaches are needed to understand microbial pathways and their functional mechanisms fully. With the rise in global population and the calamity of climate change, scientific methods to boost sustainable livelihoods and management strategies are required [107]. They emphasize the importance of future research into understudied biomes in improving our understanding of global relationships between microbial communities and ecosystem processes.

CONCLUSION

Micro and macro soil organisms are crucial to processes in the soil, like nutrient cycling. These effective procedures are essential for maintaining the quality of water, air, and habitats and for agriculture and forestry. Without a doubt, microorganisms play a significant and wide-ranging role in the biogeochemical cycle. Initially, the culturable microbes were isolated from each ecosystem and studied for characteristics that emphasized their activity. Soil enzymes function as a catalyst in the maintenance of soil health and fertility. Despite the significance of soil biological activities, little progress has been made in defining monitoring and management standards. Future research focusing on uncovering novel microbial variety in soil may be significant practices that may positively affect microbial activities for improved plant growth and render friendly biological habitats for other living beings to survive. Effective soil microbes play an important role since they are in charge of driving multiple biological transformations as well as various pools of carbon (C) and macro- and micronutrients, allowing the subsequent construction of soil-plant-microbe interaction. Microorganisms play critical roles in nitrogen, sulfur, and phosphorus cycling, as well as organic waste breakdown. The soil microbiome is critical for plant growth development and soil fertility in sustainable agriculture. The decomposition of organic matter and the operation of ecosystems allow the microbiome to contribute to nutrient cycling (nutrient recycling and resistance to biotic and abiotic stress).

ACKNOWLEDGEMENTS

The facility and support from the National Agro Foundation, Chennai, are duly acknowledged.

REFERENCES

[1]   Merchant SS, Helmann JD. Elemental Economy. Adv Microb Physiol 2012; 60: 91-210.
        [http://dx.doi.org/10.1016/B978-0-12-398264-3.00002-4] [PMID: 22633059]

[2]   Cavicchioli R, Ripple WJ, Timmis KN, *et al.* Scientists' warning to humanity: microorganisms and climate change. Nat Rev Microbiol 2019; 17(9): 569-86.
        [http://dx.doi.org/10.1038/s41579-019-0222-5] [PMID: 31213707]

[3]   Sahu N, Vasu D, Sahu A, Lal N, Singh SK. Strength of microbes in nutrient cycling: a key to soil health. Agric Important Microbes Sustain Agric Vol I Plant-soil-microbe nexus. 2017;69–86.
        [http://dx.doi.org/10.1007/978-981-10-5589-8_4]

[4]   Dinesh GK, Sinduja M, Priyanka B, *et al.* Enhancing Soil Organic Carbon Sequestration in Agriculture: Plans and Policies. Plans and Policies for Soil Organic Carbon Management in Agriculture. Springer 2022; pp. 95-121.
        [http://dx.doi.org/10.1007/978-981-19-6179-3_4]

[5]   Saccá ML, Barra Caracciolo A, Di Lenola M, Grenni P. Ecosystem services provided by soil microorganisms. Soil biological communities and ecosystem resilience. Springer 2017; pp. 9-24.
        [http://dx.doi.org/10.1007/978-3-319-63336-7_2]

[6]   Makarov MI. The role of mycorrhiza in transformation of nitrogen compounds in soil and nitrogen nutrition of plants: a review. Eurasian Soil Sci 2019; 52(2): 193-205.
        [http://dx.doi.org/10.1134/S1064229319020108]

[7]   Sokol NW, Slessarev E, Marschmann GL, *et al.* Life and death in the soil microbiome: how ecological processes influence biogeochemistry. Nat Rev Microbiol 2022; 20(7): 415-30.
        [http://dx.doi.org/10.1038/s41579-022-00695-z] [PMID: 35228712]

[8]   Burgin AJ, Yang WH, Hamilton SK, Silver WL. Beyond carbon and nitrogen: how the microbial energy economy couples elemental cycles in diverse ecosystems. Front Ecol Environ 2011; 9(1): 44-52.
        [http://dx.doi.org/10.1890/090227]

[9]   Nybo SE, Khan NE, Woolston BM, Curtis WR. Metabolic engineering in chemolithoautotrophic hosts for the production of fuels and chemicals. Metab Eng 2015; 30: 105-20.
        [http://dx.doi.org/10.1016/j.ymben.2015.04.008] [PMID: 25959019]

[10]   Hennon GMM, Dyhrman ST. Progress and promise of omics for predicting the impacts of climate change on harmful algal blooms. Harmful Algae 2020; 91: 101587.
[http://dx.doi.org/10.1016/j.hal.2019.03.005] [PMID: 32057337]

[11]   Dang H, Lovell CR. Microbial surface colonization and biofilm development in marine environments. Microbiol Mol Biol Rev 2016; 80(1): 91-138.
[http://dx.doi.org/10.1128/MMBR.00037-15] [PMID: 26700108]

[12]   Kumar S, Prasad S, Yadav KK, *et al.* Hazardous heavy metals contamination of vegetables and food chain: Role of sustainable remediation approaches - A review. Environ Res 2019; 179(Pt A): 108792.
[http://dx.doi.org/10.1016/j.envres.2019.108792] [PMID: 31610391]

[13]   Maróti G, Kereszt A, Kondorosi É, Mergaert P. Natural roles of antimicrobial peptides in microbes, plants and animals. Res Microbiol 2011; 162(4): 363-74.
[http://dx.doi.org/10.1016/j.resmic.2011.02.005] [PMID: 21320593]

[14]   Muralikrishna IV, Manickam V. Natural resource management and biodiversity conservation. Environ Manage 2017; 2017: 23-35.

[15]   Jhariya MK, Banerjee A, Meena RS. Importance of natural resources conservation: Moving toward the sustainable world. Natural Resources Conservation and Advances for Sustainability. Elsevier 2022; pp. 3-27.
[http://dx.doi.org/10.1016/B978-0-12-822976-7.00027-2]

[16]   Gavin MC, Solomon JN, Blank SG. Measuring and monitoring illegal use of natural resources. Conserv Biol 2010; 24(1): 89-100.
[http://dx.doi.org/10.1111/j.1523-1739.2009.01387.x] [PMID: 20015259]

[17]   Uzun FV. Natural Resources Management. Handbook of Research on Environmental Policies for Emergency Management and Public Safety. IGI Global 2018; pp. 1-21.
[http://dx.doi.org/10.4018/978-1-5225-3194-4.ch001]

[18]   Wassie SB. Natural resource degradation tendencies in Ethiopia: a review. Environ Syst Res 2020; 9(1): 33.
[http://dx.doi.org/10.1186/s40068-020-00194-1]

[19]   Agus C. Integrated bio-cycles system for sustainable and productive tropical natural resources management in Indonesia. Bioeconomy Sustain Dev 2020; pp. 201-16.
[http://dx.doi.org/10.1007/978-981-13-9431-7_11]

[20]   Kour D, Rana KL, Yadav AN, *et al.* Microbial biofertilizers: Bioresources and eco-friendly technologies for agricultural and environmental sustainability. Biocatal Agric Biotechnol 2020; 23: 101487.
[http://dx.doi.org/10.1016/j.bcab.2019.101487]

[21]   Bharti VS, Dotaniya ML, Shukla SP, Yadav VK. Managing soil fertility through microbes: prospects, challenges and future strategies Agro-Environmental Sustain. Manag Crop Heal 2017; Vol. 1: pp. 81-111.

[22]   Zhang L, Fang W, Li X, Lu W, Li J. Strong linkages between dissolved organic matter and the aquatic bacterial community in an urban river. Water Res 2020; 184: 116089.
[http://dx.doi.org/10.1016/j.watres.2020.116089] [PMID: 32693265]

[23]   Yadav AN, Kour D, Kaur T, *et al.* Biodiversity, and biotechnological contribution of beneficial soil microbiomes for nutrient cycling, plant growth improvement and nutrient uptake. Biocatal Agric Biotechnol 2021; 33: 102009.
[http://dx.doi.org/10.1016/j.bcab.2021.102009]

[24]   Yang T, Siddique KHM, Liu K. Cropping systems in agriculture and their impact on soil health-A review. Glob Ecol Conserv 2020; 23: e01118.
[http://dx.doi.org/10.1016/j.gecco.2020.e01118]

[25]   Lakshmi G, Okafor BN, Visconti D. Soil microarthropods and nutrient cycling. Environ Clim plant Veg growth. 2020;453–72.
[http://dx.doi.org/10.1007/978-3-030-49732-3_18]

[26]   Dubey A, Malla MA, Khan F, *et al.* Soil microbiome: a key player for conservation of soil health under changing climate. Biodivers Conserv 2019; 28(8-9): 2405-29. [Internet].
[http://dx.doi.org/10.1007/s10531-019-01760-5]

[27]   Stirling G, Hayden H, Pattison T, Stirling M. Soil health, soil biology, soilborne diseases and sustainable agriculture: A Guide. Csiro Publishing 2016.
[http://dx.doi.org/10.1071/9781486303052]

[28]   Zaramela LS, Moyne O, Kumar M, Zuniga C, Tibocha-Bonilla JD, Zengler K. The sum is greater than the parts: exploiting microbial communities to achieve complex functions. Curr Opin Biotechnol 2021; 67: 149-57.
[http://dx.doi.org/10.1016/j.copbio.2021.01.013] [PMID: 33561703]

[29]   French E, Kaplan I, Iyer-Pascuzzi A, Nakatsu CH, Enders L. Emerging strategies for precision microbiome management in diverse agroecosystems. Nat Plants 2021; 7(3): 256-67.
[http://dx.doi.org/10.1038/s41477-020-00830-9] [PMID: 33686226]

[30]   Lidbury IDEA, Raguideau S, Borsetto C, *et al.* Stimulation of distinct rhizosphere bacteria drives phosphorus and nitrogen mineralization in oilseed rape under field conditions. mSystems 2022; 7(4): e00025-22.
[http://dx.doi.org/10.1128/msystems.00025-22] [PMID: 35862821]

[31]   Aamir M, Rai KK, Zehra A, *et al.* Microbial bioformulation-based plant biostimulants: A plausible approach toward next generation of sustainable agriculture. Microbial endophytes. Elsevier 2020; pp. 195-225.
[http://dx.doi.org/10.1016/B978-0-12-819654-0.00008-9]

[32]   Hakim S, Naqqash T, Nawaz MS, *et al.* Rhizosphere engineering with plant growth-promoting microorganisms for agriculture and ecological sustainability. Front Sustain Food Syst 2021; 5: 617157.
[http://dx.doi.org/10.3389/fsufs.2021.617157]

[33]   Maitra S, Brestic M, Bhadra P, *et al.* Bioinoculants—Natural biological resources for sustainable plant production. Microorganisms 2021; 10(1): 51.
[http://dx.doi.org/10.3390/microorganisms10010051] [PMID: 35056500]

[34]   Breitkreuz C, Heintz-Buschart A, Buscot F, Wahdan SFM, Tarkka M, Reitz T. Can we estimate functionality of soil microbial communities from structure-derived predictions? A reality test in agricultural soils. Microbiol Spectr 2021; 9(1): e00278-21.
[http://dx.doi.org/10.1128/Spectrum.00278-21] [PMID: 34346741]

[35]   Smith JL, Paul EA. The significance of soil microbial biomass estimations. Soil biochemistry. Routledge 2017; pp. 357-98.
[http://dx.doi.org/10.1201/9780203739389-7]

[36]   Smith JL. Cycling of nitrogen through microbial activity. Soil biology: effects on soil quality. CRC Press 2018; pp. 91-120.

[37]   Umesha S, Singh PK, Singh RP. Microbial biotechnology and sustainable agriculture. Biotechnology for sustainable agriculture. Elsevier 2018; pp. 185-205.
[http://dx.doi.org/10.1016/B978-0-12-812160-3.00006-4]

[38]   Lamb D, Erskine PD, Parrotta JA. Restoration of degraded tropical forest landscapes. Science (80-). 2005;310(5754):1628–32.
[http://dx.doi.org/10.1126/science.1111773]

[39]   Kesavan PC, Swaminathan MS. Strategies and models for agricultural sustainability in developing Asian countries. Philos Trans R Soc Lond B Biol Sci 2008; 363(1492): 877-91.
[http://dx.doi.org/10.1098/rstb.2007.2189] [PMID: 17761471]

[40]   Glick BR. Using soil bacteria to facilitate phytoremediation. Biotechnol Adv 2010; 28(3): 367-74.
       [http://dx.doi.org/10.1016/j.biotechadv.2010.02.001] [PMID: 20149857]

[41]   Aislabie J, Deslippe JR, Dymond J. Soil microbes and their contribution to soil services. Ecosyst Serv
       New Zealand–conditions trends Manaaki Whenua Press Lincoln. New Zeal 2013; 1(12): 143-61.

[42]   Ma Y, Oliveira RS, Freitas H, Zhang C. Biochemical and molecular mechanisms of plant-microb-
       -metal interactions: relevance for phytoremediation. Front Plant Sci 2016; 7: 918.
       [http://dx.doi.org/10.3389/fpls.2016.00918] [PMID: 27446148]

[43]   Nihorimbere V, Ongena M, Smargiassi M, Thonart P. Beneficial effect of the rhizosphere microbial
       community for plant growth and health. Biotechnol Agron Soc Environ 2011; 15(2).

[44]   Goswami D, Thakker JN, Dhandhukia PC. Portraying mechanics of plant growth promoting
       rhizobacteria (PGPR): A review. Cogent Food Agric 2016; 2(1): 1127500.
       [http://dx.doi.org/10.1080/23311932.2015.1127500]

[45]   Arora NK. Plant microbes symbiosis: Applied facets. Springer 2015.
       [http://dx.doi.org/10.1007/978-81-322-2068-8]

[46]   Beneduzi A, Ambrosini A, Passaglia LMP. Plant growth-promoting rhizobacteria (PGPR): their
       potential as antagonists and biocontrol agents. Genet Mol Biol 2012; 35(4 suppl 1): 1044-51.
       [http://dx.doi.org/10.1590/S1415-47572012000600020] [PMID: 23411488]

[47]   Ahemad M, Kibret M. Mechanisms and applications of plant growth promoting rhizobacteria: current
       perspective. J King saud Univ. 2014;26(1):1–20.

[48]   Khare E, Arora NK. Physiologically stressed cells of fluorescent pseudomonas EKi as better option for
       bioformulation development for management of charcoal rot caused by Macrophomina phaseolina in
       field conditions. Curr Microbiol 2011; 62(6): 1789-93.
       [http://dx.doi.org/10.1007/s00284-011-9929-x] [PMID: 21479797]

[49]   Kang SM, Khan AL, Waqas M, *et al.* Plant growth-promoting rhizobacteria reduce adverse effects of
       salinity and osmotic stress by regulating phytohormones and antioxidants in *Cucumis sativus*. J Plant
       Interact 2014; 9(1): 673-82.
       [http://dx.doi.org/10.1080/17429145.2014.894587]

[50]   Cheng W. Rhizosphere priming effect: Its functional relationships with microbial turnover,
       evapotranspiration, and C–N budgets. Soil Biol Biochem 2009; 41(9): 1795-801.
       [http://dx.doi.org/10.1016/j.soilbio.2008.04.018]

[51]   Kuiper I, Lagendijk EL, Bloemberg GV, Lugtenberg BJJ. Rhizoremediation: a beneficial plant-
       microbe interaction. Mol Plant Microbe Interact 2004; 17(1): 6-15.
       [http://dx.doi.org/10.1094/MPMI.2004.17.1.6] [PMID: 14714863]

[52]   Singh JS, Kumar A, Rai AN, Singh DP. Cyanobacteria: a precious bio-resource in agriculture,
       ecosystem, and environmental sustainability. Front Microbiol 2016; 7: 529.
       [http://dx.doi.org/10.3389/fmicb.2016.00529] [PMID: 27148218]

[53]   Sahu D, Priyadarshani I, Rath B. Cyanobacteria–as potential biofertilizer. CIBTech J Microbiol 2012;
       1(2–3): 20-6.

[54]   Prasad RC, Prasad BN. Cyanobacteria as a source biofertilizer for sustainable agriculture in Nepal. J
       Plant Sci Bot Orient 2001; 1: 127-33.

[55]   Ribeiro SP, Barh D, Andrade BS, *et al.* Long-term unsustainable patterns of development rather than
       recent deforestation caused the emergence of Orthocoronavirinae species. Environ Microbiol 2022;
       24(10): 4714-24.
       [http://dx.doi.org/10.1111/1462-2920.16121] [PMID: 35859337]

[56]   Satyanarayana T, Johri BN, Prakash A. Microorganisms in sustainable agriculture and biotechnology.
       Springer Science & Business Media 2012.
       [http://dx.doi.org/10.1007/978-94-007-2214-9]

[57]   Gianfreda L, Rao MA. Potential of extra cellular enzymes in remediation of polluted soils: a review. Enzyme Microb Technol 2004; 35(4): 339-54.
[http://dx.doi.org/10.1016/j.enzmictec.2004.05.006]

[58]   Sinduja M, Sathya V, Maheswari M, Dinesh GK, Prasad S, Kalpana P. Groundwater quality assessment for agricultural purposes at Vellore District of Southern India: A geospatial based study. Urban Clim 2023; 47: 101368.
[http://dx.doi.org/10.1016/j.uclim.2022.101368]

[59]   Prasad S, Yadav KK, Kumar S, *et al.* Chromium contamination and effect on environmental health and its remediation: A sustainable approaches. J Environ Manage 2021; 285: 112174.
[http://dx.doi.org/10.1016/j.jenvman.2021.112174] [PMID: 33607566]

[60]   Reichman SM. Probing the plant growth-promoting and heavy metal tolerance characteristics of Bradyrhizobium japonicum CB1809. Eur J Soil Biol 2014; 63: 7-13.
[http://dx.doi.org/10.1016/j.ejsobi.2014.04.001]

[61]   Chaudhry Q, Blom-Zandstra M, Gupta SK, Joner E. Utilising the synergy between plants and rhizosphere microorganisms to enhance breakdown of organic pollutants in the environment. Environ Sci Pollut Res Int 2005; 12(1): 34-48.
[http://dx.doi.org/10.1065/espr2004.08.213] [PMID: 15768739]

[62]   Singh H. Mycoremediation: fungal bioremediation. John Wiley & Sons 2006.
[http://dx.doi.org/10.1002/0470050594]

[63]   Rani K, Zwanenburg B, Sugimoto Y, Yoneyama K, Bouwmeester HJ. Biosynthetic considerations could assist the structure elucidation of host plant produced rhizosphere signalling compounds (strigolactones) for arbuscular mycorrhizal fungi and parasitic plants. Plant Physiol Biochem 2008; 46(7): 617-26.
[http://dx.doi.org/10.1016/j.plaphy.2008.04.012] [PMID: 18514537]

[64]   Mougin C. Bioremediation and phytoremediation of industrial PAH-polluted soils. Polycycl Aromat Compd 2002; 22(5): 1011-43.
[http://dx.doi.org/10.1080/10406630214286]

[65]   Bertrand J-C, Caumette P, Lebaron P, Matheron R, Normand P, Ngando TS. Environmental microbiology: fundamentals and applications. Springer 2015.
[http://dx.doi.org/10.1007/978-94-017-9118-2]

[66]   Smith P, Cotrufo MF, Rumpel C, *et al.* Biogeochemical cycles and biodiversity as key drivers of ecosystem services provided by soils. Soil (Gottingen) 2015; 1(2): 665-85.
[http://dx.doi.org/10.5194/soil-1-665-2015]

[67]   Kristensen E, Connolly RM, Otero XL, Marchand C, Ferreira TO, Rivera-Monroy VH. Biogeochemical cycles: Global approaches and perspectives. Mangrove Ecosyst A Glob Biogeogr Perspect Struct Funct Serv 2017; pp. 163-209.

[68]   Mackey KRM, Van Mooy B, Cade-Menun BJ, Paytan A. Phosphorus dynamics in the environment Encycl Microbiol. 4th ed. Elsevier, Inc. 2019; pp. 506-19.

[69]   Brusseau ML. Ecosystems and Ecosystem Services. In: Brusseau ML, Pepper IL, Gerba CP (Eds.) Environmental and Pollution Science, 3rd edition. Academic Press 2019, pp. 89-102.
[http://dx.doi.org/10.1016/B978-0-12-814719-1.00006-9]

[70]   Ahmed N, Mahboob F, Hamid Z, *et al.* Nexus between Nuclear Energy Consumption and Carbon Footprint in Asia Pacific Region: Policy toward Environmental Sustainability. Energies 2022; 15(19): 6956.
[http://dx.doi.org/10.3390/en15196956]

[71]   Dervash MA, Yousuf A, Ozturk M, Bhat RA. Phytosequestration: Strategies for Mitigation of Aerial Carbon Dioxide and Aquatic Nutrient Pollution. Springer Nature 2023.
[http://dx.doi.org/10.1007/978-3-031-26921-9]

[72]   Wright RJ, Gibson MI, Christie-Oleza JA. Understanding microbial community dynamics to improve optimal microbiome selection. Microbiome 2019; 7(1): 85.
[http://dx.doi.org/10.1186/s40168-019-0702-x] [PMID: 31159875]

[73]   Gao J, Liu M, Shi S, *et al.* Disentangling responses of the subsurface microbiome to wetland status and implications for indicating ecosystem functions. Microorganisms 2021; 9(2): 211.
[http://dx.doi.org/10.3390/microorganisms9020211] [PMID: 33498486]

[74]   Madsen EL. Microorganisms and their roles in fundamental biogeochemical cycles. Curr Opin Biotechnol 2011; 22(3): 456-64. Available from: https://www.sciencedirect.com/science/article/pii/S095816691100022X
[http://dx.doi.org/10.1016/j.copbio.2011.01.008] [PMID: 21333523]

[75]   Takahashi S, Tomita J, Nishioka K, Hisada T, Nishijima M. Development of a prokaryotic universal primer for simultaneous analysis of Bacteria and Archaea using next-generation sequencing. PLoS One 2014; 9(8): e105592.
[http://dx.doi.org/10.1371/journal.pone.0105592] [PMID: 25144201]

[76]   Donachie SP, Fraser CJ, Hill EC, Butler MA. The Problem with Microbiome'. Diversity 2021; 13(4): 138.

[77]   Huss J. Methodology and ontology in microbiome research. Biol Theory 2014; 9(4): 392-400.
[http://dx.doi.org/10.1007/s13752-014-0187-6] [PMID: 25484632]

[78]   Doolittle WF, Zhaxybayeva O. Metagenomics and the units of biological organization. Bioscience 2010; 60(2): 102-12.
[http://dx.doi.org/10.1525/bio.2010.60.2.5]

[79]   Roux S, Hallam SJ, Woyke T, Sullivan MB. Viral dark matter and virus–host interactions resolved from publicly available microbial genomes. eLife 2015; 4: e08490.
[http://dx.doi.org/10.7554/eLife.08490] [PMID: 26200428]

[80]   Williamson KE, Radosevich M, Wommack KE. Abundance and diversity of viruses in six Delaware soils. Appl Environ Microbiol 2005; 71(6): 3119-25.
[http://dx.doi.org/10.1128/AEM.71.6.3119-3125.2005] [PMID: 15933010]

[81]   Zablocki O, Adriaenssens EM, Cowan D. Diversity and ecology of viruses in hyperarid desert soils. Appl Environ Microbiol 2016; 82(3): 770-7.
[http://dx.doi.org/10.1128/AEM.02651-15] [PMID: 26590289]

[82]   Trubl G, Jang HB, Roux S, *et al.* Soil viruses are underexplored players in ecosystem carbon processing. mSystems 2018; 3(5): 10.1128/msystems.00076-18.
[http://dx.doi.org/10.1128/msystems.00076-18] [PMID: 30320215]

[83]   Emerson JB, Roux S, Brum JR, *et al.* Host-linked soil viral ecology along a permafrost thaw gradient. Nat Microbiol 2018; 3(8): 870-80.
[http://dx.doi.org/10.1038/s41564-018-0190-y] [PMID: 30013236]

[84]   Giroux MS, Reichman JR, Langknecht T, Burgess RM, Ho KT. Environmental RNA as a Tool for Marine Community Biodiversity Assessments. Sci Rep 2022; 12(1): 17782.
[http://dx.doi.org/10.1038/s41598-022-22198-w] [PMID: 36273070]

[85]   Das, S.K., Varma, A. Role of Enzymes in Maintaining Soil Health. In: Shukla, G., Varma, A. (Eds.) Soil Enzymology. Soil Biology, vol 22. Berlin, Heidelberg: Springer 2010; pp. 25-42.
[http://dx.doi.org/10.1007/978-3-642-14225-3_2]

[86]   Burns RG, DeForest JL, Marxsen J, *et al.* Soil enzymes in a changing environment: Current knowledge and future directions. Soil Biol Biochem 2013; 58: 216-34.
[http://dx.doi.org/10.1016/j.soilbio.2012.11.009]

[87]   Tiwari SC, Tiwari BK, Mishra RR. Microbial populations, enzyme activities and nitrogen-phosphorus-potassium enrichment in earthworm casts and in the surrounding soil of a pineapple plantation. Biol

Fertil Soils 1989; 8(2): 178-82.
[http://dx.doi.org/10.1007/BF00257763]

[88]   Dick RP. Soil enzyme activities as integrative indicators of soil health. Biol Indic soil Heal. 1997;121–56.

[89]   Badalucco L, Kuikman PJ. Mineralization and immobilization in the rhizosphere. The rhizosphere. CRC Press 2000; pp. 175-212.

[90]   Allison SD, Vitousek PM. Responses of extracellular enzymes to simple and complex nutrient inputs. Soil Biol Biochem 2005; 37(5): 937-44.
[http://dx.doi.org/10.1016/j.soilbio.2004.09.014]

[91]   Pelletier A, Sygusch J. Purification and characterization of three chitosanase activities from Bacillus megaterium P1. Appl Environ Microbiol 1990; 56(4): 844-8.
[http://dx.doi.org/10.1128/aem.56.4.844-848.1990] [PMID: 16348170]

[92]   Shackle VJ, Freeman C, Reynolds B. Carbon supply and the regulation of enzyme activity in constructed wetlands. Soil Biol Biochem 2000; 32(13): 1935-40.
[http://dx.doi.org/10.1016/S0038-0717(00)00169-3]

[93]   Koroljova-Skorobogat'ko OV, Stepanova EV, Gavrilova VP, *et al.* Purification and characterization of the constitutive form of laccase from the basidiomycete *Coriolus hirsutus* and effect of inducers on laccase synthesis. Biotechnol Appl Biochem 1998; 28(1): 47-54.
[http://dx.doi.org/10.1111/j.1470-8744.1998.tb00511.x] [PMID: 9693088]

[94]   Klonowska A, Gaudin C, Fournel A, *et al.* Characterization of a low redox potential laccase from the basidiomycete C30. Eur J Biochem 2002; 269(24): 6119-25.
[http://dx.doi.org/10.1046/j.1432-1033.2002.03324.x] [PMID: 12473107]

[95]   Watts DB, Torbert HA, Feng Y, Prior SA. Soil microbial community dynamics as influenced by composted dairy manure, soil properties, and landscape position. Soil Sci 2010; 175(10): 474-86.
[http://dx.doi.org/10.1097/SS.0b013e3181f7964f]

[96]   Ross DJ. Some factors influencing the estimation of dehydrogenase activities of some soils under pasture. Soil Biol Biochem 1971; 3(2): 97-110.
[http://dx.doi.org/10.1016/0038-0717(71)90002-2]

[97]   Dick RP. Soil enzyme activities as indicators of soil quality. Defin soil Qual a Sustain Environ. 1994;35:107–24.
[http://dx.doi.org/10.2136/sssaspecpub35.c7]

[98]   Nannipieri P. GRECO S, Ceccanti B Ecological significance of the biological activity in soil. Soil Biochem 2017; pp. 293-356.

[99]   Ebhin Masto R, Chhonkar PK, Singh D, Patra AK. Changes in soil biological and biochemical characteristics in a long-term field trial on a sub-tropical inceptisol. Soil Biol Biochem 2006; 38(7): 1577-82.
[http://dx.doi.org/10.1016/j.soilbio.2005.11.012]

[100]   Frankenberger WT Jr, Dick WA. Relationships between enzyme activities and microbial growth and activity indices in soil. Soil Sci Soc Am J 1983; 47(5): 945-51.
[http://dx.doi.org/10.2136/sssaj1983.03615995004700050021x]

[101]   Myers MG, McGarity JW. The urease activity in profiles of five great soil groups from northern New South Wales. Plant Soil 1968; 28(1): 25-37.
[http://dx.doi.org/10.1007/BF01349175]

[102]   Bremner JM, Mulvaney RL Urease activity in soils. In R.G. Burns (Ed.) Soil Enzymes. New York: Academic Press 1978; pp. 149–96.

[103]   Dinesh GK, Sharma DK, Jat SL, Sri KS, Bandyopadhyay KK, Bhatia A, *et al.* Ecological Relationship of Earthworms with Soil Physicochemical Properties and Crop Yields in Conservation Agriculture.

Indian J Ecol 2022; 49(6): 2135-9.

[104] Chan SS, Khoo KS, Chew KW, Ling TC, Show PL. Recent advances biodegradation and biosorption of organic compounds from wastewater: Microalgae-bacteria consortium - A review. Bioresour Technol 2022; 344(Pt A): 126159.
[http://dx.doi.org/10.1016/j.biortech.2021.126159] [PMID: 34673198]

[105] Knelman JE, Nemergut DR. Changes in community assembly may shift the relationship between biodiversity and ecosystem function. Front Microbiol 2014; 5: 424.
[http://dx.doi.org/10.3389/fmicb.2014.00424]

[106] Prasad S, Malav LC, Choudhary J, *et al.* Soil Microbiomes for Healthy Nutrient Recycling. In: Yadav, A.N., Singh, J., Singh, C., Yadav, N. (Eds) Current Trends in Microbial Biotechnology for Sustainable Agriculture. Environmental and Microbial Biotechnology. Singapore: Springer 2021; pp. 1-21.
[http://dx.doi.org/10.1007/978-981-15-6949-4_1]

[107] Ghannam RB, Techtmann SM. Machine learning applications in microbial ecology, human microbiome studies, and environmental monitoring. Comput Struct Biotechnol J 2021; 19: 1092-107.
[http://dx.doi.org/10.1016/j.csbj.2021.01.028] [PMID: 33680353]

CHAPTER 2

Role of Microbes and Microbiomes in Bioleaching and Bioremediation for Polluted Ecosystem Restoration

Ramesh Poornima[1,*], Chidambaram Poornachandhra[1], Ganesan Karthikeyan[1], Thangaraj Gokul Kannan[1], Sangilidurai Karthika[1], Selvaraj Keerthana[1] and Periyasamy Dhevagi[1]

[1] *Department of Environmental Sciences, Tamil Nadu Agricultural University, Coimbatore, Tamil Nadu, India*

Abstract: In an environmental degradation era, improving microbial activity in sustainable mining and pollutant removal has become necessary for the green economy's future. Bioleaching (microbial leaching) is being studied intensively for metal extraction since it is a cost-effective and environmentally benign technique. Bioleaching with acidophiles involves the production of ferric (Fe III) and sulfuric acid. Cyanogenic microorganisms, in particular, can extract metal(s) by creating hydrogen cyanide. Furthermore, environmental degradation and its rehabilitation are serious issues worldwide. Hydrocarbons, pesticides, heavy metals, dyes, and other contaminants are the principal factors significantly degrading the environment. Residual pollutants might also be challenging to remove. Bioremediation is one of the most effective approaches for reducing environmental contaminants since it restores the damaged site to its original state. So yet, only a tiny number of microorganisms (culturable bacteria) have been used, leaving a vast amount of microbial diversity undiscovered. Various bioremediation approaches, such as chemotaxis, bioaugmentation, biostimulation, genetically engineered microbes, biofilm formation, and advanced omics, have been widely used to improve the microbe's metabolic activity, degradation potential of persistent pollutants and restoration of polluted habitats. Microorganisms contribute to the rehabilitation of polluted ecosystems by cleaning up trash in an ecologically friendly way and producing harmless products. This chapter addresses the critical processes in improving bioremediation and current breakthroughs in bioremediation, including bacteria and plants.

Keywords: Bioleaching, microbes, mechanism, bioremediation, heavy metals, organic pollutants.

* **Corresponding author Ramesh Poornima:** Department of Environmental Sciences, Tamil Nadu Agricultural University, Coimbatore, Tamil Nadu, India; E-mail: poornimaramesh93@gmail.com

Govindaraj Kamalam Dinesh, Shiv Prasad, Ramesh Poornima, Sangilidurai Karthika, Murugaiyan Sinduja & Velusamy Sathya (Eds.)

INTRODUCTION

Earth's crust withholds minerals that contain naturally occurring metals in its sulphide, oxide, and carbonate form. The metal-containing minerals are known as "ore". The desired metals were excavated from the ores and purified by a technique called "mineral processing". Pyrometallurgy and hydrometallurgy are the most used conventional methods in mineral processing. The former method utilizes heat to provoke chemical reactions to extract metals from the ores, which involves high cost and emission of greenhouse gases during the process. Hydrometallurgy utilizes an enormous quantity of water, which is a rising concern in today's scenario. Also, this method extracts only water-soluble metals, leaving behind the insoluble ones. Each method has its downsides, making researchers and scientists explore a more economical and eco-friendly way to leach the desired metals from their ores. This led them to discover the application of microorganisms in metal leaching. The conversion of insoluble metals to their soluble form with the help of microbes is termed "bioleaching".

The parallel term "biomining" refers to using bacteria or fungi to mobilize the metals from their solid state. A bioleaching or biomining method involves using bacteria to change insoluble metal oxides and sulfides into solvent particles that can then be recovered with the help of hydrometallurgy. This combined process is called "biohydrometallurgy" [1]. This solubilization and mobilization of metals occur through different processes like oxidation, complexation, and acidification exhibited by microorganisms. These biotechnological processes involving the interaction between microorganisms and the ores are described by the term "biometallurgy". Biomining is chiefly adapted for metals, such as cobalt, copper, uranium, gold, zinc, and nickel. All metals except uranium were extracted from insoluble sulfides, while uranium was from oxides [2].

The recent findings in the biometallurgy process could lead to more inventive options than conventional metal extraction techniques. Various microorganisms have now been documented to play an important role in the geochemical cycles involving the formation, sedimentation, and degradation of minerals. With the help of these microbes, pure metals can be separated from the metal-ore network. When these microbes are grown in the presence of minerals for a prolonged time, they develop resistance and produce bioreagents [3]. The genetic data from these microbes could be utilized to develop genetically engineered microbiomes for specific purposes. Unlike the chemical extraction method, bioleaching was utilized to extract metals from poor-quality ores and tailings for a long time at mechanical scales. They also encourage profitable metal recovery, even from metal-containing wastes [4]. The biotechnological processes by microorganisms in metal extraction can change uneconomical ore stores into economically viable resources.

Microbes are essential in the bioremediation of contaminated ecosystems due to their potential significance in biomining. Bioremediation is a technology that promotes natural environment restoration by eliminating contaminants and avoiding additional contamination. Bioremediation is more environmentally friendly and cost-effective than alternative remediation processes such as chemical and physical. Bioremediation can reduce pollutant toxicity, which uses the metabolic capacity of microorganisms to convert, mineralize, and immobilize hazardous compounds into less toxic forms. Microbes have yet to be shown to degrade some xenobiotic substances, such as strongly halogenated and nitrated aromatic compounds and a few insecticides [5]. The efficiency of microorganisms, on the other hand, is based on several factors, including concentration, the chemical type of pollutants, availability, and the physiological properties of the environment.

Consequently, the factors influencing microorganism degradation potential are nutritional needs or environmental circumstances. Furthermore, bioremediation is classified into two forms depending on removing harmful chemicals and their transportation methods: *in situ* and *ex situ*. Hence, the chapter unravels the potential role of microbes and microbiomes in biomining and bioremediation.

ROLE OF MICROBES IN BIOLEACHING

Bioleaching is a mineral and metal extraction technique from the parent ore utilizing biological procedures. This process depends on the interaction of microbes, unlike any conventional techniques which utilize ecologically harmful chemicals. Bioleaching can also be used to extract metals and minerals from low-grade ores. Microbes get energy for their growth from minerals, which is slower than the other methods. However, bioleaching is considered a green innovation that become considerably more essential in future years, as it is cost-effective [6]. Bioleaching is also called transforming solid/insoluble metal into water-soluble forms using microbes. For example, copper present as sulfide is oxidized by microbes to water-soluble copper sulfate while the remaining residue is disposed of. The microbiological oxidation of host minerals containing metal complexes of interest is depicted in bio-oxidation. Biohydrometallurgy encompasses bioleaching and biomining. Biohydrometallurgy is an interdisciplinary field combining elements of geoscience, biotechnology, mineralogy, microbiology, mining engineering, and hydrometallurgy. The treatment of metals and metal-containing materials by wet methods is referred to as hydrometallurgy, and it portrays the extraction and metal recovery from minerals by procedures in which the liquid phase plays a key role [7].

Biotechnology is important in mineral engineering to create economically viable processes for using wastes and low-quality metals through biochemical leaching technologies and metal upgradation through biobeneficiation. Because of the rapid increase in metal demand, high-grade ores are diminishing faster. In any case, there are still massive quantities of low-grade ores to be mined. Metal recovery *via* traditional techniques is expensive due to high energy and capital sources. Another crucial consideration is the cost of an abnormal state of pollution, which increases the cost of environmental protection. In this line, bioleaching is a vital mining technology that is widely available and cost-efficient. Many developing countries are currently involved in bioleaching activities. This situation is exemplified by the fact that few developing countries have large mineral reserves and bioleaching properties, making this method particularly appealing to developing countries due to its simplicity and low capital cost requirement [8].

Types of Bioleaching

Heap Bioleaching

The poor-grade wastes generated during mining activities contain a ratio of metals lingering in the wastes. The lingering metals were leached from the waste dumps loaded with 10 to 20 m high. The leachate can be collected at the bottom of the heap. Before dumping, an impenetrable base is created to collect the liquor that permeates through the load is collected and not lost to drainage. Leaching liquor is pumped to the top of the heap and allowed to penetrate the deposited material (Fig. **1**). This approach is favored because the liquor is oxygenated. However, evaporation may occur in arid places. The size and porosity distribution in a pile influence a few associated metrics. However, improving the drainage rate is usually performed by observation [9]. A few important parameters are provided below.

i. *Percolation rate: a quick exchange of reactants, i.e.,* oxygen and the products (*e.g.,* dissolvable metal particles), into the heap, occurs through a high permeation rate. This phenomenon is influenced by high space and molecule size in a heap.

ii. *Uncovered surface area:* A large exposed surface area of the material increases the amount of mineral in direct contact with the leaching liquid. A high level of fines in the landfill influences this circumstance, though it may reduce permeability.

iii. *Rock porosity:* Most essential minerals are contained within the rock, and for leaching to occur, the material must be sufficiently porous to allow microporous movement of reactants and products.

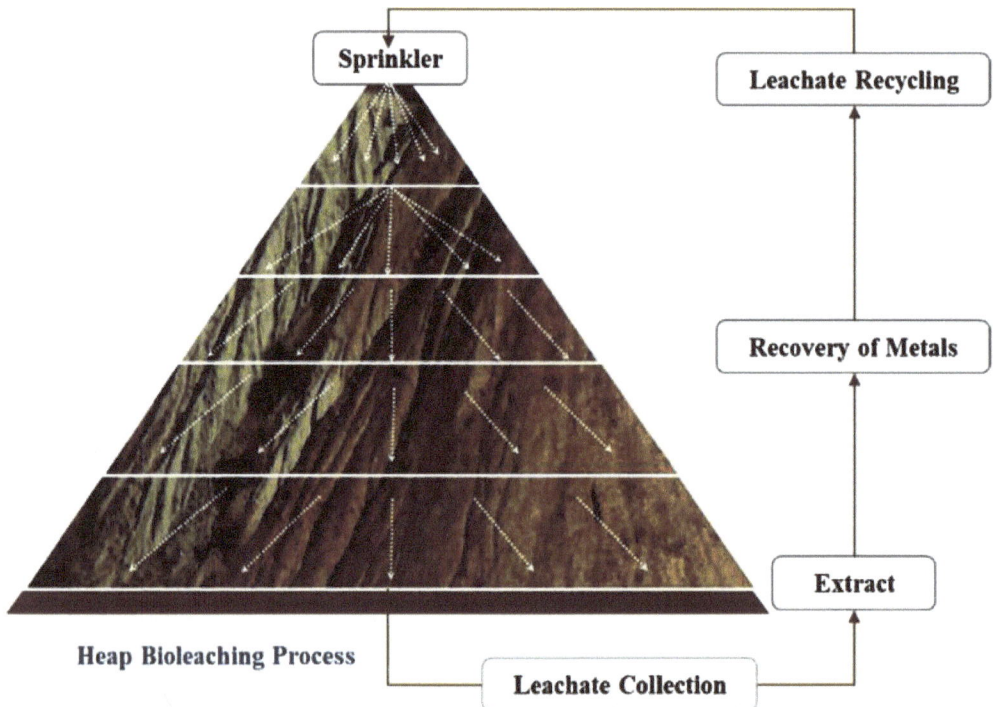

Fig. (1). Heap Bioleaching Process.

Adequate broken-down oxygen and sulfide minerals in the heap pave the way for the development of microorganisms. At the same time, supplements for microbial growth like nitrogen and phosphorous are abundant in their available form in the gangue material. The leach solution pH should be maintained at 2 to 3 by generating sulfuric acid from sulfides. For 20% of the time, a small portion of the heap should be watered so that the metal particles are washed out from the heap. In the remaining time, the solution infiltrates the pores in the rock and responds with sulfide minerals. Evaporation helps carry products back to the rock surface in the later stages of this period, where they can be washed off following the next permeation cycle. The leach solution that exits the heap is processed to recover the dissolvable metal. Due to copper leach solutions, the metal is often recovered through cementation on scrap iron. The copper precipitated from the cementation plant is handled alongside copper concentrates *via* pyrometallurgical processes. Dissolvable extraction allows for the delivery of solutions that are sufficiently focused for direct electrowinning. Following metal recovery, the solution is stored in an oxidation lake where microscopic organisms reduce ferrous particles to ferric particles. Hydrated ferric oxide is accelerated, lowering the concentration of iron in the solution and bringing the pH down [10].

Dump Bioleaching

Dump bioleaching comprises heaped-up uncrushed waste rock. Most of these dumps contain 0.1% - 0.5% copper, which is too low to recover profitably using standard methods. One of these massive landfills contains almost 10 million tonnes of waste rock. Mining operations generate massive amounts of waste minerals over a long period of time. Waste mineral dumps remain on the mining site and cannot be economically handled for metal extraction using conventional methods. In order to handle such overburden, bioleaching is an efficient option. Most of the dumps are unorganized and contain a variety of mineral particle sizes, so dump geometry should be examined for proper draining. Finger dumps have dimensions of 800 x 35 x 200 m, parallel to one another, and are more conspicuous in length than width and stature. The porousness of the landfill's fragmented rock mass is crucial because a large amount of fines would affect the penetration of the leachate. The penetration of acid solution through the mineralized particles across the mass constitutes dump leaching. By providing appropriate development conditions, local microorganisms are allowed to multiply. Metals are naturally soluble, and the profluent collected at the bottom is rich in metal and is recovered using leaching solutions.

The top dump is covered by the development of shallow, square or rectangular lakes. The lixiviant, a pH 2 sulfuric acid solution, is pumped to the lakes *via* a network of distribution lines. Segments of such lakes can be flooded by alternated wetting and rest cycles to provide continuous access to the permeating solution. Sprinklers can disperse the leach solution on the landfill's upper segments and sides. Compressed air can be used, and leach liquids can circulate through the landfill's cross-sections. Because sulfide bio-oxidation is exothermic, temperatures of 60 - 80 °C can be achieved inside the dump. For effective draining, microbial behavior at various temperatures must be understood. The most important acidophilic bacteria in the dump, *A. ferrooxidans* and *A. thiooxidans* thrive at temperatures ranging from 30 to 35 degrees Celsius. Mesophiles die from thermal shock when the temperature rises. A few thermophilic autotrophs capable of withstanding greater temperatures have been accounted for in the dumps [11]. Bioleaching with increased response energy could so continue even at higher temperatures. Biooxidation of sulfides in dump leaching activities is achieved through synergistic movement between mesophiles and thermophiles.

In situ Bioleaching

Microbial *in situ* leaching is a unique approach for extracting metal from mining ores and uneconomical subsurface metal deposits. Filtration is possible in underground mined stops. The pregnant leach solution is retrieved by pumping after leach liquids are infused and penetrated rock cavities (Fig. **2**).

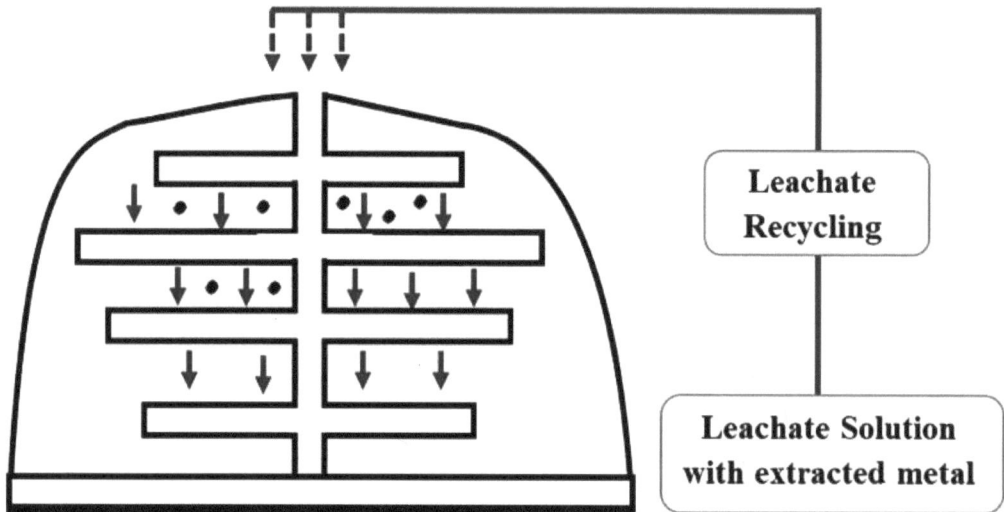

Fig. (2). *In situ* bioleaching process.

This technique has enormous promise for treating low-grade ore. The metal body is drained in place by demonstrating many ways for applying the leach liquor and retrieving the pregnant solution. Leach liquor could infiltrate from the surface or pump under strain into bore holes. The pregnant solution might be extracted from neighboring holes beneath the metal body. The porousness of the metal body is the most critical aspect influencing the efficiency of this method. The use of *in situ* draining is dependent on the ore's geographical features, as well as the hydrologic and mechanical properties of the host rock. In terms of rock porousness and availability, the possibility of *in situ* filtering must be established [12].

Stirred Tank Bioleaching

Because of mass transport constraints, the leaching rate in heaps is extremely low, with oxygen diffusion through the heap being the primary contributor. The high rates are attained when finely powdered concentrates are filtered with tiny organisms in a mixed reactor under optimal conditions. A continuous blending tank reactor with standard air circulation equipment operating at 35°C was used.

The overly expansive limit was unnecessary because the maintenance time was only one day. The filter solution was repeated to allow the copper concentration to reach a level suitable for direct electrowinning. It should be noted that the included microscopic organisms can function in solutions with up to 25,000 ppm copper [13].

Microbes involved in Bioleaching

Acidophilic bacteria or archaea dissolve minerals and metabolize iron and inorganic sulfur compounds typically found in bioleaching operations. They are frequently joined by obligate heterotrophic microbes, especially fungi and bacteria. They often grow at acidic pH levels (<3) and at a wide temperature range. The process parameters, such as oxygen availability, pH, carbon supply, temperature, and solid load, influence the variety of microorganisms in bioleaching processes [14]. At moderate temperatures, iron(II)-oxidizing mesophilic genera *viz., Acidocella, Acidithiobacillus (At.), Acidiferrobacter, Leptospirillum (L.) ferrooxidans*, and *Acidiphilium* mostly dominate bioleaching communities. At moderately high temperatures, *i.e.,* 40 to 60 °C, *Actinobacteria* (*Ferrithrix, Acidimicrobium, and Ferrimicrobium*), *Firmicutes* (*Sulfobacillus,* and *Alicyclobacillus*), *Acidithiobacillus caldus,* and *L. ferriphilum* are frequently identified during the bioleaching process. Members of the *Thermoplasmatales* belonging to archaea are also found in the moderately high-temperature range (*Ferroplasma, Acidiplasma,* and *Cuniculiplasma*). *Sulfolobales* archaea is a type of iron and sulfur oxidizing thermophilic microbe, including the genera *Sulfuracidifex, Sulfolobus, Metallosphaera, Sulfurisphaera,* and *Acidianus.* The species belonging to these genera are dominant at high temperatures (> 60 °C). Various types of microbes involved in bioleaching are given in Table **1**.

Moderately high temperature (55 °C) is optimal for the iron oxidizing archaea species *viz., Ferroplasma acidamarnus, Fp. acidiphilum, Acidiplasma cupricumulans*, and *Ap. aeolicum* [14]. *Thermoplasma acidophilum* is the first acidophilic archaea bacterium that thrives under moderate thermophile conditions. It was the first heterotrophic acidophile isolated. Later, the genera *Sulfolobus* and *Acidianus* were discovered to be acidophilic *Thermoproteota*. Golyshina *et al.* [15] recently identified a mesophilic genus, Cuniculiplasma, in the bioleaching systems, which is exclusively organotrophic. At temperatures as high as 85 °C, thermophilic archaea frequently outperform mesophiles in metal extraction efficiency. However, they are vulnerable to high pulp densities. *Acidianus* sp., including *A. brierleyi*, were the first thermophilic bacteria identified for bioleaching of sulfide minerals. *Sulfolobus metallicus*, newly renamed *Sulfuracidifex metallicus*, dominates most bioleaching processes at around 70°C [16]. *Metallosphaera sedula* and other unnamed archaea are dominant at temperatures >75 °C. Norris [17] reported that these thermophilic archaea have the potential to leach chalcopyrite even at higher temperatures (90°C). *Metallosphaera javensis*, a new species, was recently described by Hofmann *et al.* [18].

Table 1. Types of microbes involved in bioleaching.

Type	Description	Examples	References
(A). Chemolithotrophs	Flourish at higher acidic conditions (pH ≤ 2.0). Most of the preferred chemolithotrophs are reduced sulfur or ferrous oxidizers.		
Mesophiles	Grows at temperatures ranging from 28 - 37°C. They derive energy from inorganic elements like sulfur or ferrous (FE II) and use CO_2 to synthesize cell material	*Acidithiobacillus ferrooxidans, Leptospirillum ferriphilum, Leptospirillum ferrooxidan,* and *Acidithiobacillus thiooxidans*	[19, 20]
Moderate thermophiles	Grows at a higher temperature ranging from 40 - 60°C and pH ranging from 1.5 – 2.5	*Acidimicrobium sp., Ferromicrobium sp., Sulfobacillus sp., Acidithiobacillus caldus* and *Sulfobacillus thermosulfidooxidans*	[21, 22]
Thermophiles	Grows at higher temperatures ranging from 60 - 80°C and belongs to Archaea	*Sulfolobus metallicus, Sulfolobus acidocaldarius, Sulfolobus solfataricus, Sulfolobus brierley,* and *Sulfolobus Ambioalous*	[23, 24]
(B). Organotrophs	Cyanogenic microorganisms and fungi are classified as organotrophs, and when grown on organic carbon, the fungus makes organic acids, whilst the cyanogenic bacteria produce hydrogen cyanide. These end products are responsible for bioleaching.		
Fungal bioleachers	Grows at 25 - 35°C temperatures and pH from 3.0 – 7.0. Use fungal species that produce organic acids like gluconic, oxalic, and citric acids to facilitate leaching.	*Aspergillus niger, Aspergillius flavus, Penicillium simplicissimum* and *Pencillium chrysogenum*	[25, 26]
Cyanogenic bioleachers	Grows at temperatures ranging from 25 - 35°C and pH ranging from 7.0 – 11.0. They include both bacterial and fungal species	*Pseudomonas fluorescens, Pseudomonas aeruginosa, Pseudomonas putida, Chromobacterium violaceum, Bacillus megaterium, Pseudomonas chlorophis, Pseudomonas aureofaciens* and *Escherichia coli*	[27, 28]

Chemolithoautotrophic microbes such as *Leptospirillum* sp., *Ferroplasma* sp., and *Acidithiobacillus* sp. are microbes involved in metabolizing sulfur or iron compounds and facilitating carbon assimilation [14]. Most heterotrophic and mixotrophic bacteria can convert their metabolism fast or exist on extremely low concentrations of organic substances, like extracellular polymeric compounds discharged by primary producers. In addition to the sulfate and iron metabolisms found in acidophilic microbes, a few of these species have been shown to employ H_2 as donors of an electron in anaerobic conditions, either for iron (III) ions, organic molecules, or oxygen reduction [29]. The growth of several species, including *At. Ferrooxidans*, is fueled by the reduction of iron (III) ions combined with the oxidation of hydrogen, sulfur, or organic compounds. The most effective and dominating bacteria in reductive bioleaching of oxidized ores have been reported to be *At. ferroxidans* and *S. thermosulfidooxidans*, although *At. thiooxidans* that thrives at acidic pH is involved in iron (III) mineral dissolution [30].

Moreover, bioleaching bacteria use alternative phosphorous sources, such as phosphonates and scavenge remnants of inorganic phosphate to survive in habitats rich in iron but with low phosphate availability. Tolerance to various hazardous metals and metalloids is conceivable due to the defense mechanism against ROS, genetic and physiological adaptations, and specialized metal detoxifying and binding systems [31]. Furthermore, due to their proclivity to accumulate substantial quantities of inorganic polyphosphates, certain species have been regarded as PAO (polyphosphate accumulating organisms). Copper bioaccumulation in the cells of archaea bacteria and bioleaching bacteria is the cause of their resistance [32].

Bioleaching Pathways

The dissolving of metal ore or any solid substance by chemical processes catalyzed by microorganisms is known as bioleaching. It can be distinguished based on the detailed chemical reactions outlined [33].

Oxidative Bioleaching

Biomining is the industrial use of oxidative bioleaching for metal recovery from metal sulfides [34]. Archaea and iron- and sulfur-oxidizing bacteria use molecular oxygen as an electron acceptor in a series of chemical reactions known as the polysulfide and thiosulfate pathways to dissolve metal sulfides.

Acid Bioleaching

In hydrometallurgy, acid leaching is used to dissolve ores, such as oxide ores like limonitic laterites, by utilizing inorganic acids. Sulfuric acid is also produced by sulfur-oxidizing bacteria when elemental sulfur is introduced or from the sulfur moiety of the mineral being oxidized in the case of metal sulfides. Since the dissolution of metal sulfides utilize both Fe (III) ions generated by protons from acids and oxidation reactions, it is challenging to distinguish acid bioleaching and oxidative bioleaching. Heterotrophic microbes secrete organic acids, enabling their importance in the extraction of metals from ore/solids through metal complexation. This type of bioleaching is known as "heterotrophic bioleaching." However, because these bacteria are heterotrophs, this phrase is misleading. Without chemolithotrophs or sufficient carbon sources, bioleaching is not a process that heterotrophic bacteria can solely drive. Hence, heterotrophic acid bioleaching is a more appropriate description.

Reductive Bioleaching

Reductive bioleaching is the technique of dissolving any solid material by a chemical reduction aided by microorganisms. The dissimilatory reduction of Fe (III) ions is a key reaction [30]. The most well-known instance is reductive bioleaching of oxide ores, such as limonitic laterites. It is difficult to distinguish between laterite bioleaching and acid bioleaching because elemental sulfur is a source of sulfuric acid and a reducing agent [34]. Because numerous microbes, especially acidophiles, may combine Fe(III) ion reduction with hydrogen oxidation, hydrogen may be a promising reducing agent in future research.

Two distinct metal sulfide oxidation processes (mechanisms) have been hypothesized based on the existence of two distinct classes of metal sulfides. These methods can account for the presence of all inorganic sulfur compounds observed in bioleaching environments. Under natural conditions, however, chemically pure metal sulfides do not exist. Thus, mixed minerals are constantly available in practice, forming mixed reaction products and galvanic coupling [35]. The thiosulfate mechanism has been reported for S-containing pyrite, molybdenite, and tungstenite, and the polysulfide process has been characterized for the other metal sulfides. These are indirect mechanisms, while iron present in the exopolysaccharide layer of minerals in the bioleaching microbial cell plays a vital role in the dissolution of metals [36].

Bioleaching Mechanism

Acidophiles and archaea are the sovereign microbes involved in sulfide mineral–leaching. Specifically, *A. ferrooxidans* is the first sulfide-oxidizing bacteria to be isolated [37]. Three major types of mechanisms are involved in bioleaching, namely i) indirect mechanism wherein the ferrous ions in the solution are oxidized to ferric ions by the bacteria, ii) the extracellular polymeric substance (EPS) and the surface where the bacterial cells oxidize the ferrous ions, and the oxidized ferric ions in the bacterial layer leached the sulfide, iii) direct contact mechanism wherein the bacteria directly oxidizes the sulfide without utilizing the ferric ions [38].

Direct Contact Mechanism

In contact bioleaching, cells adhere to the surface of sturdy solid particles. Sessile cells release EPS (extracellular polymeric substances), which immobilize cells in biofilms. Planktonic cells are the cells floating in the fluid. Specific microbial individuals produce metabolites in a network to benefit other species. Sessile cells can be shielded by some bacteria from potentially damaging natural elements like antimicrobials. The component is mainly seen to be interacted with by microbial compounds. *A. ferrooxidans* oxidizes FeS_2 in the presence of oxygen and uses O_2 as the terminal electron acceptor [39]. No experimental data have been found to describe the process of contact leaching, in which microorganisms breakdown the metal sulfide bond in metal sulfides.

$$2\ FeS_2 + 7O_2 + 2H_2O\ \ Iron\ or\ sulphur\ oxidizer\ \rightarrow\ \ 2FeSO_4 + H_2SO_4\ \ \textbf{(1)}$$

The adsorption density of *A. ferrooxidans* hinges on various factors like pH, the residence time of interaction, surface area, cell density, and mineral composition. Beyond neutral pH, the density of adsorption drops follows the order below [40].

Pyrite > Chalcopyrite > Arsenopyrite

Indirect Mechanism

The indirect bioleaching method does not require physical contact between solids and bacteria. The three approaches are (i) secretion of natural and inorganic acids, (ii) redox responses, and (iii) emission of complex agents for ligand formation used to dissolve metals from intractable materials using tiny organisms such as bacteria and fungi.

Sulfuric acid is the most common inorganic acid used in chemical leaching. Several *Acidithiobacillus* bacteria provide it during bioleaching. Ferric particles are also a solid oxidizing agent reduced during metal sulfide solubilization. The most essential ligands generated by many microscopic organisms are oxalate, citrate succinate, and malonate. This form of bioleaching does not require contact, and the development of biofilm on sturdy particles can enhance the bioleaching procedure since biofilms provide a diffusional obstacle that encourages targeted metabolites, such as natural acids, to accumulate on molecule surfaces [41].

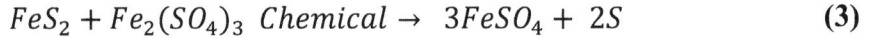

$$4\,FeSO_4 + O_2 + 2H_2SO_4\ Iron\ oxidizer \rightarrow\ 2Fe_2(SO_4)_3 + 2H_2O \qquad (2)$$

$$FeS_2 + Fe_2(SO_4)_3\ Chemical \rightarrow\ 3FeSO_4 + 2S \qquad (3)$$

The last mechanism of sulfide oxidation is also termed "Galvanic interaction" [42]. Furthermore, electrostatic, hydrophobic, and chemical forces also involve the attachment of the bacteria to the surface of the sulfide minerals. The former two reactions (electrostatic and hydrophobic) may be notable during the initial stages of cell adhesion and are reversible. However, the latter (chemical forces) interaction involves the formation of polymeric bridges and is irreversible. The electrochemical reactions are as follows:

Anodic reaction

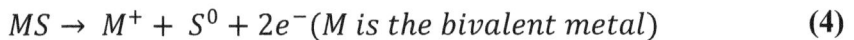

$$MS \rightarrow M^+ + S^0 + 2e^-(M\ is\ the\ bivalent\ metal) \qquad (4)$$

Cathodic reduction of oxygen

$$O_2 + 4H^+ + 4e^+ \rightarrow\ 2H_2O \qquad (5)$$

Polysulfate and Thiosulfate Mechanism

Recently, the disintegration of metal sulfides was categorized based on how well they dissolve in acids. Both of these are acid-insoluble metal sulfides (thiosulfate system) and acid-dissolvable metal sulfides (polysulfide component), respectively (Fig. **3**). In the polysulfide system, the accumulated actions of H^+ and Fe^{3+} particles break apart several metal sulfides, including ZnS, $CuFeS_2$, $FeAsS$, MnS_2, and PbS. The complex interactions between the molecules of metal and sulfur are ruptured as a result of the proton attack, and an $H_2S_2^+$ radical is created and transformed into H_2S_2. It is also oxidized to natural sulfur by higher polysulfides and polysulfide radicals. As a result, this component is referred to as the polysulfide mechanism. Metal sulfides, such as MoS_2 and FeS_2, are oxidized only by Fe^{3+} in the thiosulfate process [43].

The primary mechanism of sulfide leaching is the ferric ions secreted by the extracellular polymeric substances with glucuronic acid [42]. The chemistry involved depends on the mineral's structure and the metal sulfide solubility. The ferric ions are regenerated by the bacteria and are concentrated on the cell wall, facilitating the dissolution of metals. Two major pathways are involved based on the metal solubility and electronic configuration [44]. (a) Thiosulphate pathway wherein the ferric ions strike the surface of the mineral (FeS_2, MoS_2, WS_2) (b) Polysulphide pathway wherein the ferric ions/ the proton attacks the surface (ZnS, $CuFeS_2$, PbS, MnS_2).

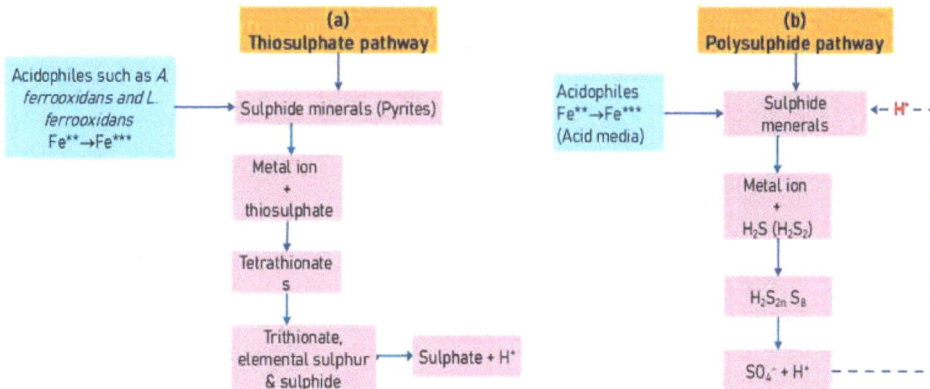

Fig. (3). Sulphate pathways involved in bioleaching a) thiosulphate and b) polysulphide.

Metal Microbe Interaction – Cell Attachment

Leaching bacteria appear to develop, adhering to the surfaces of mineral sulfides in general. Certain microorganisms with significant attachment capabilities to mineral surfaces can attach up to 80–90% of the planktonic cells in less than a day when the surface area is not constrained. Cell adhesion is influenced by the strain, the cultivation environment, and, in some situations, the presence of extra, noteworthy colonizers (such as iron oxidizers) on mineral surfaces [45]. The process of attachment involves extracellular polymeric substances (EPS), which govern microbial cells. Surface contact, which is mediated by cell adhesion, increases EPS formation. Cell attachment to mineral ores is brought on by electrostatic interactions between negatively charged pyrite surfaces and positively charged cell surfaces [35]. The attachment of *A. ferriphilus* R1 cells to mineral sulfides is assumed to be mostly unaffected by hydrophobic interactions, despite their apparent involvement [46]. It is unknown and needs to be clarified how leaching bacteria adjust the composition and quantity of their EPS to a new growth substrate (planktonic cells grown on soluble substrates like iron(II) sulfate seldom ever produce EPS). Future research will need to address the attachment sites on mineral surfaces and how cells recognize/sense them.

Bioleaching of Metals

Copper: Chalcopyrite, the most bounteous copper mineral, is exceptionally stubborn and not promptly amiable to mesophilic bio-oxidation. Chalcopyrite metals are being attempted to be bioleached, with a focus on using thermophiles. The microorganism that is most frequently thought of is *A. ferrooxidans*. In several bioreactor forms, *Leptospirillum ferriphilum* frequently outperforms *A. ferrooxidans*. Studies on *L. ferrooxidans* and *A. thiooxidans* in copper-contaminated environments revealed highly acidic levels of bioleaching. At higher temperatures (40–80 °C), both extreme and moderate thermophiles grow enormously. In a few bioreactors, *Acidithiobacillus caldus* is the actual sulfur oxidizer. *Sulfolobus* species require temperatures more than 60 °C. The use of heap and mixed tank bioleaching processes for copper extraction from blended oxide-sulfides, as well as chalcopyrite metals and their concentrates is becoming more and more popular. Heap bioleaching is generally utilized among various other strategies. The copper bioleaching by heap method is a practical innovation.

Cobalt: When microbes act on cobalt sulfide, it dissolves. Utilizing *A. ferrooxidans, A. thiooxidans, L. ferrooxidans*, thermophilic acidophiles, cobalt sulfide (CoS), and cobaltite (CoAsS) can be biodissolved. Business-scale cobalt bioleaching initially tried to use cobalt-containing pyrite in blended reactors. The cobalt is primarily scattered in pyrite, siegenite [$(Co, Ni)_3S_4$], pentlandite [$(Ni, Fe, Co)_9S_8$] and bravoite [$(Fe, Ni, Co)S_2$].

Nickel: Manufactured, exceptionally unadulterated NiS was bio-leached utilizing *A. ferrooxidans* to recuperate 92% - 98% nickel. In bioleaching of nickel sulfide utilizing nickel-resistant *A. ferrooxidans*, 43% recuperation was accounted for. Because nickel sulfides are sensitive to temperature, utilizing all thermophiles enhances energy extraction. *A. ferrooxidans* was used in bioleaching experiments using a copper-nickel-press sulfide metal. Nickel disintegration was found to be more rapid and to always occur before copper disintegration. Nickel disintegration was just a few times higher than copper from copper-nickel sulfide ores [42]. Thermophiles can produce nickel resilience, such as *Sulfobacillus* isolated from nickel mineral sources [47]. Bioheap draining was considered the most financial choice for nickel. *A. caldus, Sulfobacillus* spp., and *Acidimicrobium* were used to bioleach nickel sulfide assemblies at higher temperatures of up to 55°C.

Uranium: Uranium can be bioleached utilizing heap, dump, vat, or tank bioleaching techniques. The variables influencing uranium bioleaching are metal mineralogy and qualities, pH, temperature, and microbial culture. Considerations for uranium bioleaching frequently employ enhanced, pure strains of *A. ferrooxidans* that oxidize iron and sulfur, or mixed societies of *A. ferrooxidans, A. thiooxidans,* or *L. ferrooxidans*. *A. ferrooxidans* or *L. ferrooxidans* are responsible for producing the oxidant as a lixiviant in the previous bio-oxidation response, which pursues the indirect path. The microscopic organisms could likewise straightforwardly oxidize U(IV) to U(VI), encouraging simple corrosive disintegration through enzymatic activity [48].

Zinc: Combining *A. ferrooxidans* with *A. thiooxidans* results in a faster rate of zinc disintegration from zinc sulfide minerals than alone. Zinc disintegration benefits from the addition of a suitable concentration of ferric iron; greater ferric concentrations slow down zinc disintegration due to the formation of jarosites. Utilizing iron-oxidizing microorganisms, ferric sulfate is bio-oxidized from the zinc leach solution with ferrous particles and then used once more in the leaching process. For PbeZn concentrates containing 40% Zn and 20% Pb, the indirect bioleaching method has been investigated using an integrated pilot plant. It generally involves the leaching of ferric, bio-oxidation of ferrous particles to recover the lixicant, precipitation of Fe, extraction of solvent, and electrowinning for zinc.

Recovery of Metals From Acid Mine Drainage and Bioleaching of Industrial Waste

Known as secondary or urban mining, biohydrometallurgy offers techniques for recovering metals from various forms of residues and wastes produced by mining and industrial activities. By recovering remaining precious metals that were not commercially relevant at the time of tailings deposition, reprocessing mine waste offers the advantage of decreasing the waste's environmental impact. Several case studies have explored the bioleaching of tailings at the laboratory scale. Cu, Co, and Ni extraction rates from sulfide tailings [49] and laterite tailings [50] were high. Indium and other trace metals have also been a target [51]. Economic factors for recycling mine tailings are uncommon but essential for application [52].

Additionally, a stirred-tank reactor was used commercially to recover Co from pyrite concentrate at the Kilembe mine in Kasese, Uganda [53]. Using ferric iron or sulfuric acid-producing litho-autotrophic acidophiles (like *Acidithiobacillus*) and organic acids (or cyanide-excreting heterotrophs), industrial, metal-rich residues like fly ash from coal combustion or sludge, slag, electronic scrap, and spent catalysts have frequently been bioleached at pilot scale [54]. As a result, biotechnology-based recycling of metals will gain attraction in the future, owing to economic and environmental factors such as carbon footprint and energy usage. Biotechnology tools make metal extraction from industrial wastewaters, acid mine drainage, and metal-rich process fluids possible. Examples include bioaccumulation, biosorption, bioelectrochemistry, biomineralization, and bioprecipitation. Metal sorption on biomolecules or biomass is referred to as biosorption. Bioaccumulation occurs when live cells absorb metals into the cell's core. There are numerous publications and patents from mostly laboratories and a few pilot-scale studies that discuss the biosorption of various metals [55]. However, there are no known uses for industrial biosorption. Based on the solubility of metal complexes, the microbially influenced chemical precipitation of metals (bioprecipitation) appears more promising for application.

According to Reichel *et al.* [56], one example of this is the pilot-scale removal of iron from acid mine drainage *via* microbial iron (II) oxidation to iron (III) hydroxides like schwertmannite. While pure metal fractions can be obtained by varying the pH of the polymetallic solutions during the precipitation process, nickel, copper, cobalt, and zinc can be extracted chemically using hydrogen sulfide as metal sulfides. Microbial sulfur or sulfate reduction with sulfate-reducing bacteria (SRB), which can be acidophiles or neutrophils [57], can also create hydrogen sulfide. There are commercial uses. Acid mine drainage treatment using acidophilic sulfate reducers can recover metals as metal sulfide precipitates, eliminates sulfate, and raises pH.

As previously mentioned, biomineralization—the biotechnological production of minerals from metals in solution—is usually linked to bioprecipitation. Examples include metal sulfide bio precipitation and schwertmannite. Additionally, certain microorganisms have demonstrated the capacity to produce pure metal or metal complex nanoparticles. Nanoparticles can occur intra – (within cells) or extracellular (cell surfaces), and successful laboratory-scale experiments have produced Ag, Au, Pd, Pt, Te, Se, Si, Ti, and Zr nanoparticles [58]. In recent years, bioelectrochemistry has been widely employed in the laboratory for the extraction of metal from solutions. The essential premise is that heterotrophic bacteria adhering to an anode (electroactive biofilm) oxidize organic molecules in a bioelectric system (electro-bioreactor). As evidenced by the separation of Pb, Cu, Zn, and Cd from diluted solutions, the electrons generated at the cathode are used for the electrochemical reduction of metal cations to pure metals in elemental form. A progressive adjustment of the electrochemical potential was used to separate the pure metals. Cr, Co, Hg, Se, and Ag are among the other metals produced this way [59].

ROLE OF MICROBES IN BIOREMEDIATION OF POLLUTED HABITATS

Bioremediation is the use of living organisms, usually bacteria, that have the ability to degrade harmful environmental pollutants or convert them to a less toxic form. Based on the utilization of various organisms, several specialized words, such as phyto- and myco-remediation, are used to describe bioremediation. Depending on the concentration and metabolic requirements of the contaminants, different bacteria may digest the different toxins. Microorganisms are expected to be vital in bioremediation and have the greatest potential worldwide. Microbes degrade contaminants through enzymatic activity, utilize them as the sole carbon source and energy source for reproduction and growth. The required conditions must be met by organisms utilized in bioremediation [60]: (a) in a contaminated environment, the organism must be able to live and exhibit biological activity; (b) the contaminant, which is frequently adsorbed to solid surfaces or is insoluble in watery environments, should be accessible to the organism; (c) bioremediation requires efficient enzymes in the organisms; (d) the substrate must have access to the active site of the enzymes involved in bioremediation; (e) there must be close intra- or extracellular contact in the enzymatic and contaminant systems, (f) having the potential to live and thrive in contaminated ecosystems.

The majority of the microorganisms found in polluted locations are Pseudomonas species. Pseudomonas putida received the first patent for petroleum breakdown as a bioremediation agent in 1974 [61]. The mechanism of microbes involved in remediating inorganic pollutants like heavy metals is given in Fig. (**4**).

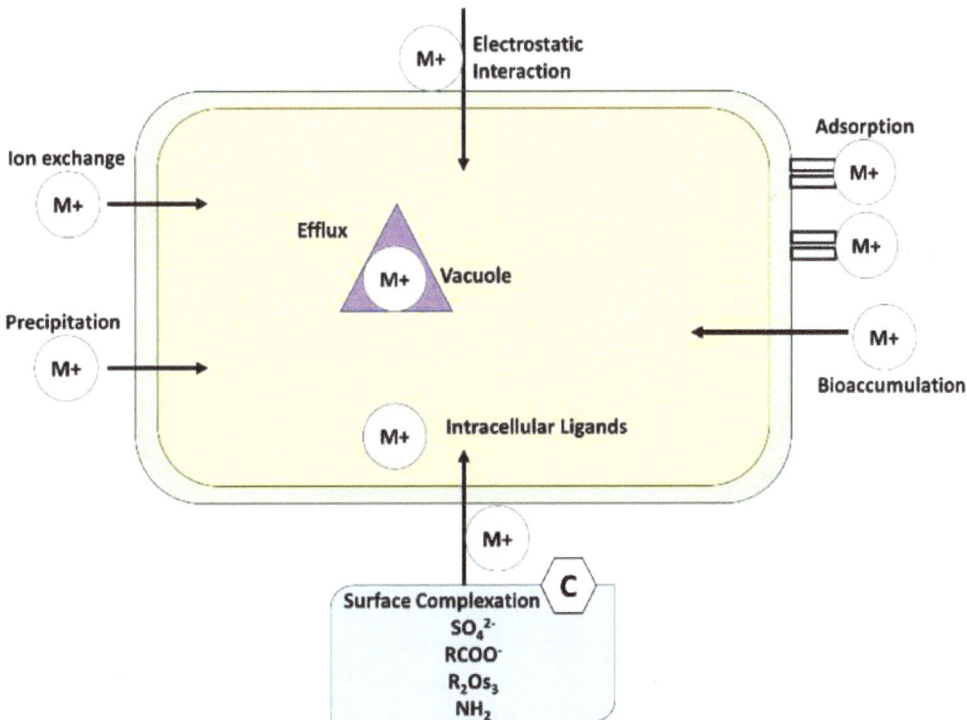

Fig. (4). Mechanism involved in remediating heavy metals by microbes.

Furthermore, while considering organic pollutants, oil slicks and petroleum can be biodegraded with the help of *Mycobacteria, Corynebacteria, Pseudomonas*, and yeasts [62]. *Pseudomonas, Sphingomonas, Alcaligens, Mycobacterium,* and *Rhodococcus* are among the aerobic bacteria recognized as hydrocarbon and pesticide degraders. Aerobic bacteria are more commonly utilized in the process than anaerobic bacteria. Chloroform, polychlorinated biphenyls (PCBs), trichloroethylene (TCE), and dechlorinate are all known to be degraded by anaerobic bacteria in sediments. A ligninolytic fungus called *Phanaerochaete chrysosporium* is useful in the bioremediation of several refractory toxic environmental contaminants, such as DDT, TNT, polymers, and a number of high molecular weight aromatics.

Furthermore, methylotrophs have been linked to the bioremediation of various contaminants, including chlorinated aliphatic trichloroethylene and 1,2-dichloroethane. *Geobacter metallireducens* can transform metals into non-toxic forms [63]. Microbial consortiums can break down the bulk of harmful compounds that individual microorganisms cannot mineralize independently. The microbial consortium destroys these contaminants in stages, utilizing synergy and co-metabolism. The various microbes in the remediation of organic and inorganic pollutants are given in Table **2**.

Table 2. Microbes involved in bioremediation of organic and inorganic pollutants.

Pollutant	Microorganisms	References
Organic Pollutants		
1,1,1-Trichloroethane	*Butane-utilizing enrichment culture*	[64]
Chloroethenes	*The consortium that contains Dehalococcoides*	[65]
Benzene, anthracene, hydrocarbons PCB's	*Pseudomonas spp.*	[66]
Benzene, toluene, and o-xylene (BTX)	*Pseudomonas putida MHF 7109*	
Halogenated hydrocarbons, phenoxy acetates	*Bacillus spp.*	
Glyphosate	*Pseudomonas puteda, P. aeruginosa and Acinetobacter faecalis*	[67]
Atrazine	*Acinetobacter spp., Enterobacter spp., Bacillus spp., Providencia sp, and Pseudomonas spp.*	[68]
Inorganic pollutants		
Lead, nickel and mercury	*Saccharomyces cerevisiae*	[69]
Fe^{2+}, Zn^{2+}, Pb^{2+}, Mn^{2+} and Cu^2	*Pseudomonas fluorescens and Pseudomonas aeruginosa*	[70]
Copper, cobalt, lead and chromium	*Lysinibacillus sphaericus CBAM5*	[71]

(Table 2) cont.....

Pollutant	Microorganisms	References
Cadmium	*Aspergillus versicolor, Paecilomyces sp., A. fumigatus, Paecilomyces sp., Microsporum sp., Terichoderma sp., Cladosporium sp*	[72]
Uranium, Copper, Nickel, Chromium	*Aeromonas sp., Pseudomonas aeruginosa*	[73]

Bioremediation Techniques

Bioremediation has been classified into *in situ* and *ex situ* based on transporting and removing harmful chemicals.

In situ Remediation Technique

In situ treatments entail treating pollutants at their source with little disruption. It not only saves on transportation costs, but it also uses risk-free microorganisms to eliminate chemical pollutants. Since it alleviates the soiled region without needing excavation, it also lessens interruptions. Additionally, it is a safer way to get rid of dangerous compounds. Different *in situ* bioremediation techniques have been used to successfully treat sites contaminated with dyes, chlorinated solvents, hydrocarbons, and heavy metals [74]. The intended *in situ* bioremediation process is improved, whereas intrinsic or spontaneous attenuation is left alone. Biosparging, bioventing, and bioaugmentation are the most often employed *in situ* methods.

Bioventing is a popular *in-situ* treatment method that involves injecting aeration and nutrients into polluted soil to increase the development and activity of the indigenous microbial community. This method decreases pollutant volatilization and discharges into the environment by requiring a low airflow rate and enough oxygen for biodegradation. When the water table is far buried beneath the soil's surface, it is frequently employed to eliminate simple hydrocarbons and is more effective. The pace of pollution removal may vary from site to site due to changes in soil texture and hydrocarbon structure. By injecting pressurized air under the water table, the concentration of groundwater oxygen is raised, resulting in an improved rate of pollutant biodegradation by native bacteria. It promotes aerobic decomposition as well as volatilization. In order to minimize the migration of volatile elements into the environment, pressure control is critical during the biosparging process. Pollutant biodegradability and soil permeability are two main characteristics that influence biosparging effectiveness. Biosparging is more common in areas with a high concentration of medium-weight petroleum compounds, such as jet fuel or diesel.

Sometimes, it takes longer for heavier materials, like lubricating oil, to biodegrade than for lighter ones, like petrol. But these places can also make use of biosparging [75]. In some cases, bioaugmentation—the act of introducing efficient pollutant-degrading strains with a high rate of degradation or are very well suited to thrive under existing conditions—can increase the rate of biodegradation. The outcome will be limited, and the treatment will be ineffective if the introduced organism(s) are not widely distributed throughout the matrix and cannot compete with the native microbial population. In a natural setting, bioaugmentation is renowned for isolating, characterizing, and standardizing microbes to digest contaminants.

Ex situ Remediation Technique

Ex situ bioremediation is performed elsewhere rather than at the sites. It entails decontaminating contaminated sites and transporting contaminated groundwater or soil to a treatment facility. When carried out in the absence of a natural environment, this procedure is independent of environmental conditions, allowing the treatment plan for target pollutants to be modified utilizing physico-chemical approaches before or during degradation. *Ex-situ* bioremediation can be classified as either a solid phase treatment or a slurry phase treatment, depending on the physical state of the pollutant during bioremediation.

Solid-phase bioremediation is an *ex-situ* method that excavates polluted soil and piles it. It also includes organic trash such as leaves, animal manure, agricultural waste, and home, industrial, and municipal garbage. Bacterial growth is transported by pipelines installed into the piles. Air movement through the pipes is required for ventilation and microbial respiration. When compared to slurry-phase procedures, solid-phase systems needed significantly more space and cleaning time. Biopiles, windrows, land farming, and composting are a few examples of solid-phase treatment techniques. The slurry-phase bioremediation process is quicker than the other types of treatment. The bioreactor creates the ideal environment for microorganisms to break down contaminants in the soil by mixing contaminated soil with water, nutrients, and oxygen. In this process, contaminated soil must be separated from stones and debris. The concentration of contaminants determines the additional water concentration, the rate of biodegradation, and the physicochemical parameters of the soil. After this process is completed, the dirt is collected and dried using vacuum filters, pressure filters, and centrifuges. The next step is to dispose of the dirt and treat the resulting fluids.

The process of composting involves gathering impure soil and non-hazardous organic materials, such as manure or vegetal waste. This boosts the required temperature and encourages efficient microbial growth [76]. It effectively removes TNT, RDX, and PAH. Biopiles combine land cultivation with composting; designed cells are created as aerated composting heaps, and the process is continued by adding compost to the contaminated soil. In a more sophisticated kind of land farming, biopiles are used to manage the physical losses of pollutants through volatilization and leaching as well as to remove petroleum hydrocarbons [76]. In bioreactors, slurry phase treatment consists of a solid-liquid combination, which is far quicker than other treatment methods. In order to generate a stable interaction between soil toxins and indigenous microorganisms, impure soil is mixed with water and other additives in a vast reservoir known as a bioreactor. Furthermore, the appropriate environment for bacterial pollutant breakdown is preserved by supplying oxygen and nutrients and regulating the atmosphere inside the bioreactor. After treatment, water is separated from solid wastes, which are then either deposited or subjected to additional disinfection if contaminants are still present.

Factors Influencing Bioremediation

Biotic Factors

At the contaminated site, a variety of native life forms are biotic variables that have a substantial impact on the microbial degradation processes. These characteristics influence the bioremediation process by lowering the microbial mobility and survival activities associated with degradation. These changes are brought on by protozoan grazing, other eukaryotic interactions, and microbial competition for scarce carbon sources. The amount of pollutants, microbes that break down materials, and enzymes produced by each cell (catalyst) determine the pace of deterioration. Maintaining microbial growth, which requires adequate oxygen and nutrients in usable form, is crucial for successful and lasting bioremediation. Some research has demonstrated that techniques such as repeating bioaugmentation, repeated inoculation, and pre-induction can help to avoid a decrease in effective microbial growth [77].

Abiotic Factors

Temperature, aeration, and resistance of ambient pH by modifying the redox state and electrokinetic of the contaminated soil may also affect the degradation rate of pollutants [78]. The numerous abiotic elements of the site have a considerable impact on *in situ* bioremediation. Microbial treatment necessitates a relatively limited set of physicochemical conditions for successful pollutant degradation, and deviations in these ideal parameters almost always result in decreased degradation efficiency. The temperature significantly impacts the activity of microbes, eventually governing the degradation rate by controlling the pace of reactions catalyzed by enzymes. The rate of biodegradation slows with decreasing temperature, and at very low temperatures, the rate of biodegradation is extremely low.

In contrast, a rise in soil temperature leads to an increase in the metabolic activity of microbes, which leads to an enhanced rate of biodegradation. An extremely high temperature, on the other hand, might be detrimental to some microbes. Furthermore, a wide pH range promotes biodegradation, although, in terrestrial and aquatic systems, a pH range of 6.5-8.5 is thought to be optimal for the process to proceed efficiently. In addition, it significantly impacts nutrient availability and solubility, thereby affecting biological activity. For instance, 6.5 is the ideal pH for phosphorus solubility; henceforth, the solubility decreases below this pH. Various nutrients, including calcium, nitrogen, sulfur, magnesium, phosphorus, potassium, copper, zinc, manganese, trace elements, and iron, are also essential for cell development. Nutrient sufficiency is required for effective microbial decomposition. Nitrogen and phosphorous are commonly lacking in polluted environments and are given to the bioremediation system in usable forms (ammonia for nitrogen, phosphate for phosphorous). Water aids in the ejection of metabolic waste from the cell and distributes nutrients and organic components throughout the microbial system. Since it relates to oxygen exchange, the amount of water in the soil pores is also considerably affected. Additionally, abundance of water is harmful because it clogs the soil and prevents oxygen exchange in the soil (unless anaerobic conditions are required), lowering the rate of biodegradation [78]. According to research, these parameters will likely influence the *in situ* bioremediation capability through various pathways.

Methods to Enhance Bioremediation

Several methods can be used to improve the bioremediation process. The following efficient processes to enhance bioremediation are shown in Fig. (**5**).

Chemotaxis

Microorganisms exhibit a wide range of behavioral modifications that improve bioremediation *in situ* treatment. Chemotaxis is widely recognized with these chemotactic characteristics, and prevailing microbes may readily move toward contaminants. Chemotaxis is thus regarded vital for *in situ* bioremediation since it considerably improves the contaminant's bioavailability, eventually improving its performance. In a variety of aquatic environments, a wild bacterial strain degraded naphthalene faster than its immobile (non-chemotactic) mutant [79]. Chemotaxis has been shown to considerably increase non-aqueous phase liquid parameters like the rate of mass movement and the breakdown of hydrophobic contaminants [79]. In contaminated soil, bacteria can easily access hydrophobic organic pollutants linked to non-aqueous phase liquids and form biofilms on the surfaces they are related to. Chemotaxis aids in abiotic surface attachment, which increases biofilm growth and aids bacterial absorption of nutrients (contaminants). Several bacterial strains from different taxonomic groupings have shown chemotaxis of different contaminants, including explosives, nitroaromatic compounds (NACs), petroleum-coupled hydrocarbons, polycyclic aromatic hydrocarbons, and their associated metabolic intermediates [80].

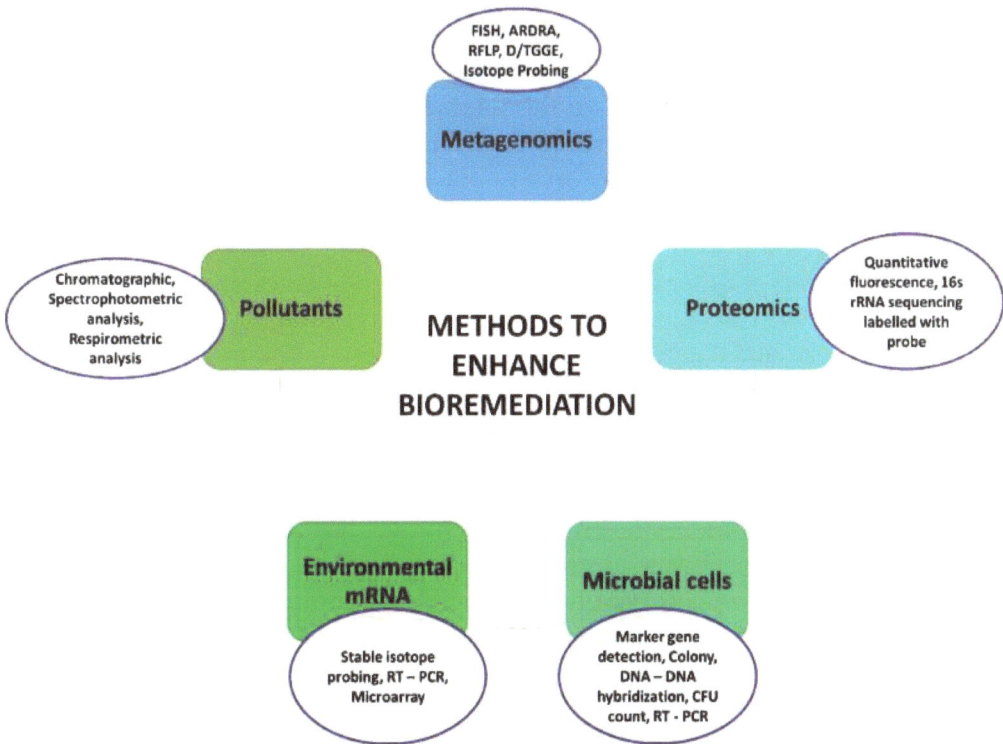

Fig. (5). Methods to enhance bioremediation processes.

Initially, bacterial chemotaxis was categorized according to the chemoattractant used in the process. Chemotaxis was then divided according to metabolism, further dividing chemotactic reactions into metabolism - (1) dependent and (2) independent. Furthermore, bacteria may travel towards electron acceptors/donors as part of metabolism-dependent chemotaxis. Investigating and knowing mechanisms can help enhance chemotactic behavior and create bioremediation technologies. Improving *in situ* bioremediation demands additional improvements in quantitative and qualitative tests. The molecular processes underlying the control of chemotactic responses must also be investigated. Transcriptomics and whole-genome sequencing studies have shown more genetic components related to chemotaxis than previously thought [81]. Phosphorylation and dephosphorylation have been shown in studies to help aid bacterial chemotaxis. Thus, allowing non-chemotactic strains to move by genetic alteration may aid in furthering our understanding of the regulatory mechanisms involved in chemotaxis and, as a result, improve the *in situ* bioremediation process.

Biosurfactants

Adaptations such as biofilm development or biosurfactant production can be exploited to improve *in situ* bioremediation. By overcoming the hydrophobic organic pollutants' bioaccessibility to prospective biodegrading bacteria, a surfactant application can improve the interaction between microorganisms and contaminants. The biodegradation of the chlorinated insecticide endosulfan has been discovered to be aided by biosurfactants produced by *Bacillus subtilis* MTCC1427. According to Odukkathil and Vasudevan [82], this sped up the slow microbial degradation of resistant endosulfan by as much as 30%. Additionally, the bioremediation of other contaminants, such as n-alkanes and PAHs, has been accelerated by biosurfactants. Microorganisms form a slimy coating under typical environmental circumstances due to diverse bacteria sticking to a substrate or matrix. A microbial film is a slimy covering that is revealed to be a helpful characteristic of bacteria in the environment. Microbial biofilms have already been proven advantageous in pathogenesis, with their vital role in bacterial survival despite a well-documented host immune response. Biofilm reactors improved the biodegradation kinetics of polychlorinated hydrocarbons and chlorinated aliphatic pollutants despite earlier research showing the importance of biofilm in promoting *in situ* bioremediation [83]. In order to optimize the *in situ* bioremediation process, more research on the influence of biofilm on microbial degradation is needed.

Genetically Engineered Microbes (GEMs)

Mixing genetic components from various species in one recipient strain to produce bacteria with higher degrading capacity may be a better alternative for bioremediation. To accomplish this, a number of aspects of designing compatible GEMs should be optimized, including creating new metabolic pathways, enlarging the range of existing pathways, preventing substrate diversion into harmful intermediates or ineffective routes, increasing substrate flux through pathways to prevent the accumulation of obstructing intermediates, boosting catabolic activities, raising the bioavailability of hydrophobic pollutants, and enhancing efficiency. The availability of a microbe's whole genome and multiple catabolic operons enables the production of GEMs by mixing genes and promoters, increasing their *in situ* potential. Numerous GEMs have been successfully generated and have proven *in vitro* bioremediation. Use of these GEMs *in situ* has minimal influence since it is associated with the risk of horizontal gene transfer, which results in uncontrolled diffusion throughout the environment [84]. As a result, GEMs may be cause for alarm. Scientists have attempted to prevent GEM survival in the environment for decades by establishing the restricted suicidal arrangement called 'bacterial confinement systems,' which is characterized by many catabolic killer genes [85].

Killer gene pairs are linked to several bacteriophages, bacterial chromosomes, and plasmids. The anti-killing method involves the production of two separate genes: the poison and a specialized antidote. The killer toxin is somewhat stable over time, but the anti-killing toxin is extremely unstable. Antidotes (antisense protein or RNA) prevent or counteract their corresponding toxins [86]. Different breakdown rates of the toxin and the antidote dictate the molecular process of toxin inactivation in cells without plasmid. Several lethal genes, including *hok, sac, colE3, Relf*, and *nuc*, have been related to the evolution of various confinement systems. Molin invented the first containment system [87]. Later, the researchers modified the suicidal system theory to create a 'fail-safe' mechanism for releasing GEMs into the atmosphere for successful bioremediation.

Additionally, utilizing GEMs in bioreactors is anticipated to be more effective because there is no rivalry with native microbes, and they are maintained under controlled temperature and growth conditions. To assess GEM's effectiveness and risk elements once implemented into natural ecosystems, field studies are also required. Biotechnologists should constantly take ethical considerations into account before using such bioremediation technologies. The organism should only be limited to the region of action, and its altered genetic make-up should be considered as a whole.

Omics Approaches

An effective bioremediation procedure demands the integration of several multidimensional elements that will allow us to understand better and anticipate the destiny of pollution. Metagenomics, transcriptomics, proteomics, and other molecular methods demonstrate their utility in *in-situ* bioremediation, increasing our understanding of precise processes. These methods aid in collecting data on biodegrading metabolites, proteins, and genes. Furthermore, developing next-generation sequencing (NGS) has paved the way for extensive proteomic, metagenomic, and bioinformatic research of eco-friendly bacteria, resulting in a clear understanding of the biodegradation processes.

The systematic examination of microbial populations at the genome level is known as metagenomics. A fundamental barrier to culture-dependent systems is the fact that more than 99% of microorganisms (dwelling in various natural environments) are either difficult to cultivate or uncultivable [88]. The bioremediation process is aided by this method's ability to examine a sample's full genome sequence (genomics) and large samples of genomes directly from the environment (metagenomics). Several research studies have been undertaken to explain how proteomics aids bioremediation. Furthermore, various unculturable microorganisms may be employed to generate metagenomics libraries that can subsequently be transferred to culturable bacteria to find desired degradative genes and their properties [89]. Metagenome identification and selection from contaminated environments are crucial in metagenomics. Metagenomic research produces more promising outcomes with a greater degradation ratio than prior bioremediation procedures.

Metagenomics is an effective method for examining the genetic makeup of bacteria that live in a variety of habitats, but it plays a very small part in revealing gene expression and activity. The improved metatranscriptomics and metaproteomics techniques make it possible to predict the practical performance of microbial consortia. Because they confirm gene activity within a particular environmental context, metatranscriptomics investigations are important in environmental restoration research. On the other hand, the combined effect of different omic-based techniques and environmental proteomics can produce far better findings. Proteomics is used in studies that add to the knowledge of adaption processes, emphasiing the extremophilic environment. Some proteomics investigations have indicated that it can significantly increase the bioremediation rate. Similar to this, cisdichloroethene (cDCE), a potential carcinogen, was found to upregulate gene expression in the JS666 strain of *Polaromonas* sp [90]. On the other hand, the transcriptome of the *Geobacter uraniireducens* strain, which was growing in uranium-polluted underground sediments, was discovered using microarray analysis of the entire genome [91].

Several other studies have examined the role and effectiveness of proteomics and metatranscriptomics in microbial-mediated bioremediation. Proteomics is frequently utilized to advance understanding of the physiology of microbial response to various temperatures, xenobiotics, and a variety of other stress-causing agents because it is expected to help research the discrepancy in the physiology of microorganisms during bioremediation. The usefulness of metagenomics has now been established by its considerable contribution to the evaluation of the functional range of microbial consortiums. Using metabolome-dependent approaches, it is feasible to build models that predict microbial behavior from environmental samples. Metabolomics is a relatively recent approach in omics research that includes examining the metabolite profiles of cells under controlled settings. When microbial cells are exposed to diverse environmental conditions, they release a variety of low molecular weight metabolites, which may be functionally quantified *via* metabolomic research. Through metagenomic techniques, we have gained a better understanding of the physiological potential of bacteria. Researchers used metabolomics to investigate the significance of aerobic processes in biodegradation, finding that the process involves over 4776 compounds [92]. That indicates the significant metabolic variability of the contaminated location. Several other studies have proved the application of metabolomics in investigating environmental pollutant biodegradation [93].

Nonetheless, detailed research that can illustrate the consequences of multiple distinct processes occurring concurrently in a microbiological system is required. That would aid in visualizing the performance of bacteria in a specific environmental situation. The improvement of *in silico* investigations is making these descriptions possible. Thus, bioinformatics may aid in the discovery of previously unknown microbial communities. Furthermore, some of the selected genes/transcripts and enzymes may be assessed *in vitro* to determine their optimal performance, which can then be employed to improve the potential of *in situ* bioremediation approaches.

FUTURE PROSPECTS

Bioleaching techniques are proliferating, and the endeavors are committed to the research center scale, with a few investigations concentrated on pilot-scale improvement. The presence of hydrocarbons in the bioreactors is another method of testing the current bioleaching activities. Inadvertently introducing these complexes into the production process can inhibit bacterial motility. Likewise, a few researchers utilize frontline innovation to comprehend the water system and bio-load colonization. Then again, research on bioleaching from polymetallic sources and uncommon components is additionally part of the scientific talk.

Research should focus on the bioleaching of silicate, carbonate, and oxide metals, even though basic mechanical improvements have been made in examining sulfide mineral interactions with microbes. A few lab-scale studies focused on these minerals were conducted, and a methodical study is urgently required to understand the fundamentals of microbial mineral interactions. Our mineral resources are becoming harder to extract and process, rarer, and of lower quality. It takes a lot of energy to coordinate the liberation of the mineral grains. By specifically concentrating on the exposed mineral grains, the naturally considerate process of bioleaching will play a significant part in making mineral extraction possible. A few private organizations have started conducting research to develop certain proteins for use in bioleaching. As these particular catalysts target particular minerals and solubilize them in the fluid condition from which the metal may be extracted, this may provide superior openings. Even if the information is available, very little of it has been disseminated, and the research pertaining to the advancement of bioleaching with chemicals is cloaked. The development of innovation toward these sources is normal for the time being. New procedures should target growing new catalysts that can enhance the associations of microorganisms with minerals while quickening the energy to beat the disadvantage of the bioleaching process. Mechanical development should advance in two ways: process and hardware improvement. Significant testing must be done to make bioleaching amiable to mineral extractions.

Though bioremediation is an environmentally friendly and cost-effective method of restoring degraded soil's physicochemical and biological properties [94], more research and ethical issues must be addressed before using improved and efficient GEMs. Furthermore, additional research is needed to establish the catabolic capacity of pollution-degrading microorganisms, individually and in combination. This will advance our understanding of the microbial consortia-mediated cleaning of polluted sites. Many conservative bioremediation techniques involve a sizeable amount of volatile hydrocarbons that do not decompose and can be volatilized to the atmosphere. As a result, standard procedures necessitate hydrocarbon and waste treatment, which also damages the volatile organic portion.

It is critical to enhance our knowledge of the interaction between microbial communities and the contaminated environment to fully use the potential of recognized and new species. The goal of future studies should be to combine all omics-based studies with computational methodologies, allowing researchers to establish a clear and complete understanding of microbial-mediated bioremediation processes. Without a doubt, the efficient use of bioremediation technologies will be made possible by the systematic deployment of microbial consortia with support from and comprehension of clearly characterized molecular and biochemical pathways. Although the applications of these techniques are still in their infancy, the development of omics approaches has led to the emergence of huge amounts of data that needs to be properly structured in a database. The successful monitoring of the target organism and the eradication of the pollutants will be aided by the application of omics approaches in the study of microbial molecular activity, leading to the treatment of hazardous pollutants.

CONCLUSION

Bioleaching is an innovation that is evolving quickly. It is utilized worldwide to recover metals from relinquished old mine squanders and expel noxious synthetic substances from old mine regions. Bioleaching can be considered the quickening of a characteristic ecologic process utilizing microbes that leach or discrete sulfide minerals from metal. However, the procedure is moderate, yet it has turned out to be practical, and as of now, organizations are utilizing bioleaching to extract copper from poor-grade minerals. Improvement is vital concerning both specialized and organic aspects. It incorporates expanding the filtering rate and the microorganisms' resilience to substantial metals. Hereditary change of bioleaching microscopic organisms, regardless of whether by transformation and choice or by hereditary designing, will bring results more rapidly than regular methods like screening and modification, and meanwhile, significant advancement has been made in the improvement of a hereditary framework for *T. ferrooxidans*. Furthermore, understanding the hazardous effects of various chemical pollutants has led to advancements in research efforts to develop remediation procedures that might be used to eradicate contaminants. Bioremediation is one such technology that is deemed safe, affordable, and ecologically friendly, and it eliminates contaminants by hastening the natural biodegradation process. As a result, an urgent need is to improve the understanding of microbial communities and their reactions to environmental contaminants to increase beneficial organisms' pollutant degrading capability. Without question, bioremediation is paving the way for a cleaner, healthier planet that would be more sustainable.

ACKNOWLEDGEMENTS

The authors would like to thank Tamil Nadu Agricultural University, Coimbatore, Tamil Nadu, India, for their support.

REFERENCES

[1]   Dunbar WS. Biotechnology and the mine of tomorrow. Trends Biotechnol 2017; 35(1): 79-89.
[http://dx.doi.org/10.1016/j.tibtech.2016.07.004] [PMID: 27612568]

[2]   Johnson DB, Bryan CG, Schlömann M, Roberto FF. Biomining technologies: extracting and recovering metals from ores and wastes. Springer 2022.

[3]   Banerjee I, Burrell B, Reed C, West AC, Banta S. Metals and minerals as a biotechnology feedstock: engineering biomining microbiology for bioenergy applications. Curr Opin Biotechnol 2017; 45: 144-55.
[http://dx.doi.org/10.1016/j.copbio.2017.03.009] [PMID: 28371651]

[4]   Senthil Kumar P, Yaashikaa PR, Baskar G. Biomining of natural resources. Waste Bioremediation 2018; pp. 313-42.

[5]   Gangola S, Joshi S, Kumar S, Pandey SC. Comparative analysis of fungal and bacterial enzymes in biodegradation of xenobiotic compounds Smart bioremediation Technol. Elsevier 2019; pp. 169-89.

[6]   Zhuang WQ, Fitts JP, Ajo-Franklin CM, Maes S, Alvarez-Cohen L, Hennebel T. Recovery of critical metals using biometallurgy. Curr Opin Biotechnol 2015; 33: 327-35.
[http://dx.doi.org/10.1016/j.copbio.2015.03.019] [PMID: 25912797]

[7]   Kaksonen AH, Mudunuru BM, Hackl R. The role of microorganisms in gold processing and recovery—A review. Hydrometallurgy 2014; 142: 70-83.

[http://dx.doi.org/10.1016/j.hydromet.2013.11.008]

[8]   Watling H. Review of biohydrometallurgical metals extraction from polymetallic mineral resources. Minerals (Basel) 2014; 5(1): 1-60.
[http://dx.doi.org/10.3390/min5010001]

[9]   Ma L, Wang X, Feng X, *et al.* Co-culture microorganisms with different initial proportions reveal the mechanism of chalcopyrite bioleaching coupling with microbial community succession. Bioresour Technol 2017; 223: 121-30.
[http://dx.doi.org/10.1016/j.biortech.2016.10.056] [PMID: 27788425]

[10]   Petersen J. Heap leaching as a key technology for recovery of values from low-grade ores – A brief overview. Hydrometallurgy 2016; 165: 206-12.
[http://dx.doi.org/10.1016/j.hydromet.2015.09.001]

[11]   Ghorbani Y, Franzidis J-P, Petersen J. Heap leaching technology—current state, innovations, and future directions: a review. Miner Process Extr Metall Rev 2016; 37: 73-119.

[12]   Vereecken J, Groudev SN, Groudeva VI. Recovery of copper by dump leaching with use of bacteria and cementation at the Vlaikov Vrah mine, Bulgaria. EMC'91 Non-Ferrous Metall Futur 1991:109–17.

[13]   O'Gorman G, von Michaelis H, Olson GJ. Novel *in-situ* metal and mineral extraction technology. US: Little Bear Laboratories, Inc., Golden, Colorado, USA, pp. 62.
[http://dx.doi.org/10.2172/835781]

[14]   Johnson DB, Quatrini R. Acidophile microbiology in space and time. Curr Issues Mol Biol 2020; 39: 63-76.
[http://dx.doi.org/10.21775/cimb.039.063] [PMID: 32083998]

[15]   Golyshina OV, Lünsdorf H, Kublanov IV, Goldenstein NI, Hinrichs KU, Golyshin PN. The novel extremely acidophilic, cell-wall-deficient archaeon *Cuniculiplasma divulgatum* gen. nov., sp. nov. represents a new family, Cuniculiplasmataceae fam. nov., of the order Thermoplasmatales. Int J Syst Evol Microbiol 2016; 66(1): 332-40.
[http://dx.doi.org/10.1099/ijsem.0.000725] [PMID: 26518885]

[16]   Itoh T, Miura T, Sakai HD, Kato S, Ohkuma M, Takashina T. *Sulfuracidifex tepidarius* gen. nov., sp. nov. and transfer of *Sulfolobus metallicus* Huber and Stetter 1992 to the genus *Sulfuracidifex* as *Sulfuracidifex metallicus* comb. nov. Int J Syst Evol Microbiol 2020; 70(3): 1837-42.
[http://dx.doi.org/10.1099/ijsem.0.003981] [PMID: 31958046]

[17]   Norris PR. Acidophile Diversity in Mineral Sulfide Oxidation. In: Rawlings DE, Johnson, DB (Eds.) Biomining. Berlin, Heidelberg: Springer 2007; pp. 199-216.

[18]   Hofmann M, Norris PR, Malik L, *et al. Metallosphaera javensis* sp. nov., a novel species of thermoacidophilic archaea, isolated from a volcanic area. Int J Syst Evol Microbiol 2022; 72(10): 5536.
[http://dx.doi.org/10.1099/ijsem.0.005536] [PMID: 36251741]

[19]   Hong Y, Valix M. Bioleaching of electronic waste using acidophilic sulfur oxidising bacteria. J Clean Prod 2014; 65: 465-72.
[http://dx.doi.org/10.1016/j.jclepro.2013.08.043]

[20]   Chen S, Yang Y, Liu C, Dong F, Liu B. Column bioleaching copper and its kinetics of waste printed circuit boards (WPCBs) by Acidithiobacillus ferrooxidans. Chemosphere 2015; 141: 162-8.
[http://dx.doi.org/10.1016/j.chemosphere.2015.06.082] [PMID: 26196406]

[21]   Cheng Y, Guo Z, Liu X, *et al.* The bioleaching feasibility for Pb/Zn smelting slag and community characteristics of indigenous moderate-thermophilic bacteria. Bioresour Technol 2009; 100(10): 2737-40.
[http://dx.doi.org/10.1016/j.biortech.2008.12.038] [PMID: 19171481]

[22]   Behrad Vakylabad A. A comparison of bioleaching ability of mesophilic and moderately thermophilic

culture on copper bioleaching from flotation concentrate and smelter dust. Int J Miner Process 2011; 101(1-4): 94-9.
[http://dx.doi.org/10.1016/j.minpro.2011.09.003]

[23]   Kim HJ, Lee S, Kim J, Mitchell RJ, Lee JH. Environmentally friendly pretreatment of plant biomass by planetary and attrition milling. Bioresour Technol 2013; 144: 50-6.
[http://dx.doi.org/10.1016/j.biortech.2013.06.090] [PMID: 23867527]

[24]   Roshani M, Shojaosadati SA, Safdari SJ, Vasheghani-Farahani E, Mirjalili K, Manafi Z. Bioleaching of molybdenum by two new thermophilic strains isolated and characterized. Iran J Chem Chem Eng 2017; 36: 183-94.

[25]   Amiri F, Mousavi SM, Yaghmaei S. Enhancement of bioleaching of a spent Ni/Mo hydroprocessing catalyst by Penicillium simplicissimum. Separ Purif Tech 2011; 80(3): 566-76.
[http://dx.doi.org/10.1016/j.seppur.2011.06.012]

[26]   Ilyas S, Chi R, Lee J. Fungal bioleaching of metals from mine tailing. Miner Process Extr Metall Rev 2013; 34(3): 185-94.
[http://dx.doi.org/10.1080/08827508.2011.623751]

[27]   Natarajan G, Ting YP. Pretreatment of e-waste and mutation of alkali-tolerant cyanogenic bacteria promote gold biorecovery. Bioresour Technol 2014; 152: 80-5.
[http://dx.doi.org/10.1016/j.biortech.2013.10.108] [PMID: 24291311]

[28]   Valix M. Bioleaching of electronic waste: milestones and challenges Curr Dev Biotechnol Bioeng. Elsevier 2017; pp. 407-42.
[http://dx.doi.org/10.1016/B978-0-444-63664-5.00018-6]

[29]   Kucera J, Lochman J, Bouchal P, *et al.* A model of aerobic and anaerobic metabolism of hydrogen in the extremophile Acidithiobacillus ferrooxidans. Front Microbiol 2020; 11: 610836.
[http://dx.doi.org/10.3389/fmicb.2020.610836] [PMID: 33329503]

[30]   Malik L, Hedrich S. Ferric iron reduction in extreme acidophiles. Front Microbiol 2022; 12: 818414.
[http://dx.doi.org/10.3389/fmicb.2021.818414] [PMID: 35095822]

[31]   Martínez-Bussenius C, Navarro CA, Jerez CA. Microbial copper resistance: importance in biohydrometallurgy. Microb Biotechnol 2017; 10(2): 279-95.
[http://dx.doi.org/10.1111/1751-7915.12450] [PMID: 27790868]

[32]   Orell A, Navarro CA, Rivero M, Aguilar JS, Jerez CA. Inorganic polyphosphates in extremophiles and their possible functions. Extremophiles 2012; 16(4): 573-83.
[http://dx.doi.org/10.1007/s00792-012-0457-9] [PMID: 22585316]

[33]   Glombitza F, Reichel S. Metal-containing residues from industry and in the environment: Geobiotechnological urban mining. Geobiotechnology I Met Issues 2014; pp. 49-107.

[34]   Roberto FF, Schippers A. Progress in bioleaching: part B, applications of microbial processes by the minerals industries. Appl Microbiol Biotechnol 2022; 106(18): 5913-28.
[http://dx.doi.org/10.1007/s00253-022-12085-9] [PMID: 36038754]

[35]   Dong B, Jia Y, Zhao H, *et al.* Evidence of weak interaction between ferric iron and extracellular polymeric substances of Acidithiobacillus ferrooxidans. Hydrometallurgy 2022; 209: 105817.
[http://dx.doi.org/10.1016/j.hydromet.2022.105817]

[36]   Schippers A. Biogeochemistry of metal sulfide oxidation in mining environments, sediments, and soils. Spec Pap Soc Am 2004; pp. 49-62.
[http://dx.doi.org/10.1130/0-8137-2379-5.49]

[37]   Sand W, Gehrke T. Analysis and function of the EPS from the strong acidophile Thiobacillus ferrooxidans. Microb Extracell Polym Subst Charact Struct Funct 1999; pp. 127-41.
[http://dx.doi.org/10.1007/978-3-642-60147-7_7]

[38]   Silverman MP. Mechanism of bacterial pyrite oxidation. J Bacteriol 1967; 94(4): 1046-51.

[http://dx.doi.org/10.1128/jb.94.4.1046-1051.1967] [PMID: 6051342]

[39]   Jin J, Liu GL, Shi SY, Cong W. Studies on the performance of a rotating drum bioreactor for bioleaching processes — Oxygen transfer, solids distribution and power consumption. Hydrometallurgy 2010; 103(1-4): 30-4.
[http://dx.doi.org/10.1016/j.hydromet.2010.02.013]

[40]   Chandraprabha MN, Modak JM, Natarajan KA. Effect of LPS removal on electrophoretic softness of Acidithiobacillus ferrooxidans cells. Adv. Mater. Res., vol. 20, Trans Tech Publ; 2007, p. 341–4.
[http://dx.doi.org/10.4028/0-87849-452-9.341]

[41]   Li Y, Jia R, Al-Mahamedh HH, Xu D, Gu T. Enhanced biocide mitigation of field biofilm consortia by a mixture of D-amino acids. Front Microbiol 2016; 7: 896.
[http://dx.doi.org/10.3389/fmicb.2016.00896] [PMID: 27379039]

[42]   Natarajan KA, Iwasaki I. Role of galvanic interactions in the bioleaching of Duluth gabbro copper-nickel sulfides. Sep Sci Technol 1983; 18(12-13): 1095-111.
[http://dx.doi.org/10.1080/01496398308059919]

[43]   Mishra D, Kim DJ, Ahn JG, Rhee YH. Bioleaching: A microbial process of metal recovery; A review. Met Mater Int 2005; 11(3): 249-56.
[http://dx.doi.org/10.1007/BF03027450]

[44]   Schippers A, Sand W. Bacterial leaching of metal sulfides proceeds by two indirect mechanisms *via* thiosulfate or *via* polysulfides and sulfur. Appl Environ Microbiol 1999; 65(1): 319-21.
[http://dx.doi.org/10.1128/AEM.65.1.319-321.1999] [PMID: 9872800]

[45]   Bellenberg S, Díaz M, Noël N, *et al.* Biofilm formation, communication and interactions of leaching bacteria during colonization of pyrite and sulfur surfaces. Res Microbiol 2014; 165(9): 773-81.
[http://dx.doi.org/10.1016/j.resmic.2014.08.006] [PMID: 25172572]

[46]   Sampson MI, Phillips CV, Blake RC II. Influence of the attachment of acidophilic bacteria during the oxidation of mineral sulfides. Miner Eng 2000; 13(4): 373-89.
[http://dx.doi.org/10.1016/S0892-6875(00)00020-0]

[47]   Chen G, Yang H, Li H, Tong L. Recovery of cobalt as cobalt oxalate from cobalt tailings using moderately thermophilic bioleaching technology and selective sequential extraction. Minerals (Basel) 2016; 6(3): 67.
[http://dx.doi.org/10.3390/min6030067]

[48]   Groudev S, Spasova I, Nicolova M, Georgiev P. *In situ* bioremediation of contaminated soils in uranium deposits. Hydrometallurgy 2010; 104(3-4): 518-23.
[http://dx.doi.org/10.1016/j.hydromet.2010.02.027]

[49]   Lorenzo-Tallafigo J, Iglesias-González N, Romero-García A, *et al.* The reprocessing of hydrometallurgical sulphidic tailings by bioleaching: The extraction of metals and the use of biogenic liquors. Miner Eng 2022; 176: 107343.
[http://dx.doi.org/10.1016/j.mineng.2021.107343]

[50]   Hosseini Nasab M, Noaparast M, Abdollahi H, Amoozegar MA. Kinetics of two-step bioleaching of Ni and Co from iron rich-laterite using supernatant metabolites produced by Salinivibrio kushneri as halophilic bacterium. Hydrometallurgy 2020; 195: 105387.
[http://dx.doi.org/10.1016/j.hydromet.2020.105387]

[51]   Martin M, Janneck E, Kermer R, Patzig A, Reichel S. Recovery of indium from sphalerite ore and flotation tailings by bioleaching and subsequent precipitation processes. Miner Eng 2015; 75: 94-9.
[http://dx.doi.org/10.1016/j.mineng.2014.11.015]

[52]   Drobe M, Haubrich F, Gajardo M, Marbler H. Processing tests, adjusted cost models and the economies of reprocessing copper mine tailings in Chile. Metals (Basel) 2021; 11(1): 103.
[http://dx.doi.org/10.3390/met11010103]

[53]   Morin DHR, d'Hugues P. Bioleaching of a cobalt-containing pyrite in stirred reactors: a case study

from laboratory scale to industrial application. Biomining 2007; pp. 35-55.

[54]   Faramarzi MA, Mogharabi-Manzari M, Brandl H. Bioleaching of metals from wastes and low-grade sources by HCN-forming microorganisms. Hydrometallurgy 2020; 191: 105228.
[http://dx.doi.org/10.1016/j.hydromet.2019.105228]

[55]   de Freitas GR, da Silva MGC, Vieira MGA. Biosorption technology for removal of toxic metals: a review of commercial biosorbents and patents. Environ Sci Pollut Res Int 2019; 26(19): 19097-118.
[http://dx.doi.org/10.1007/s11356-019-05330-8] [PMID: 31104247]

[56]   Reichel S, Janneck E, Burghardt D, *et al.* Microbial production of schwertmannite: development from microbial fundamentals to marketable products. Solid State Phenom., vol. 262, Trans Tech Publ; 2017, p. 568–72.
[http://dx.doi.org/10.4028/www.scientific.net/SSP.262.568]

[57]   Kaksonen AH, Deng X, Bohu T, *et al.* Prospective directions for biohydrometallurgy. Hydrometallurgy 2020; 195: 105376.
[http://dx.doi.org/10.1016/j.hydromet.2020.105376]

[58]   Saitoh N, Fujimori R, Nakatani M, Yoshihara D, Nomura T, Konishi Y. Microbial recovery of gold from neutral and acidic solutions by the baker's yeast Saccharomyces cerevisiae. Hydrometallurgy 2018; 181: 29-34.
[http://dx.doi.org/10.1016/j.hydromet.2018.08.011]

[59]   Nancharaiah YV, Venkata Mohan S, Lens PNL. Metals removal and recovery in bioelectrochemical systems: A review. Bioresour Technol 2015; 195: 102-14.
[http://dx.doi.org/10.1016/j.biortech.2015.06.058] [PMID: 26116446]

[60]   Alexander M. Biodegradation and bioremediation. San Diego, CA, USA: Academic Press 1999.

[61]   Prescott LM, Harley JP, Klein DA. Isolation of pure cultures Microbiol. 5th ed. New York: McGraw-Hill Companies 2002; pp. 106-10.

[62]   Sardrood BP, Goltapeh EM, Varma A. An introduction to bioremediation Fungi as bioremediators. Springer 2012; pp. 3-27.

[63]   Kumar A, Bisht BS, Joshi VD, Dhewa T. Review on bioremediation of polluted environment: a management tool. Int J Environ Sci 2011; 1: 1079-93.

[64]   Jitnuyanont P, Sayavedra-Soto LA, Semprini L. Bioaugmentation of butane-utilizing microorganisms to promote cometabolism of 1,1,1-trichloroethane in groundwater microcosms. Biodegradation 2001; 12(1): 11-22.
[http://dx.doi.org/10.1023/A:1011933731496] [PMID: 11693291]

[65]   Adamson DT, McDade JM, Hughes JB. Inoculation of a DNAPL source zone to initiate reductive dechlorination of PCE. Environ Sci Technol 2003; 37(11): 2525-33.
[http://dx.doi.org/10.1021/es020236y] [PMID: 12831039]

[66]   Cybulski Z, Dziurla E, Kaczorek E, Olszanowski A. The influence of emulsifiers on hydrocarbon biodegradation by Pseudomonadacea and Bacillacea strains. Spill Sci Technol Bull 2003; 8(5-6): 503-7.
[http://dx.doi.org/10.1016/S1353-2561(03)00068-9]

[67]   Olawale AK, Akintobi OA. Biodegradation of glyphosate pesticide by bacteria isolated from agricultural soil. Rep Opinion 2011; 3: 124-8.

[68]   El-Bestawy E, Sabir J, Mansy AH, Zabermawi N. Comparison among the efficiency of different bioremediation technologies of Atrazine-contaminated soils. J Bioremediat Biodegrad 2014; 5: 237.

[69]   Infante J C, De Arco RD, Angulo ME. Removal of lead, mercury and nickel using the yeast Saccharomyces cerevisiae. Rev Mvz Cordoba 2014; 19(2): 4141-9.
[http://dx.doi.org/10.21897/rmvz.107]

[70]   Tigini V, Prigione V, Giansanti P, Mangiavillano A, Pannocchia A, Varese GC. Fungal biosorption, an

innovative treatment for the decolourisation and detoxification of textile effluents. Water 2010; 2(3): 550-65.
[http://dx.doi.org/10.3390/w2030550]

[71]   Peña-Montenegro TD, Lozano L, Dussán J. Genome sequence and description of the mosquitocidal and heavy metal tolerant strain Lysinibacillus sphaericus CBAM5. Stand Genomic Sci 2015; 10(1): 2.
[http://dx.doi.org/10.1186/1944-3277-10-2] [PMID: 25685257]

[72]   Mohammadian Fazli M, Soleimani N, Mehrasbi M, Darabian S, Mohammadi J, Ramazani A. Highly cadmium tolerant fungi: their tolerance and removal potential. J Environ Health Sci Eng 2015; 13(1): 19.
[http://dx.doi.org/10.1186/s40201-015-0176-0] [PMID: 25806110]

[73]   Sinha A, Sinha R, Khare SK. Heavy metal bioremediation and nanoparticle synthesis by metallophiles. Geomicrobiol Biogeochem 2014; pp. 101-18.
[http://dx.doi.org/10.1007/978-3-642-41837-2_6]

[74]   Pande V, Pandey SC, Joshi T, Sati D, Gangola S, Kumar S, *et al.* Biodegradation of toxic dyes: a comparative study of enzyme action in a microbial system Smart bioremediation Technol. Elsevier 2019; pp. 255-87.

[75]   Singh SP, Tiwari G. Application of bioremediation on solid waste management: a review. J Bioremed Biodegr 2014; p. 5.

[76]   Shinde S. Bioremediation. An overview. Recent Res Sci Technol 2013; 5: 67-72.

[77]   Lima D, Viana P, André S, *et al.* Evaluating a bioremediation tool for atrazine contaminated soils in open soil microcosms: The effectiveness of bioaugmentation and biostimulation approaches. Chemosphere 2009; 74(2): 187-92.
[http://dx.doi.org/10.1016/j.chemosphere.2008.09.083] [PMID: 19004466]

[78]   Malik A. Environmental Microbiology: Bioremediation. Dipetik Novemb 2006; 26: 2015.

[79]   Law AMJ, Aitken MD. Bacterial chemotaxis to naphthalene desorbing from a nonaqueous liquid. Appl Environ Microbiol 2003; 69(10): 5968-73.
[http://dx.doi.org/10.1128/AEM.69.10.5968-5973.2003] [PMID: 14532051]

[80]   Gordillo F, Chávez FP, Jerez CA. Motility and chemotaxis of Pseudomonas sp. B4 towards polychlorobiphenyls and chlorobenzoates. FEMS Microbiol Ecol 2007; 60(2): 322-8.
[http://dx.doi.org/10.1111/j.1574-6941.2007.00293.x] [PMID: 17374130]

[81]   Lange C, Zaigler A, Hammelmann M, *et al.* Genome-wide analysis of growth phase-dependent translational and transcriptional regulation in halophilic archaea. BMC Genomics 2007; 8(1): 415.
[http://dx.doi.org/10.1186/1471-2164-8-415] [PMID: 17997854]

[82]   Odukkathil G, Vasudevan N. Enhanced biodegradation of endosulfan and its major metabolite endosulfate by a biosurfactant producing bacterium. J Environ Sci Health B 2013; 48(6): 462-9.
[http://dx.doi.org/10.1080/03601234.2013.761873] [PMID: 23452211]

[83]   Arora PK, Sasikala C, Ramana CV. Degradation of chlorinated nitroaromatic compounds. Appl Microbiol Biotechnol 2012; 93(6): 2265-77.
[http://dx.doi.org/10.1007/s00253-012-3927-1] [PMID: 22331236]

[84]   Naik MG, Duraphe MD. Review paper on-Parameters affecting bioremediation. Adv Res Pharm Biol 2012; p. 2.

[85]   Torres B, Jaenecke S, Timmis KN, García JL, Díaz E. A dual lethal system to enhance containment of recombinant micro-organisms. Microbiology (Reading) 2003; 149(12): 3595-601.
[http://dx.doi.org/10.1099/mic.0.26618-0] [PMID: 14663091]

[86]   Petersen J. Phylogeny and compatibility: plasmid classification in the genomics era. Arch Microbiol 2011; 193(5): 313-21.
[http://dx.doi.org/10.1007/s00203-011-0686-9] [PMID: 21374058]

[87] Molin S, Klemm P, Poulsen LK, Biehl H, Gerdes K, Andersson P. Conditional suicide system for containment of bacteria and plasmids. Bio/Technology 1987; 5: 1315-8.

[88] Bursle E, Robson J. Non-culture methods for detecting infection. Aust Prescr 2016; 39(5): 171-5.
[http://dx.doi.org/10.18773/austprescr.2016.059] [PMID: 27789929]

[89] Jaiswal S, Singh DK, Shukla P. Gene editing and systems biology tools for pesticide bioremediation: a review. Front Microbiol 2019; 10: 87.
[http://dx.doi.org/10.3389/fmicb.2019.00087] [PMID: 30853940]

[90] Jennings LK, Chartrand MMG, Lacrampe-Couloume G, Lollar BS, Spain JC, Gossett JM. Proteomic and transcriptomic analyses reveal genes upregulated by cis-dichloroethene in Polaromonas sp. strain JS666. Appl Environ Microbiol 2009; 75(11): 3733-44.
[http://dx.doi.org/10.1128/AEM.00031-09] [PMID: 19363075]

[91] Holmes DE, O'Neil RA, Chavan MA, *et al.* Transcriptome of *Geobacter uraniireducens* growing in uranium-contaminated subsurface sediments. ISME J 2009; 3(2): 216-30.
[http://dx.doi.org/10.1038/ismej.2008.89] [PMID: 18843300]

[92] Dong X, Greening C, Rattray JE, *et al.* Metabolic potential of uncultured bacteria and archaea associated with petroleum seepage in deep-sea sediments. Nat Commun 2019; 10(1): 1816.
[http://dx.doi.org/10.1038/s41467-019-09747-0] [PMID: 31000700]

[93] Brune KD, Bayer TS. Engineering microbial consortia to enhance biomining and bioremediation. Front Microbiol 2012; 3: 203.
[http://dx.doi.org/10.3389/fmicb.2012.00203] [PMID: 22679443]

[94] Maestroni B, Cannavan A. Integrated analytical approaches for pesticide management. Academic Press, London, UK 2018.

Role of Microbes in the Production of Renewable Energy

Role of Microbes and Microbiomes in Microbial Fuel Cells: A Novel Tool for a Clean and Green Environment

Sagia Sajish[1,*], **Karthika Ponnusamy**[2] and **B.N. Brunda**[1]

¹ Division of Microbiology, Indian Agricultural Research Institute, New Delhi, India

² Department of Microbiology, College of Basic Science & Humanities, Chaudhary Charan Singh Haryana Agricultural University, Haryana, India

Abstract: Over the recent decades, there has been a tremendous need to develop alternative, sustainable, clean, and renewable energy resources. This demand is attributed to the exhaustion of fossil fuel reserves and the associated economic risks, the impact of fossil fuel use on the environment, and the associated global warming. Bioelectrochemical systems (BES), which use biological entities to generate electricity, are promising alternative clean renewable energy. Microbial fuel cell (MFC), a type of BES, exploits the potential of electro-active microorganisms for extracellular electron transfer to generate electricity. In an MFC, microbes oxidize the organic substrates fed into the anode chamber into electrons, protons, and CO_2. The electrons flow through the connected external load/circuit towards the cathode, creating the potential difference across the electrode and subsequent current output. A terminal electron acceptor at the cathode accepts the electrons and protons. In addition to electricity generation, MFC has extended applications in wastewater treatment, heavy metal remediation, bioremediation of environmental pollutants, biosensors for monitoring the environment, *etc*. This chapter will help understand the basic principle of an MFC and the role of microbes in a microbial fuel cell, genetic engineering, biofilm engineering approaches, and electrode engineering approaches for increasing the overall efficiency of an MFC for its practical implementation.

Keywords: Bioelectrochemical systems, bioremediation, biosensors, climate change, electroactive microorganisms, microbial fuel cells, wastewater treatment.

INTRODUCTION

The depletion of fossil fuel reserves is considered to be the major driver of unsustainability, which puts us and our future generations at risk of environmental

---
* **Corresponding author Sagia Sajish:** Division of Microbiology, Indian Agricultural Research Institute, New Delhi, India; E-mail: sagiagri001@gmail.com

Govindaraj Kamalam Dinesh, Shiv Prasad, Ramesh Poornima, Sangilidurai Karthika, Murugaiyan Sinduja & Velusamy Sathya (Eds.)

and economic security. However, extraction and use of fossil fuels for energy pose environmental risks with higher greenhouse gas emissions and associated global warming. Therefore, the global energy crisis, exhaustion of fossil fuel reserves, global warming, and the associated climate change have necessitated the search for alternative, sustainable, clean, and renewable energy sources [1]. The quest for such a clean and renewable energy source has been augmented in recent decades with promising results. Research and developments have been made in renewable energy sources like solar energy, wind power, geothermal energy, conversion of biomass to energy, *etc.*

According to the Renewables 2022 report by the International Energy Agency (IEA), global renewable energy capacity is estimated to surge by 70% (2400 GW) between the 2022 and 2027 forecast period. This is an 85% increase from the last five years' rate, which is mainly attributed to fossil fuel reserves depletion and the global energy crisis, making renewable energy resources an economically/environmentally viable energy resource [2]. However, no such individual renewable energy can compete and replace the use of fossil fuels currently; the combination of these sources is an alternative area to be investigated [3]. One such promising technology is the bioelectrochemical system. The bioelectrochemical system combines biochemical pathways (biological metabolism) and electrochemical techniques to generate electricity.

Bioelectrochemical systems (BES) are of two types: Microbial fuel cells (MFC) and microbial electrolysis cells (MEC). In microbial fuel cells, electroactive microorganisms convert chemical energy stored in organic/inorganic substrates into electrical energy. In Microbial Electrolysis Cell, biomass produces hydrogen using an applied external potential. A classic MFC comprises an anode and a cathode isolated by a proton exchange membrane, which prevents oxygen diffusion from the cathode to the anode, allowing only protons (H^+) to pass through it. In an anodic chamber, the microbial consortia oxidize the substrates (e.g., Glucose) [4]. Microbial fuel cells from various sources can be used with complex substrates like lignocellulosic biomass [5] and wastewater [6].

Thus, MFC is a reliable technology for waste to electricity production and subsequent decrease in the total amount of carbon dioxide liberated into the atmosphere compared to fossil fuels. Apart from bioelectricity generation, microbial fuel cells are also used in wastewater treatment, bioremediation of pollutants, recovery of heavy metals, desalination process, as a biosensor, *etc.* The ability to use various substrates and ambient operating conditions makes it more promising than conventional wastewater treatment methods and bioremediation. This chapter signifies the role of microbial fuel cells as a clean, renewable, and sustainable energy source. Special emphasis has been given to the role of electroactive microorganisms in MFC and strategies for efficient biofilm development on the anode surface and future perspective.

MICROBIAL FUEL CELL - HISTORY AND FUNDAMENTALS

Luigi Galvani, who coined the term animal electricity by first discovering the movement of a dead frog's muscles upon the strike by an electrical pulse, is considered the first electrochemist [7]. The first successful fuel cell with a 5KW system (hydrogen-oxygen fuel cell) was developed in 1959 by Francis Bacon [8]. Since then, diverse types of fuel cells have been developed and are categorized according to the electrolyte used. Subsequently, all these developments paved the way for the development of Biological fuel cells (BFs), wherein biochemical pathways are used to perform redox reactions for electricity generation. Biological fuel cells are classified into two types- enzymatic fuel cells (EFCs) [9] and microbial fuel cells (MFCs) [4]. Enzymes are used in EFCs, and live microorganisms are used in MFCs. The electrochemical catalytic efficiency of enzymes is superior to that of microbes but is unstable and less durable in contrast to living microbes. The use of microorganisms (*Saccharomyces cerevisiae*) for the generation of electricity was first performed at the beginning of the twentieth century (1911) by M.C. Potter [10]. It was in 1931 when Branet Cohen constructed microbial fuel cells that produce 35 volts and 2mA current when connected in series [11]. The practical application of MFC was demonstrated by using a benthic MFC to power a meteorological buoy for remote monitoring [12].

Configuration of MFC

A typical microbial fuel cell (MFC) consists of an anodic and cathode chamber partitioned by a proton/cation exchange membrane. On the anode of MFC, the proliferating microorganisms use their metabolic pathways to tap the chemical energy stored in the organic substrates supplied in the anode chamber. The oxidation of organic matter in the anode generates electron proton and carbon dioxide. The electron produced in the anode flows through an external circuit connected to a resistor or a load towards the cathode, generating a potential difference across the electrodes. As an electron moves towards the cathode, by convention, current flows from the positive towards the negative terminal. A proton moves across the proton exchange membrane towards the cathode for each electron that moves through the circuit. Oxygen is commonly used as an oxidant in the cathode of MFC, which serves as the terminal electron acceptor for the incoming protons and electrons from the anode generating water. Metal oxidants like chromium, cadmium, and copper can also accept electrons [4].

Typically, MFCs operate as close systems wherein the anode is completely maintained in an anaerobic condition. Such a condition is made to sustain in the anode so as to enable the efficient proliferation of anaerobic exo-electrogenic bacteria that can transfer the electrons from the substrate to the anode. Various substrates like simple sugars, lignocellulosic biomass [5], wastewaters [6], activated sludge [13], and soil sediments [14] are found to be used as substrates in anode chamber [15]. The Proton exchange membrane (PEM) used in a typical MFC allows the passage of protons across to the cathode chamber. The best PEM should have high conductivity to cations, selective permeability for protons, durability, no detrimental effect on microbes, and low internal resistance [16]. In addition, an MFC can generate energy that can be theoretically estimated from the potential difference between the electron donor and electron acceptor redox reactions [17].

Double chamber MFC: This is the simplest MFC design in which one bottle, cube, or chamber is made as an anode and another bottle as a cathode partitioned by a proton/cation exchange membrane (Fig. **1**). Oxidation reaction occurs in the anode, and reduction reaction takes place at the cathode [18].

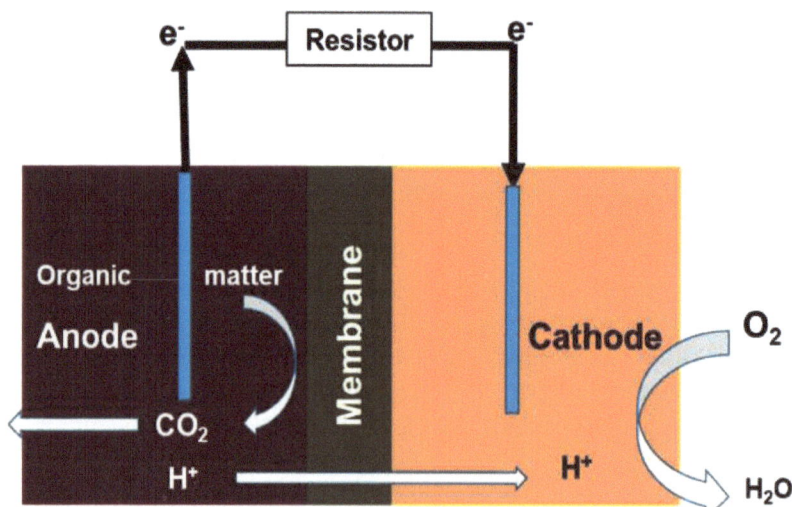

Fig. (1). Schematic representation of dual chamber MFC.

Single chamber MFC: This model of single chamber MFC was introduced first by Doo Hyun Park and J. Gregory Zeikus [19]. It is made of a single chamber where anode and cathode electrodes will be separated by a proton exchange membrane (PEM). The space between the anode and cathode gets reduced compared to dual chamber MFC, thus reducing the internal ohmic resistance and subsequently increasing the current output. Here, the cathode is exposed to air, thus eliminating the need for a catholyte [20].

Upflow MFC: This is a cylindrical-shaped fuel cell with an anode chamber at the bottom and a cathode chamber at the top with a layer of glass wool or glass beads between the chambers. No separate catholyte and anolyte are used in this system. The substrate provided at the bottom moves upward and leaves at the top from the cathode. This technology is amenable to scaling up and appropriate for wastewater treatment plants. However, the energy required to pump the substrates upwards in this model is a major drawback [21].

Stacked MFC: The stacked MFC model is the combination of several microbial fuel cells that are connected in parallel or series. The primary purpose of this model is to increase the overall power output of MFCs. However, the final voltage and power will not precisely be the total voltage of the individual MFCs connected together. The total current produced is observed to be more when connected in parallel than in series [22].

BIO-ELECTROCHEMICALLY ACTIVE MICROORGANISMS

Microorganisms with the ability of extracellular electron transfer (EET) are known as exoelectrogens or bio-electrochemically active microorganisms [23]. Most microorganisms depend on the oxidation of organic matter for their energy needs by means of respiration. Electrons derived from the organic matter in cellular respiration are carried by the electron carriers in the electron transport chain (ETC), transporting protons across the membrane, thus creating a proton motive force. This proton motive force (PMF) is used for ATP synthesis, and the electron is accepted by a terminal electron acceptor like oxygen in case of aerobic respiration or other soluble substances like iron and manganese oxides in anaerobic respiration [24].

In the case of exoelectrogens, electron obtained from the oxidation of organic matter is passed on to an external terminal electron acceptor outside the cell. In the case of a microbial fuel cell, the anode acts as the terminal electron acceptor. This potential of exoelectrogens is used in microbial fuel cells for bio-electricity generation, and their proliferation inside the MFC anodic chamber is considered the primary determinant factor in bioelectricity generation. Anaerobic sludge from industrial and municipal wastewater was found to be a rich source of exo-electrogenic bacteria. These exo-electrogens are primarily bacteria and evolved in conditions like anaerobic environments wherein metal oxides are used as terminal electron acceptors [17].

Depending on the type of anaerobic respiration employed by these bacteria, they are categorized as metal-reducing bacteria like *Geobacter, Shewanella* and *Geothrix*, sulfate-reducing bacteria like *Desulfuromonas*, nitrate-reducing bacteria such as *Pseudomonas* and *Ochrobactrum*, fermentative bacteria including

Clostridium and *E.coli*, purple sulfur and purple non-sulfur photosynthetic bacteria, *etc* [17]. Elucidation of the metabolic activities and the mechanism of electron transport helps scale up overall development of MFC technology.

Mechanism of Electron Transfer in MFC

Exo-electrogenic bacteria are found to use one or more than one of the following three types of mechanisms to transfer electrons to the anode [25], (i) Direct contact using outer membrane cytochromes and nanowires, (ii) Chemical redox mediators or electron shuttles and (iii) Conductive biofilms

Direct electron transfer (DET) is mediated by the outer membrane cytochrome (OMC) proteins present on the bacterial surface. These cytochromes are heme-containing proteins located on the bacteria's outer cellular membrane and are part of the electron transport chain (ETC) [25]. DET was first discovered in *Shewanella,* which was later named a cable bacterium [26]. Other genera that are capable of DET include *Geobacter, Geothrix, Rhodoferax,* and *Geoalkalibacter.* DET, however, requires physical contact between the cytochromes and the MFC anode. DET can also occur through microbial nanowires, conductive appendages that transfer an electron from the bacterial surface to the anode. Nanowires are nothing but pili appendages and are present on the outer membrane of bacteria [25]. It was seen that nanowire-mediated electron transfer also occurs cell by cell, covering a distance of almost 1 cm [27]. These appendages also help in the formation of thick biofilm on the anode surface.

Electron shuttles are mediator compounds that help to transfer the electrons from the bacterial surface to the anode. These mediators can be externally supplied (exogenous mediators) or synthesized by the bacteria itself and secreted outside (endogenous mediators). Synthetic exogenous redox mediators were first used by the researchers Allen and Bennetto [28]. The attributes of the best exogenous mediators are solubility, stability, reusability, environmentally friendly, and appropriate redox potential to shuttle the electron between bacteria and anode.

Some of the synthetic compounds used as mediators include methylene blue, indophenol, 2,6-dichlorophenol, thionin, benzyl viologen, neutral red, and safranine-O [29]. In addition, some of the exoelectrogenic bacteria secrete Flavin and phenazine-derived compounds into the anodic media as self or endogenous mediators [25]. Two exclusively studied bacteria capable of secreting self-mediators are *Pseudomonas aeruginosa* and *Shewanella oneidensis.* Most of the exogenous mediators are toxic and highly expensive. Therefore, exploring electroactive microorganisms capable of producing endogenous mediators will help develop MFC as a sustainable technology.

Exoelectrogens with the ability to form nanowires and Exo Polymeric Substances (EPS) develop as thick conductive biofilms on the anode surface [25, 30]. Biofilm (with high conductivity) formation is found to enhance the power density of the MFC. Electrons are passed between the adjacent cells in the network and ultimately reach the anode. In the case of mixed cultures, direct interspecies electron transfer (DIET) has also been observed [31].

ELECTROACTIVE MICROBIAL GENERA IN MICROBIAL FUEL CELLS

Microbial fuel cells (MFCs) are bio-electrochemical systems consisting of electrochemically active microbes and a biocatalyst mimicking the bacterial interaction for treating wastewater and generation of electricity through the oxidation of organics [4, 15, 32, 33]. Bacteria, Archaea, and Eukarya are the three domains that constitute all life. Yeasts (Eukarya) can be isolated from highly diversified MFCs, but bacteria and archaea predominate (Table **1**) [34].

Table 1. Electrochemically active microbes.

Sl. No.	Microbes	Substrate	References
1.	*Bacillus*	Glucose	[38]
2.	*Enterobacter*	Wastewater	[39]
3.	*Geobacter*	Glucose	[40]
4.	*Proteus*	Sucrose, Glucose	[41]
5.	*Shewanella*	Lactase, pyruvate, acetate, glucose	[42]
6.	*Proteus mirabilis*	Glucose	[43, 44]
7.	*Erwinia dissolved*	Glucose	[44]
8.	*Aeromonas hydrophila*	Acetate	[44, 45]
9.	*Geobacter metallireducens*	Acetate	[43, 44]
10.	*Shewanella putrefacien*	Lactase, pyruvate, acetate, glucose	[44]
11.	*S. oneidensis MR-1*	Lactase	[44, 46]
12.	*G. sulfurreducens*	Acetate	[44]
13.	*Rhodoferax ferrireducens*	Glucose	[44, 47]
14.	*Lactobacillus plantarum*	Glucose	[44]
15.	*Klebsiella pneumoniae*	Glucose	[48]
16.	*Gammaproteo* and *shewanellaaffinis* (KMM3586)	Cyctenin	[49, 50]
17.	*Escherichia coli*	Sewage sludge, Glucose	[51 - 53]
18.	*Deltaproteo bacterium*	Marine sediment	[49, 50]

(Table 1) cont.....

Sl. No.	Microbes	Substrate	References
19.	*Acidithiobacillus ferrooxidans*	Potato wastewater	[54]
20.	*Geothrix fermentans*	Lactate	[55]
21.	*Klebsiella pneumoniae strain Gut-S2, Enterococcus avium strain Gut-S*	Acetate or lactate	[56]
22.	*Pseudomonas aeruginosa RA5, Micrococcus luteus RA2, Diaphorobacter oryzae* RA3, *Cloacibacterium normanense RA1*	Cellulose	[57]
23.	*Aeromonas jandaei SCS5*	Acetate	[58]
24.	*Raoultella electrica*	Glucose	[59]
25.	*Tolumonas osonensis OCF 7*	Glucose	[60]
26.	*Citrobacter sp. SX-1*	Acetate	[61]
27.	*Comamonas denitrificans DX-4*	Acetate	[62]
28.	*Brevibacteria sp. (2 strains)*	Glucose/wastewater organics (EDs)	[63]
29.	*Ochrobactrum anthropi YZ-1*	Acetate	[64]
30.	*Pseudomonas otitidis* AATB4	Municipal wastewater	[65]
31.	*Spirulina*	Wastewater	[66]
32.	*Saccharomyces sp.*	Glucose	[67]
33.	*Pseudomonas methanica*	Methane	[68]
34.	*Clostridium butyricum*	Glucose, Starch	[69]
35.	*Acidithiobacillus* and *Ferroplasma*	Mining wastewater	[70]
36.	*Pseudomonas* and *Klebsiella*	Palm oil effluent	[71]
37.	*Bacillus circulans*	Chitin biomass	[72]
38.	*Paenibacillus lautus*	Glucose	[73]
39.	*Desulfuromonas acetoxidans*	Sodium acetate	[74]

Mixed culture MFCs provide a more steady electricity supply for a long-time than pure microbial culture [35]. This is due to the synergistic interaction between the microbial populations, where they evolve cooperatively to escape pollutant toxicity rather than competing with each other [36]. For example, *Shewanella* and *Geobacter* are well-investigated genera because they produce more electricity and generate hydrogen in MFC systems and MFCs, respectively. Therefore, they are reactively, biologically stable, and economically [37]. But apart from that, many other microbial genera are used in the microbial fuel cell.

FACTORS AFFECTING THE DEVELOPMENT OF ANODE BIOFILM

The electrodes (anode and cathode) are separated from each other by a proton/cation exchange membrane. The electrochemically active/ exo-electrogenic microbes develop on the anode, whereas the cathode is abiotic. The anode is where limiting mechanisms for electrochemical reactions operate [75]. Organic material degradation yields electrons, which flow to the cathode side of the circuit, where the microbes serve as the biocatalyst. An electrical current is generated by this electron transport from the anode across the membrane to the cathode [76]. Among the many variables influencing the performance of the MFC, the following seem to be the most significant:

 i. Substrate;
 ii. Microorganisms and their metabolic pathways;
 iii. Electron transfer mechanism in an anodic chamber;
 iv. Electrode material and membrane type;
 v. Role of QS signals in the formation of electroactive biofilms
 vi. Operating conditions of MFC such as temperature, pH, and salinity;
 vii. Geometric design of the fuel cell [15, 77, 78].

Due to their metabolism and the mediators they utilize to transfer electrons to the anode, choosing microorganisms in the anodic chamber is crucial [78]. The type of inoculum and substrate loaded into the anode chambers are critical biological factors influencing the overall output of MFCs.

Substrate

Substrates used for MFCs vary widely from simplest to complex organic compounds. Pure substrates, including cellulose, glucose, glycerol, butyrate, lactate, acetate, proteins, cysteine, and glycine, have rarely been employed. High-performance substrates include glucose and acetate. Acetate is a non-fermentable substrate that serves as a suitable electron source for dissimilatory iron-reducing bacteria, producing power output up to 66% greater than butyrate and other substrates [79]. Wastewater is a rich media that can be processed by MFCs, among other substrates. Organic substances like sea debris and marine soil also contain highly complex and diverse components that can produce power [74, 80]. Fermenting bacteria efficiently break down fermentable organic substrates and supply anode-respiring microbes with biodegradable components, facilitating carbon reduction in anodic communities [81]. Each substrate generates a specific amount of energy, determined by the MFC design and other factors that cannot be evaluated. Substrate concentration is another factor that needs to be considered [82].

Microorganisms and their Metabolism

Microorganisms that produce multiple enzymes are ideal; however, purified microbes can be employed in defined reactions with multiple substrates (or mixed substrates) [83]. MFC uses mediator-dependent and mediator-free microorganisms. *Proteus vulgaris, Saccharomyces cerevisiae, Proteus mirabilis, E. coli, Pseudomonas fluorescens*, and A*ctinobacillus succinogenes* need mediators. *D. desulfuricans, Shewanella putrefaciens, Geobacter sp.*, and *Rhodoferax ferrireducens* do not require mediators and have acquired attention recently [15]. Stress resistance and nutrient adaptation render mixed microbial cultures better than pure cultures [84]. Due to their high columbic efficiency and durability, exoelectrogenic bacteria in wastewater are preferable for MFCs that can oxidize organic matter, convey electrons to the anode, and conserve energy for maintenance [85].

Cell potential is mainly determined by microorganism metabolism and consequent anode potential. MFCs are limited by microbial catabolism [86]. Oxidation of organic molecules provides heterotrophic organisms energy. Two important metabolic routes, respiratory chain and fermentation, occur in the anodic chamber due to the exogenous oxidants or external terminal electron acceptors [4, 44, 87]. Aerobic respiration generates the most energy, although facultative or obligate anaerobes in MFCs use inorganic chemicals as terminal acceptors for anaerobic respiration. Sulfate, fumarate, nitrate, and metal ions ($Fe3^+$) constitute electron acceptors [87].

Electron Transfer Mechanism

Electron shuttle compounds or redox active mediators should transfer anodic chamber electrons to the anode. Certain MFC microorganisms have a non-conductive outer lipid membrane with peptidoglycans and lipopolysaccharides that retards the flow of electrons to the anode [44, 88]. Mediators include dyes like methyl viologen, methylene blue, humic acid neutral red, and thionine, which must be able to cross the membrane of the bacteria to reach the reductive species and get reduced during microbial metabolism. Mediator and reductive metabolite redox potentials should match for effective electron transfer [89]. However, most mediators are expensive, hazardous, water-soluble phenolic compounds that result in non-ecofriendly and cost-ineffective, challenging MFC commercialization [90]. This can be solved by exploiting electrogenesis bacteria that convey electrons from substrate to anode using microbial metabolites, nanowires, and endogenous mediators.

Electrode Material and Membrane

MFC anodes are biofilm reactors; therefore, they should have conductivity, high specific surface area, biocompatibility, high porosity, Stability, and low fouling and corrosion. On account of its toxicity to bacteria, copper is not a helpful anode material. Carbon has low electron conductivity but is excellent for bacterial adhesion, and carbon felt, fabric, foam, paper, and fibers are available [4, 91]. MFC electrode material with higher porosity and mechanical strength is carbon clothes, but the high cost tends to make it expensive [92]. Graphene-based electrodes have enhanced active microbe-electrode interaction, better electron transfer rate, and high specific surface area [93].

MFC performance can be improved by increasing the anode electrode nitrogen content through ammonia, heat, or diazonium treatment, which increases bacterium adhesion without affecting electron transfer or microbial survival [94 - 96]. Nafion is considered to be the most popular proton exchange membrane due to its strong electrical conductivity and selective proton permeability. Modern composite membranes are found to be cheaper and more effective than Nafion. Activated carbon nanofibre/Nafion nanocomposite membranes, disulphonated poly(arylene ether sulphone), earthen pot, and sulfonated poly(ether ether ketone) (SPEEK) in poly(ether sulphone) (PES) membranes at varied SPEEK compositions are some of the newly tested separators in MFCs.

Role of Quorom Sensing (QS) Signals in the Formation of Electroactive Biofilms (EABs)

QS signals regulate biofilm formation and EABs, which are the biological functions of exoelectrogens in nature [97]. Due to its confined architecture and spatial boundaries, EPS around microbes was extensively researched [98]. EPS surfactants solubilize non-degradable organic contaminants [99]. Quorum sensing (QS) signals generated, secreted, and absorbed by cells can influence biofilm dynamics. QS signals accelerate EAB maturation by enriching them, increasing EPS secretion, improving EET by decreasing charge transfer resistance, and enhancing substrate breakdown and electricity generation.

QS signals speed up EAB evolution, start-up time, and MFC voltage output. In addition, QS signals diminish EAB charge transfer resistance, increasing exoelectrogens and extracellular electron transfer (EET) at the EAB-anode interface. They can also improve bioelectrochemical systems' (BESs) electricity output [100]. For example, short-chain '3-oxoC6-HSL' AHL addition decreases hydrogen scavengers and enhances electrochemically active bacteria, which increases hydrogen yield and electron recovery [101].

Operating Conditions

The microbial growth is mainly affected by salinity, pH, ionic strength, and temperature, which affects MFC performance. MFCs operate under standard pH conditions, temperature, and pressure. Bacteria thrive best at neutral pH, but some wastewaters have very low pH, which inhibits MFC effectiveness. An anodic pH of 7.5 was optimal for energy generation and organic matter elimination. The internal resistance of the MFC is higher than that of chemical fuel cells that use acid or alkaline electrolytes due to neutral pH's low proton concentration. Ionic strength can be adjusted to increase solution conductivity and lower internal resistance without affecting pH. High salinity also enhances conductivity and power production, where proton transport increases with conductivity, lowering internal system resistance. Temperature affects reaction rates of oxygen catalyzed by platinum on the cathode, bacterial kinetics, and proton mass transfer across the liquid, affecting MFC performance. MFC studies are carried out at 20–35°C. For repeatable power cycles and performance, MFCs at 4–30°C need a longer startup time [78].

Design of the MFC

System design is a significant barrier to MFC power densities. Typical MFCs have two "H"-shaped chambers but produce poor power densities. Two-chamber MFC output power depends on the material, anode/cathode construction, surface area, and PEM. The two-chamber MFCs are usually preferred for concept testing despite their complexity. MFC anodic chambers must be sealed to prevent oxygen diffusion, which degrades MFC performance [102].

BIOFILM ENGINEERING

Microbial biofilm is defined as a composite aggregated mass of microbial communities that self-immobilize on solid substrates by excreting protective and adhesive matrix and composed primarily of polysaccharides, with smaller amounts of glycolipids, proteins, and glycoproteins, a negligible proportion of nucleotides, and rarely, some metals [103] where the complexity of microbial biofilms has increased in the process of evolution. The biofilm density naturally increases with the time of the microbial culture. The inactive or dead cells in the biofilm matrix develop an inner covering, which precludes the biofilms from being electrochemically active [104]. MFCs derive power from biofilm-growing microorganisms (exoelectrogens) on the anode surface, immersed in an electrolytic solution at a given substrate concentration, where they transfer electrons to the anode. The electrons flow *via* the external circuit/load towards the cathode, where they get reduced by the terminal electron acceptor. The bulk electrolytic solution, which is constant in substrate concentration, pH, temperature, and pressure, along with the biofilm enveloping the electrode, constitutes the anodic chamber. The microbial population in biofilm is assumed to be uniform and constant, and the unique source of electricity is biofilm [105].

Exoelectrogens develop into a conductive biofilm covering the anode where the biofilm development is unique, and the thickness varies from 30–50 μm. Adhesins and extracellular matrix components stimulate a single bacterial cell to form biofilm on an electrode. Eventually, essential proteins promote biofilm formation, including nanowires (pili) and outer membrane c-type cytochromes (OMC c-Cyts) like OmcZ and OmcS [106]. It has been observed in *Shewanella* that electrons are exchanged indirectly in developing biofilm, which leads to more power generation [37]. Extracellular polysaccharides (EPS) have a variety of distinct roles in different cell types, including the electrochemical properties of the biofilm. The pleiotropic effects due to the differences in the composition of EPS subsequently affect the surface charge and change surface attachment. The concept of a conductive biofilm was first stated in research on *G. sulfurreducens* current-producing biofilms [107] and was later confirmed by direct measurements.

Geobacter biofilms can be transistors or supercapacitors and have conductivities comparable to conductive polymers. These species' pili have a metallic-like conductivity that allows long-range electron transfer (>1 cm) [27]. Studies with *Geobacter sulfurreducens* also illustrated that EPS is a point of attachment for peripheral redox proteins. This enables the electron transfer to distant acceptors in multicellular communities. The mutant strain failed to generate electrogenic biofilms on electrodes, lacking an exopolysaccharide matrix coding gene. That confirmed that in addition to biofilm formation, EPS is necessary for electron transfer [108].

Shewanella oneidensis MR-1, a cell surface polysaccharide, enables cells to adhere to graphite anodes and generate current in MFCs by restricting outer membrane cytochromes from hitting anodes and directing EET through them. Therefore, cell surface engineering can enhance bacterial MFC system current [109], and it is a strategy for maintaining microbes active or viable, which are required for the production of electroactive biofilms, a possible way to enhance the energy output of MFCs. Constructing anodic biofilm with an improved microbial population profile enhances catalytic current. For example, for a flavin-mediated electron transfer, a synthetic riboflavin pathway from *Bacillus subtilis* was expressed into *E. coli* to overproduce flavins, and a hydrophobic *S. oneidensis* strain CP2-1-S1 was used as the exoelectrogen to improve its adherence to the carbon electrode. *S. oneidensis* was better at attaching to the anode than recombinant *E. coli* due to its hydrophobic interactions and overproduction of flavins. Current generation increased from 0.19 to 1.84 A/m^2 at 0 V *vs* SHE [110].

Surface chemistry and morphology affect biofilm-anode electrode interaction. Changes in surface chemistry, like hydrophilicity/hydrophobicity, surface charge, oxygen or nitrogen functional groups, and immobilized mediators, can promote bacterial-level adhesion. Fed-batch is the most common approach for developing electroactive (EA) biofilms. Fed-batch is usually done by the addition of fresh substrate dose when the energy output drops below a particular baseline or by replacing the complete portion of the solution. In some circumstances, inoculum must be added several times in the initial batch, but the fresh medium is added in subsequent batches. Maintenance of substrate concentration above a threshold to avoid current reduction improves performance. Chronoamperometry is usually done at 0.6 V to 0.5 V *vs.* SHE to develop EA biofilms. Electrode potential variations or a preliminary phase at an open circuit before polarisation potential has rarely been used to generate EA biofilms.

Bioanodes made from garden compost (0.04 V *vs* SHE) or coastal sediments (0.14 V *vs.* SHE) had far lower microbial colonization at open circuits than those made from arctic soils (0.1 V *vs* SHE). Transplanting electroactive (EA) biofilms on porous or solid conductive support allows for biofilm generation (electrodes, metallic particles). This scheme minimizes biofilm microbial diversity by removing non-electroactive or non-attached microorganisms (planktonic bacteria) [111]. The electroactive biofilm transplanting technique consists of successive operations of EA biofilm transplanting, which involves scraping the biofilm on the anode of the MFC and its inoculation into a new MFC. Electrostatic interactions between nutrient buffer organic compounds with negative charges and microbial debris also produce biofilms [112].

Quorum sensing plays a vital role in the biofilm formation of microbial communities. The anode or biofilm enhancement chamber accepts QS signaling molecules. QS-based regulatory strategies are divided into enrichment and inhibition methods due to their effects, which help improve the overall bioreactor's performance. The QS enrichment can be done by adding direct QS signaling molecules or adding QS enhancers such as boron, and inhibition methods reduces the excess QS signaling molecules [101]. Therefore, the biofilm of the microbes used in the anode compartment of the microbial fuel cell is engineered with the use of genetically modified micro-organisms which can be used as mixed cultures rather than pure culture, optimizing the operating conditions of the anode chamber, using the suitable substrate and regulating the QS system of the microbial community. Also, gene editing, omic-based methodologies, metabolic engineering and bioinformatics have opened new biofilm-related research areas in microbial fuel cells.

ANODE -THE HEART OF MFC

The anode in microbial fuel cells is the site for microbial attachment and biofilm formation and is a recipient of electrons. Therefore, selecting an anode material significantly improves the electron transfer rate, microbial adhesion, electrochemical efficiency, and electrical output. There are two approaches for developing anode material for an effective MFC- creating or selecting new anode materials and modifying material surfaces. The basic properties required for an efficient anode material include high conductivity, low resistance, low impedance, increased surface area, bio-compatibility, stability, durability, accessibility, and cost of the material. Several MFC materials were used as anodes, including carbon-based materials, natural polymers, conducting polymers, metal and metal oxides, *etc* [113].

Carbon-based Anode Materials

Carbon-based materials are commonly used as anode because of their large surface area, low porosity, high conductivity, high thermal and chemical stability, biocompatibility, easy accessibility, and reasonable electron transfer rate. Carbon materials used as anode were activated carbon cloth, carbon nanotubes (CNT), granular activated carbon, carbon rods, carbon brushes, carbon fiber, graphite felt, and graphite oxide. Graphene, which has better biocompatibility and mechanical properties than graphite, is one of the emerging and promising carbon anode materials for MFC [114]. Modifying carbon anode material by embellishing carbon cloth with nitrogen-doped CNT (NCNT) increased the biocompatibility of the unmodified carbon cloth or CNT. The modified NCNT resembled bamboo, increased current density, anode potential, and decreased internal resistance [115].

Metal/Metal Oxides Based Electrodes

Several metals like silver, nickel, gold, aluminum, stainless steel, copper, molybdenum, cobalt, titanium, and iron used in MFC yield better current density than conventional graphite electrodes. Though the metals used have high conductivity, their use in MFC is limited due to their low biocompatibility [116]. Silver, gold and copper provide better current intensities among the metals used due to their high compatibility. Thermal oxygenation can improve the overall functional qualities of metal electrodes [115]. Several transformation and surface modification strategies, such as chemical oxidation, gas diffusion, ammonia treatment, and binder-assisted pasting with carbon-based materials, were used to improve the disadvantages of metal-based electrodes. Screen printing of graphene multiwalled carbon nanotube (CNT) composited with SDS (Sodium dodecyl sulfate) onto stainless steel metal electrodes revealed better biocompatibility and hydrophilicity [117].

Natural Waste-derived Anode

Easy availability, affordability, and recyclable nature make fabricating anode material with natural materials like biomass wastes advantageous and promising. Anode materials fabricated with these low-cost natural materials have a large surface area with a microporous structure, facilitating improved biofilm attachment and development [118]. An interesting example is an anode based on layered corrugated carbon (LCC) made from low-cost materials, which results in increased current densities [119]. Several naturally derived materials like *Hibiscus cannabinus* stem, corn straw, wild mushroom, king mushroom, bamboo charcoal, waste paper, forest residues, chestnut shells, silk cocoon, and coffee biomass have been used after carbonization to fabricate anode materials [120, 121].

Electrode Modification to Promote Anode Biofilm Development

The anode surface can be modified to enable effective biofilm attachment and to facilitate extra-cellular electron transport through several approaches like carbon metal coating, functional group modification, deposition of metal or metal oxides, *etc*. Functional group modification proved significant among these approaches to enhance biofilm development, biofilm stability, and energy production and was also found to be inexpensive. The surface of the anode material can be modified to alter its functional groups, surface charge, surface wetness, and surface roughness to enhance the biofilm attachment. Biofilm development occurs through various phases initiated by unstable adsorption followed by stable adsorption, microcolony formation, expansion and maturation of biofilm, aging, and dissociation of biofilm [122]. Adhesion of bacteria, which is the initial critical step and subsequent immobilization, is facilitated by various physicochemical interactions between the material surface and the outer envelope of bacteria. The roughness of the surface plays a crucial role in the entrapment of the bacteria, whereas other surface properties like surface charge and functional groups are essential for its adhesion. Most bacteria possess a net negative charge on their outer envelope or different polymeric substances, and modifying the anode material surface to have a more positively charged group aids in enhanced bacterial adhesion. Positive charges can be imparted to the electrode material through ammonia, thermal, and acid treatment [123].

Anode materials with exposed C-, O-, S-, N-, C=O, and C-N functional groups are found to have efficient chemical bonding with bacterial outer envelope components and reduce the startup time of MFC. Coating or doping anode materials with such functional group-containing compounds enhances the affinity between anode and bacteria [124]. Anode materials with rough surfaces have high porosity and enhanced sites for bacterial adhesion. Various treatments like chemical oxidation, thermal treatment, and acid treatment increase the roughness of the electrode material [125].

STRAIN IMPROVEMENT FOR IMPROVED MFC PERFORMANCE

MFC Performance Improvements through Microbial Modifications

The low rate of extracellular transfer from the exo-electrogens to the anodic surface remains a crucial tailback hampering the application of MFC in practical use. Although investigations on changing the microbial composition in an anodic biofilm have been carried out extensively to improve the overall MFC performance, genetic/chemical modification of exo-electrogens was found to be a promising tool [126 - 128].

Chemical Modification of Microbial Cells

Secretion of endogenous redox mediators from the cytoplasm of exo-electrogens into the anode media and the attachment of the electrogenic bacteria to the anode surface is a critical process ruling the extracellular electron transfer. Both the above-mentioned processes are hindered by the presence of a non-conductive lipopolysaccharide layer (LPS) in the outer membrane of Gram-negative bacteria. Chemical treatment of the cells for rough membrane surface, larger pore size and channels on the cell membrane was found to facilitate the enhanced diffusion of redox mediators and biofilm attachment [129, 130]. Lowering of phosphate concentration in the anodic chamber and the addition of trace heavy metal ions was also found to increase the concentration of redox mediators [131, 132].

Genetic Modification of Exoelectrogens

Precendently, chemical mutagenesis created several mutations in diverse microorganisms. Although the chosen phenotypes were desirable from the scientist's viewpoint, there was no conclusive proof that these organisms might cause disease in people, animals, or plants. Owing to the availability of whole genome sequencing and knock-in/allelic exchange technologies in most compelling electrogenic bacteria, genetic engineering/modification of such microorganisms seems to be a promising method to enhance the MFC output. A growing number of MFC research teams have done this using various genetic techniques, which include,

a. Electron mediator synthesis pathways can be altered,
b. Gene regulatory circuits can be rewired to affect electron transfer pathways,
c. Cofactors can be changed, and
d. Other parameters can be altered [133, 134]

The electron shuttles (also known as electron mediators) are secondary metabolites released by electrochemically active microorganisms. They typically consist of many redox-active substances. Under various development conditions, many microorganisms depend on various electron mediators (such as flavins, phenazines, and quinine/quinone compounds) [135]. The capacity of *P. aeruginosa* to manufacture its own redox mediators, such as pyorubrin (aeruginosin A) and PYO (phenazine-1-carboxamide), has made it a favorite among the various exoelectrogens among MFC researchers. PYO is a soluble phenazine-based blue colored redox mediator and the manipulation of its production pathway has been found to increase the efficiency of extracellular electron transfer in *P.aeruginosa*. A fourfold rise in the electricity yield of MFCs was reported with a modified overexpression strain of *P.aeruginosa* phzM (methyltransferase encoding gene) compared to wild-type strain [136]. PYO demethylase (PodA), an enzyme, has recently been found to inhibit *P. aeruginosa* biofilm development, an exciting indication that PYO is essential for biofilm establishment [137]. These changes helped to increase the MFC power output and extracellular electron transfer efficiency. In order to improve the bioelectricity production of MFCs, it may be adequate to manipulate the routes for the synthesis of electron mediators.

Through modification of the metabolic pathways of electrogenic microorganisms directly or indirectly, genetically modified bacteria could also be used to increase extracellular electron transfer efficiency [109]. A study was performed by altering the electron transport channels on *P. aeruginosa,* which can utilize a variety of electron mediators. According to the study, the wild strain uses a non-phenazine redox mediator as the predominant one, and when the strains's quorum sensing circuit (rhl QS) was overexpressed, phenazine mediators like phenazine-carboxylate and PYO predominated. Replacement of non-phenazine redox mediators with phenazine compounds resulted in a 1.6-fold rise in the power output [135].

The majority of metabolic and biosynthetic reactions depend on cofactors, which also serve as redox-active electron shuttles. Cofactors like $NADP^+$ (nicotinamide adenine dinucleotide phosphate) or NAD^+ (nicotinamide adenine dinucleotide) and their respective reduced forms, NADPH or NADH, play a significant role in catabolic and anabolic reactions; also, as redox mediators in an MFC's electron transfer [138]. Modifying the NADH to NAD^+ ratio was reported to increase the

MFC performance. Lactate production pathway abrogation in *E. coli* led to an increase in the cell's [NADH]/[NAD$^+$] ratio and subsequent higher power output in MFCs [132]. As of now, there has not been much effort made to improve the NAD$^+$ pool directly to change the NADH to NAD$^+$ ratio. The overall NAD$^+$ levels rose when the pncB gene, which is connected to the NAD$^+$ salvage pathway, was overexpressed [139]. It was shown that the nadE (NAD$^+$ synthetase gene) overexpression increased the NAD$^+$/NADH cofactor pool, which led to an approximately three-fold increase in power production compared to the wildtype strain. The increase in power performance was also explained by the nadE overexpression strain's reduced electron transfer resistance, which was about a factor of two, and increased electrochemical efficiency compared to the wild strain. Additionally, nadE overexpression boosted the concentration of the redox electron mediator (PYO) by almost 1.5 times, which helped to improve electrochemical performance [140]. The modification of endogenous redox-active mediators/cofactors may thus be one more successful strategy to boost the electricity production of MFCs, according to these findings.

Exo-electrogenic microorganisms are genetically engineered to enhance the number of electrons released and to improve the rate of extracellular electron transfer so as to improve the overall bioelectrochemical activity. This is a new and largely unexplored field of study that will produce fascinating discoveries in the future [134, 136].

Selection and Modification of Exo-electrogenic Strains

Cell Cultures

Electrogenic microorganisms like *Geobacter* and *Shewanella* are frequently employed in MFCs. MFC systems' functioning has also involved several yeast strains, including *Saccharomyces cerevisiae, Candida melibiosica, and Kluyveromyces marxianus.* Promising strains for producing energy include proteobacteria, cyanobacteria, and archaebacteria. However, eukaryotic algae can function in the anode and cathode as both electron generators and acceptors. MFCs receive both pure and mixed cultures as anodic inoculations. Due to their clear-cut metabolic pathways, pure cultures may convert substrates into electricity more effectively. The concentration and purity of the substrate are also more stringently required. Therefore, the capacity of pure cultures to use complex substrates for energy generation may be restricted by their ability to utilize only specific substrates.

At the moment, pure cultures are primarily used for studies on the electron transport mechanism and effectiveness of bio-electricity generation. However, to increase the volumetric power density and lower the startup time of MFCs, a pure culture-based bioaugmentation technique was developed. Because they are more able to adapt to challenging substrates, mixed cultures are ideal for scaling up MFC systems. In addition, the mixed culture's synergistic effect from the multiple strains may also help MFC systems function effectively [141].

The mixed culture most commonly utilized in MFC systems is activated sludge. Based on their distinct functionalities, the co-culture of designated strains may improve the MFC's performance. *Pseudomonas aeruginosa* and *Enterobacter aerogenes* were combined in a co-culture system created by Schmitz and Rosenbaum. The redox electron mediator made by *Pseudomonas aeruginosa* can increase the anode's ability to transport electrons more effectively. Under optimal oxygen conditions, this co-culture system can enhance electrical current generation by about 400% [142].

Strain Modification through Genetic Engineering

Even if several wild-type strains produce bio-electricity, increasing the strains' electrochemical efficiency using the proper techniques is still essential. Several physical and chemical techniques could be used to improve the capacity of strains to produce electricity. Another effective method for enhancing strains' electrochemical activity is genetic editing. The electro-shuttle route, metabolic activity, and substrate utilization are affected mainly by gene change. Therefore, an efficient way to increase the efficiency of MFC systems is by increasing genetic engineering-based extracellular electron transfer (EET).

A strain of *Shewanella oneidensis* was created with a gene cluster for flavin production. The power density of the MFC was found to rise by 110% with the use of this strain. Additionally, improving intracellular electron flow and EET efficiency can be achieved by increasing the intracellular NADH to NAD^+ ratio [143]. Metabolic engineering was used in a study to reverse methanogenesis for the generation of electricity from methane, whereas, a synthetic microbial consortium was used in the anode chamber. The consortium consists of engineered *Methanosarcina acetivorans* producing methyl-coenzyme M reductase to generate acetate from methane, *Paracoccus denitrificans* secreting redox mediators, and *Geobacter sulfurreducens* to consume acetate and generating electrons with subsequent energy generation in MFC [144]. An engineered *Shewanella oneidensis* strain was created to produce energy directly from xylose. It is capable of $2.1 \ mW/m^2$ of maximum power density. In laboratory-scale MFC systems, strain modification through genetic engineering methods has so far produced the required outcomes. However, much progress has not been made in scaling up MFC systems [145].

Experimental results with the flagellin gene of bacteria showed that both the wild-type strain K-12 and the FliC-deficient *E. coli* strain (fliC) were capable of generating power in microbial fuel cells (MFCs). The biofilm of the mutated strain fliC generated a 193% greater power density compared to the wild-type strain despite having comparable growth curves. The mutated strain was also found to have enhanced biofilm formation compared to the wild-type strain [146].

SIGNIFICANCE OF MFC FOR A CLEAN AND GREEN ENVIRONMENT

The microbial fuel cell is a distinguished alternative and sustainable clean energy technology with several applications such as bioelectricity generation, biodegradation, waste water treatment, biosensors, heavy metal recovery, toxic detection, bio-remediation, desalination, biosensors, *etc*.

Microbial Fuel Cell in Bio-energy Generation

Since Potter discovered the ability of bacteria to generate bio-electricity in 1911, microbial fuel cell technology has progressively advanced as sustainable and renewable energy technology [10]. Initially, simple monomeric sugar and organic acids like glucose, xylose, maltose, sucrose, trehalose, fructose, acetate, succinate, propionate, lactate, butyrate, malate, *etc*., had been used as a substrate for bioelectricity generation in MFC. Later, with advancements in architecture and understanding of the microbial role in MFC, complex substrates like cellulose, pectin, chitin, and wastewater from different sources have been used as substrates. Although MFC seems to be a promising alternative technology for renewable and clean energy, its generated power is unsuitable for practical use due to its low power density, low voltage output, and high internal resistance. The Power density of a typical MFC varies from 1 to 2000 mW m^{-1}, and the maximum theoretical voltage across the electrode in an MFC is 1.14V (the required operating voltage for low-power electronic devices like LEDs is 2 to 5 V) [147]. Several MFCs can be stacked (connected in parallel or series) to overcome the problem of low voltage and power density. A stacked Microbial Fuel Cell was demonstrated to power LED bulbs in outdoor public areas [148]. However, voltage reversal and consequent overall voltage decay are still problematic in stacked microbial fuel cells (MFC). Therefore, the current research on MFC is focused on formulating ways to amplify the power density produced by microbial fuel cells to use it for large-scale industrial use.

Microbial Fuel Cells for Wastewater Treatment

As mentioned above, substrates used during the initial years of microbial fuel cells are simple sugars like glucose. The use of wastewater as a substrate in MFC began in 2004 [149]. The conventional wastewater treatment methods, namely aerated lagoons, activated sludge, and trickling filters, consume vast amounts of energy. On the other hand, wastewater has 3-10 times the energy necessary to treat itself [150]. Currently exploited technology for tapping the energy stored in the wastewater is anaerobic digestion, in which chemical energy in the wastewater is converted into biogas, which can be further combusted to produce electricity. However, conventional wastewater technologies cost about 3% of the total global electricity demand. In addition, the disposal of the effluent/sludge demands 50% of the total treatment cost [151, 152]. Microbial fuel cell technology is a novel and promising technology that can overcome the abovementioned complications of conventional wastewater treatment methods. MFC translates the chemical energy conserved in the organic content of wastewater into bio-electricity, thereby reducing the COD content and generating electricity simultaneously. Also, MFC produces less sludge than traditional activated sludge and anaerobic digesters [153]. Therefore, microbial fuel cells are a promising technology for a sustainable and clean environment. Wastewater collected from various sources like municipalities and industries can be used as substrates for MFC (Table **2**).

Table 2. Wastewater sources used as a substrate for MFC.

Wastewater source	COD removal (%)	Power density	References
Domestic waste water	25.8	422 mW/m^2	[6]
Municipal wastewater	79	116 mW	[157]
Yeast industry waste water	73-92	51.02 mW/m^2	[158]
Pharmaceutical waste water	80.55	2.01 W/m^3	[159]
Electroplating industry wastewater	87	260 mW/m^2	[160]
Brewery industrial wastewater	79–83	0.8 W/m^3	[161]
Seafood industrial wastewater	52	530 $\pm$ 15 mW/m^2	[162]
Azo dye-containing wastewater	94.04	148.29 mW/m^3	[163]
Swine wastewater treatment	72	33.3 mW/m^3	[164]
Fish market wastewater	90	420 mW/m^2	[165]

The composition of wastewater influences the efficiency and overall performance of a MFC. The energy output with simple and pure sugars is higher than with waste water as substrate. This can be attributed to various complex polymeric sugar components, toxic chemicals, high ammonia concentration, production of volatile acids, *etc* [154]. Accordingly, the construction of MFC depends on the type and source of wastewater. For instance, wastewater from the food and brewery industry contains rich sources of nutrients and organic matter, whereas wastewater from a distillery unit will be rich in salt and other minerals. The total power density produced from microbial fuel cells with wastewater as substrate ranges from 10-50 mW/m^2 [155]. Besides electricity generation, microbial fuel cells are also employed to remove heavy metals and nitrates from wastewater and to produce other value-added chemicals like hydrogen [156] and methane [155]. Therefore, scaling up the MFC system is essential for real-world wastewater treatment applications. Scaling up can be ensured in two ways- enlarging reactor size and stalking several MFCs in series or parallel.

Microbial Fuel Cells for Bioremediation

Bioremediation is the technology wherein microorganisms neutralize or remove toxic environmental pollutants. It involves the degradation and conversion of pollutants into less toxic forms or even their eradication [166]. The microbial fuel cell is one of the recently developed potential tools for bioremediation that uses bacteria to remove/degrade various environmental pollutants. Energy production, pollutant removal, and less sludge production make MFC an exceptional technology compared to conventional technologies. Microbial fuel cells have been used by several researchers for the bioremediation of several classes of pollutants, including antibiotics, synthetic dyes, ethyl acetate, polycyclic aromatic compounds, phenolic compounds, xenobiotics, perchlorate, pesticides, sulfur-containing compounds, emerging contaminants and trace organic compounds (Table **3**) [167]. Various researchers have also carried out the cathodic reduction of pollutants for reductive dechlorination of pollutants like Tetrachloroethane (PCE), Trichloroethane (TCE), *etc* [168]. Incorporation of microbial fuel cells with phytoremediation in the form of constructed wetland microbial fuel cells was shown to improve bioremediation efficiency [169].

Table 3. Use of MFC for bioremediation of different classes of pollutants.

Pharmaceuticals	Bioremediation	References
Sulphamethoxazole	85% degradation within 12 hours	[170]
Paracetamol	70% degradation within 9 hours	[171]
Phenolic compounds		
p-Nitrophenol (PNP)	64.69% degradation	[172]
2,4- Dichloro Phenol	60% degradation	[173]
Polycyclic aromatic hydrocarbons (PAHs)		
Naphthalene,	76.9% degradation	[174]
Acenaphthene	52.5% degradation	
Phenanthrene	36.8% degradation	
Phenanthrene	97% degradation	[175]

(Table 3) cont.....

Pharmaceuticals	Bioremediation	References
Pesticides		
DDE (2,2-bis (p-chlorophenyl)-1,1-dichloroethylene	39% degradation within two months	[176]
Pyraclostrobin	1.7 mg/L/h removal rate	[177]

Microbial Fuel Cells as Biosensors

Sensors are devices that can detect the properties of circumstances arising around them and that information into signals that can be decoded. Biosensors use biological entities like microorganisms, enzymes, DNA, *etc.*, along with physical transducers that transform the biological response from the former part into a thermal, optical, or electrical signal. BOD measurement, a commonly used indicator for assessing water quality, is a time-consuming method that takes around five days [178]. MFC, as a biosensor alternative to conventional laboratory BOD measurement, was first proposed by Karube I in 1997 using *Clostridium butyricum* in the anode [179]. A linear correlation between BOD and current output has been observed, thus confirming the viability of MFC biosensors for BOD measurements. The impact of pollutants or toxicants in the environment will be reflected in the metabolic efficiency and pathways of the microorganisms occupying that environment. The presence of pollutants/toxicants affects the proliferation of microorganisms inside the anodic chamber of MFC, subsequently affecting the performance and current output [180]. As the level and toxicity of the toxicants increase, the current generation will decrease accordingly. Based on the analyte detected, MFCs are classified as antibiotics biosensors, acidic toxicity biosensors, organic toxicants biosensors, and heavy metal biosensors [181]. MFC can also be used as a power source for other commercial sensors by including a convertor and capacitor to elevate the MFC voltage [182]. The use of specific microbial cultures or genetically engineered microorganisms for specific pollutants/toxicants helps to detect and monitor specific pollutants in the environment. Compared to battery-based commercial biosensors, MFC-based biosensors have long-term stable use and require low maintenance [183].

FUTURE PERSPECTIVES

Microbial fuel cells that can translate the chemical energy trapped in various substrates into electrical energy seem to be a promising alternative, sustainable, clean, and renewable energy source. Electro active microbial proliferation and the electron flux between them and the electrode are the primary determinant factor in the overall output of a microbial fuel cell. Accordingly, investigations into the biochemical pathway and their metabolomics and molecular mechanisms underlying the electron flux will help us design efficient microbial consortia for effective electron transfer and power output. The main challenge lies in scaling up of MFC to be used in large-scale operations. Exploration of cheap electrode materials and their modifications to allow conductive biofilm formation and the amenability to scaling up is the immediate necessity for the practical use of MFC for real-world problems. Future research shall be advanced by the hand-in-hand collaboration of biologists, physicists, and chemists to form conductive biofilms on cost-effective, sustainable electrode materials with the potential of generating bio-electricity for practical use.

CONCLUSION

Bio-electrochemical systems and the developing Microbial fuel cells (MFC) are promising clean and renewable energy technologies. With efficient COD removal and less sludge removal, incorporating MFC into wastewater treatment plants and traditional methods has proven to be efficient in simultaneous wastewater treatment and energy generation. Besides these, MFC is also helpful in the bioremediation of various environmental pollutants and heavy metal recovery, as well as being a durable and low-maintenance biosensor for monitoring various toxicants/pollutants. Although there is potentialy for being a promising technology, the power output is not up to the level for practical applications. Research is advancing to increase the current output and to scale up practical applications. This chapter discusses a fundamental understanding of microbial fuel systems and the factors affecting their performance. Particular emphasis has been placed on electroactive biofilms, biofilm engineering approaches, electrode modification for conductive biofilm development, and strain development approaches. With the growing pressure on our environment from using fossil fuels, climate change and the demand for renewable energy sources will encourage advanced development of MFC technology for its fruitful practical implementations.

REFERENCES

[1] Mufutau Opeyemi B. Path to sustainable energy consumption: The possibility of substituting renewable energy for non-renewable energy. Energy 2021; 228: 120519.
[http://dx.doi.org/10.1016/j.energy.2021.120519]

[2] International Energy Agency. 2022. Available from: https://iea.blob.core.windows.net/assets/ ada7af90-e280-46c4-a577-df2e4fb44254/Renewables2022.pdf

[3] Alcayde A, Montoya FG, Baños R, Perea-Moreno AJ, Manzano-Agugliaro F. Analysis of research topics and scientific collaborations in renewable energy using community detection. Sustainability 2018; 10(12): 4510.
[http://dx.doi.org/10.3390/su10124510]

[4] Logan BE, Hamelers B, Rozendal R, *et al*. Microbial fuel cells: methodology and technology. Environ Sci Technol 2006; 40(17): 5181-92.
[http://dx.doi.org/10.1021/es0605016] [PMID: 16999087]

[5] Shrivastava A, Sharma RK. Lignocellulosic biomass based microbial fuel cells: Performance and applications. J Clean Prod 2022; 361: 132269.
[http://dx.doi.org/10.1016/j.jclepro.2022.132269]

[6] Ahn Y, Logan BE. Effectiveness of domestic wastewater treatment using microbial fuel cells at ambient and mesophilic temperatures. Bioresour Technol 2010; 101(2): 469-75.
[http://dx.doi.org/10.1016/j.biortech.2009.07.039] [PMID: 19734045]

[7] Galvani L, Aldini J. De viribus electricitatis. W.: Junk 1792.

[8] Carrette L, Friedrich KA, Stimming U. Fuel cells-fundamentals and applications. Fuel Cells (Weinh) 2001; 1.

[9] Luckarift HR, Atanassov PB, Johnson GR, Eds. Enzymatic fuel cells: From fundamentals to applications. John Wiley & Sons 2014.
[http://dx.doi.org/10.1002/9781118869796]

[10] Potter MC. Electrical effects accompanying the decomposition of organic compounds. Proc R Soc Lond, B 1911; 84(571): 260-76.
[http://dx.doi.org/10.1098/rspb.1911.0073]

[11] Cohen B. The bacterial culture as an electrical half-cell. J Bacteriol 1931; 21(1): 18-9.

[12] Tender LM, Gray SA, Groveman E, *et al.* The first demonstration of a microbial fuel cell as a viable power supply: Powering a meteorological buoy. J Power Sources 2008; 179(2): 571-5.
[http://dx.doi.org/10.1016/j.jpowsour.2007.12.123]

[13] Rashid N, Cui YF, Saif Ur Rehman M, Han JI. Enhanced electricity generation by using algae biomass and activated sludge in microbial fuel cell. Sci Total Environ 2013; 456-457: 91-4.
[http://dx.doi.org/10.1016/j.scitotenv.2013.03.067] [PMID: 23584037]

[14] Bose D, Santra M, Sanka RVSP, Krishnakumar B. Bioremediation analysis of sediment-microbial fuel cells for energy recovery from microbial activity in soil. Int J Energy Res 2021; 45(4): 6436-45.
[http://dx.doi.org/10.1002/er.6163]

[15] Pant D, Van Bogaert G, Diels L, Vanbroekhoven K. A review of the substrates used in microbial fuel cells (MFCs) for sustainable energy production. Bioresour Technol 2010; 101(6): 1533-43.
[http://dx.doi.org/10.1016/j.biortech.2009.10.017] [PMID: 19892549]

[16] Kim JR, Cheng S, Oh SE, Logan BE. Power generation using different cation, anion, and ultrafiltration membranes in microbial fuel cells. Environ Sci Technol 2007; 41(3): 1004-9.
[http://dx.doi.org/10.1021/es062202m] [PMID: 17328216]

[17] Guang L, Koomson DA, Jingyu H, Ewusi-Mensah D, Miwornunyuie N. Performance of exoelectrogenic bacteria used in microbial desalination cell technology. Int J Environ Res Public Health 2020; 17(3): 1121.
[http://dx.doi.org/10.3390/ijerph17031121] [PMID: 32050646]

[18] Kumar R, Singh L, Zularisam AW. Exoelectrogens: Recent advances in molecular drivers involved in extracellular electron transfer and strategies used to improve it for microbial fuel cell applications. Renew Sustain Energy Rev 2016; 56: 1322-36.
[http://dx.doi.org/10.1016/j.rser.2015.12.029]

[19] Kumar R, Singh L, Zularisam AW. Microbial fuel cells: Types and applications. Waste biomass management–A holistic approach. 2017:367-84.

[20] Rabaey K, Boon N, Siciliano SD, Verhaege M, Verstraete W. Biofuel cells select for microbial consortia that self-mediate electron transfer. Appl Environ Microbiol 2004; 70(9): 5373-82.
[http://dx.doi.org/10.1128/AEM.70.9.5373-5382.2004] [PMID: 15345423]

[21] He Z, Wagner N, Minteer SD, Angenent LT. An upflow microbial fuel cell with an interior cathode: assessment of the internal resistance by impedance spectroscopy. Environ Sci Technol 2006; 40(17): 5212-7.

[http://dx.doi.org/10.1021/es060394f] [PMID: 16999091]

[22]   Aelterman P, Rabaey K, Pham HT, Boon N, Verstraete W. Continuous electricity generation at high voltages and currents using stacked microbial fuel cells. Environ Sci Technol 2006; 40(10): 3388-94. [http://dx.doi.org/10.1021/es0525511] [PMID: 16749711]

[23]   Ortega-Martínez AC, Juárez-López K, Solorza-Feria O, *et al.* Analysis of microbial diversity of inocula used in a five-face parallelepiped and standard microbial fuel cells. Int J Hydrogen Energy 2013; 38(28): 12589-99. [http://dx.doi.org/10.1016/j.ijhydene.2013.02.023]

[24]   Haddock BA, Jones CW. Bacterial respiration. Bacteriol Rev 1977; 41(1): 47-99. [http://dx.doi.org/10.1128/br.41.1.47-99.1977] [PMID: 140652]

[25]   Yang Y, Xu M, Guo J, Sun G. Bacterial extracellular electron transfer in bioelectrochemical systems. Process Biochem 2012; 47(12): 1707-14. [http://dx.doi.org/10.1016/j.procbio.2012.07.032]

[26]   Kim BH, Kim HJ, Hyun MS, Park DH. Direct electrode reaction of Fe (III)-reducing bacterium, Shewanella putrefaciens. J Microbiol Biotechnol 1999; 9(2): 127-31.

[27]   Malvankar NS, Vargas M, Nevin KP, *et al.* Tunable metallic-like conductivity in microbial nanowire networks. Nat Nanotechnol 2011; 6(9): 573-9. [http://dx.doi.org/10.1038/nnano.2011.119] [PMID: 21822253]

[28]   Allen RM, Bennetto HP. Microbial fuel-cells. Appl Biochem Biotechnol 1993; 39-40(1): 27-40. [http://dx.doi.org/10.1007/BF02918975]

[29]   Babanova S, Hubenova Y, Mitov M. Influence of artificial mediators on yeast-based fuel cell performance. J Biosci Bioeng 2011; 112(4): 379-87. [http://dx.doi.org/10.1016/j.jbiosc.2011.06.008] [PMID: 21782506]

[30]   Torres CI, Marcus AK, Lee HS, Parameswaran P, Krajmalnik-Brown R, Rittmann BE. A kinetic perspective on extracellular electron transfer by anode-respiring bacteria. FEMS Microbiol Rev 2010; 34(1): 3-17. [http://dx.doi.org/10.1111/j.1574-6976.2009.00191.x] [PMID: 19895647]

[31]   Chen S, Rotaru AE, Liu F, *et al.* Carbon cloth stimulates direct interspecies electron transfer in syntrophic co-cultures. Bioresource technology. 2014 Dec 1;173: 82-6.32.

[32]   Rabaey K, Verstraete W. Microbial fuel cells: novel biotechnology for energy generation. TRENDS in Biotechnology. 2005 Jun 1;23(6):291-8.

[33]   Rozendal RA, Leone E, Keller J, Rabaey K. Efficient hydrogen peroxide generation from organic matter in a bioelectrochemical system. Electrochem Commun 2009; 11(9): 1752-5. [http://dx.doi.org/10.1016/j.elecom.2009.07.008]

[34]   Greenman J, Gajda I, You J, *et al.* Microbial fuel cells and their electrified biofilms 2021. Available from: https://www.sciencedirect.com/science/article/pii/S2590207521000150 [http://dx.doi.org/10.1016/j.bioflm.2021.100057]

[35]   Ishii S, Suzuki S, Yamanaka Y, Wu A, Nealson KH, Bretschger O. Population dynamics of electrogenic microbial communities in microbial fuel cells started with three different inoculum sources. Bioelectrochemistry 2017; 117: 74-82. [http://dx.doi.org/10.1016/j.bioelechem.2017.06.003] [PMID: 28641173]

[36]   Chen BY, Chang JS. Assessment upon species evolution of mixed consortia for azo dye decolorization. J Chin Inst Chem Eng 2007; 38(3-4): 259-66. [http://dx.doi.org/10.1016/j.jcice.2007.04.002]

[37]   Obileke K, Onyeaka H, Meyer EL, Nwokolo N. Microbial fuel cells, a renewable energy technology for bio-electricity generation: A mini-review. Electrochem Commun 2021; 125: 107003. [http://dx.doi.org/10.1016/j.elecom.2021.107003]

[38]   Nimje VR, Chen CY, Chen CC, *et al.* Stable and high energy generation by a strain of *Bacillus subtilis* in a microbial fuel cell. J Power Sources 2009; 190(2): 258-63.
[http://dx.doi.org/10.1016/j.jpowsour.2009.01.019]

[39]   Rezaei F, Xing D, Wagner R, Regan JM, Richard TL, Logan BE. Simultaneous cellulose degradation and electricity production by *Enterobacter cloacae* in a microbial fuel cell. Appl Environ Microbiol 2009; 75(11): 3673-8.
[http://dx.doi.org/10.1128/AEM.02600-08] [PMID: 19346362]

[40]   Richter H, McCarthy K, Nevin KP, Johnson JP, Rotello VM, Lovley DR. Electricity generation by *Geobacter sulfurreducens* attached to gold electrodes. Langmuir 2008; 24(8): 4376-9.
[http://dx.doi.org/10.1021/la703469y] [PMID: 18303924]

[41]   Chen K-C, Huang W-T, Wu J-Y, Houng J-Y. Microbial decolorization of azo dyes by *Proteus mirabilis*. J Ind Microbiol Biotechnol 1999; 23(1): 686-90.
[http://dx.doi.org/10.1038/sj.jim.2900689] [PMID: 10455502]

[42]   Watson VJ, Logan BE. Power production in MFCs inoculated with *Shewanella oneidensis* MR-1 or mixed cultures. Biotechnol Bioeng 2010; 105(3): 489-98.
[http://dx.doi.org/10.1002/bit.22556] [PMID: 19787640]

[43]   Choi Y, Jung E, Kim S, Jung S. Membrane fluidity sensing microbial fuel cell. Bioelectrochemistry 2003; 59(1-2): 121-7.
[http://dx.doi.org/10.1016/S1567-5394(03)00018-5] [PMID: 12699828]

[44]   Du Z, Li H, Gu T. A state of the art review on microbial fuel cells: A promising technology for wastewater treatment and bioenergy. Biotechnol Adv 2007; 25(5): 464-82.
[http://dx.doi.org/10.1016/j.biotechadv.2007.05.004] [PMID: 17582720]

[45]   Pham CA, Jung SJ, Phung NT, *et al.* A novel electrochemically active and Fe(III)-reducing bacterium phylogenetically related to *Aeromonas hydrophila*, isolated from a microbial fuel cell. FEMS Microbiol Lett 2003; 223(1): 129-34.
[http://dx.doi.org/10.1016/S0378-1097(03)00354-9] [PMID: 12799011]

[46]   Schröder U, Nießen J, Scholz F. A generation of microbial fuel cells with current outputs boosted by more than one order of magnitude. Angew Chem Int Ed 2003; 42(25): 2880-3. Available from: https://www.scopus.com/inward/record.uri?eid=2-s2.0-17744405443&doi=10.1002%2Fanie.200350918&partnerID=40&md5=b3a829a2a01635ed721a489c092743a0
[http://dx.doi.org/10.1002/anie.200350918] [PMID: 12833347]

[47]   Chaudhuri SK, Lovley DR. Electricity generation by direct oxidation of glucose in mediatorless microbial fuel cells. Nat Biotechnol 2003; 21(10): 1229-32.
[http://dx.doi.org/10.1038/nbt867] [PMID: 12960964]

[48]   Rhoads A, Beyenal H, Lewandowski Z. Microbial fuel cell using anaerobic respiration as an anodic reaction and biomineralized manganese as a cathodic reactant. Environ Sci Technol 2005; 39(12): 4666-71.
[http://dx.doi.org/10.1021/es048386r] [PMID: 16047807]

[49]   Logan BE, Murano C, Scott K, Gray ND, Head IM. Electricity generation from cysteine in a microbial fuel cell. Water Res 2005; 39(5): 942-52.
[http://dx.doi.org/10.1016/j.watres.2004.11.019] [PMID: 15743641]

[50]   Zhou M, Jin T, Wu Z, Chi M, Gu T. Microbial fuel cells for bioenergy and bioproducts. Sustainable Bioenergy and Bioproducts: Value Added Engineering Applications. 2012:131-71.
[http://dx.doi.org/10.1007/978-1-4471-2324-8_8]

[51]   Nevin KP, Richter H, Covalla SF, *et al.* Power output and columbic efficiencies from biofilms of *Geobacter sulfurreducens* comparable to mixed community microbial fuel cells. Environ Microbiol 2008; 10(10): 2505-14.
[http://dx.doi.org/10.1111/j.1462-2920.2008.01675.x] [PMID: 18564184]

[52]   Park DH, Zeikus JG. Improved fuel cell and electrode designs for producing electricity from microbial degradation. Biotechnol Bioeng 2003; 81(3): 348-55.
[http://dx.doi.org/10.1002/bit.10501] [PMID: 12474258]

[53]   Zou Y, Xiang C, Yang L, Sun LX, Xu F, Cao Z. A mediatorless microbial fuel cell using polypyrrole coated carbon nanotubes composite as anode material. Int J Hydrogen Energy 2008; 33(18): 4856-62.
[http://dx.doi.org/10.1016/j.ijhydene.2008.06.061]

[54]   Li Z, Haynes R, Sato E, Shields MS, Fujita Y, Sato C. Microbial community analysis of a single chamber microbial fuel cell using potato wastewater. Water Environ Res 2014; 86(4): 324-30. Available from: https://www.scopus.com/inward/record.uri?eid=2-s2.0-84901818356&doi=10.2175%2F106143013X13751480308641&partnerID=40&md5=8cd342a7dea6d5fe5f39d7a8c246e321
[http://dx.doi.org/10.2175/106143013X13751480308641] [PMID: 24851328]

[55]   Mehta-Kolte MG, Bond DR. *Geothrix fermentans* secretes two different redox-active compounds to utilize electron acceptors across a wide range of redox potentials. Appl Environ Microbiol 2012; 78(19): 6987-95. Available from: https://www.scopus.com/inward/record.uri?eid=2-s2.0-84868307698&doi=10.1128%2FAEM.01460-12&partnerID=40&md5=cffd987bec8bfc26258061babccd37b0
[http://dx.doi.org/10.1128/AEM.01460-12] [PMID: 22843516]

[56]   Naradasu D, Miran W, Sakamoto M, Okamoto A. Isolation and characterization of human gut bacteria capable of extracellular electron transport by electrochemical techniques. Front Microbiol 2019; 9: 3267.
[http://dx.doi.org/10.3389/fmicb.2018.03267] [PMID: 30697198]

[57]   Aparna PP, Meignanalakshmi S. Comparison of power generation of electrochemically active bacteria isolated from the biofilm of single chambered multi-electrode microbial fuel cell developed using *Capra hircus* rumen fluid. Energy Sources A Recovery Util Environ Effects 2016; 38(7): 982-8.
[http://dx.doi.org/10.1080/15567036.2013.835363]

[58]   Sharma SCD, Feng C, Li J, *et al.* Electrochemical characterization of a novel exoelectrogenic bacterium strain SCS5, isolated from a mediator-less microbial fuel cell and phylogenetically related to *Aeromonas jandaei*. Microbes Environ 2016; 31(3): 213-25.
[http://dx.doi.org/10.1264/jsme2.ME15185] [PMID: 27396922]

[59]   Kimura ZI, Chung KM, Itoh H, Hiraishi A, Okabe S. Raoultella electrica sp. nov., isolated from anodic biofilms of a glucose-fed microbial fuel cell. International Journal of Systematic and Evolutionary Microbiology. 2014 Apr;64(Pt_4):1384-8.

[60]   Luo J, Yang J, He H, *et al.* A new electrochemically active bacterium phylogenetically related to *Tolumonas osonensis* and power performance in MFCs. Bioresour Technol 2013; 139: 141-8.
[http://dx.doi.org/10.1016/j.biortech.2013.04.031] [PMID: 23651598]

[61]   Xu S, Liu H. New exoelectrogen Citrobacter sp. SX-1 isolated from a microbial fuel cell. J Appl Microbiol 2011; 111(5): 1108-15.
[http://dx.doi.org/10.1111/j.1365-2672.2011.05129.x] [PMID: 21854512]

[62]   Xing D, Cheng S, Logan BE, Regan JM. Isolation of the exoelectrogenic denitrifying bacterium *Comamonas denitrificans* based on dilution to extinction. Appl Microbiol Biotechnol 2010; 85(5): 1575-87.
[http://dx.doi.org/10.1007/s00253-009-2240-0] [PMID: 19779712]

[63]   Feng Y, Lee H, Wang X, Liu Y. Electricity generation in microbial fuel cells at different temperature and isolation of electrogenic bacteria. 2009 Asia-Pacific Power and Energy Engineering Conference, Wuhan, China, 2009, pp. 1-5.
[http://dx.doi.org/10.1109/APPEEC.2009.4918327]

[64]   Zuo Y, Xing D, Regan JM, Logan BE. Isolation of the exoelectrogenic bacterium *Ochrobactrum anthropi* YZ-1 by using a U-tube microbial fuel cell. Appl Environ Microbiol 2008; 74(10): 3130-7.
[http://dx.doi.org/10.1128/AEM.02732-07] [PMID: 18359834]

[65] Wang J, Ren K, Zhu Y, Huang J, Liu S. A Review of Recent Advances in Microbial Fuel Cells: Preparation, Operation, and Application. BioTech (Basel) 2022; 11(4): 44.
[http://dx.doi.org/10.3390/biotech11040044] [PMID: 36278556]

[66] Colombo A, Marzorati S, Lucchini G, Cristiani P, Pant D, Schievano A. Assisting cultivation of photosynthetic microorganisms by microbial fuel cells to enhance nutrients recovery from wastewater. Bioresour Technol 2017; 237: 240-8.
[http://dx.doi.org/10.1016/j.biortech.2017.03.038] [PMID: 28341382]

[67] Sreelekshmy BR. Exploration of electrochemcially active bacterial strains for microbial fuel cells: an innovation in bioelectricity generation. J Pure Appl Microbiol 2020; 14(1): 103-22.
[http://dx.doi.org/10.22207/JPAM.14.1.12]

[68] Bennetto HP, Delaney GM, Mason JR, Roller SD, Stirling JL, Thurston CF. The sucrose fuel cell: Efficient biomass conversion using a microbial catalyst. Biotechnol Lett 1985; 7(10): 699-704.
[http://dx.doi.org/10.1007/BF01032279]

[69] Niessen J, Schröder U, Scholz F. Exploiting complex carbohydrates for microbial electricity generation? a bacterial fuel cell operating on starch. Electrochem Commun 2004; 6(9): 955-8.
[http://dx.doi.org/10.1016/j.elecom.2004.07.010]

[70] Ni G, Christel S, Roman P, Wong ZL, Bijmans MFM, Dopson M. Electricity generation from an inorganic sulfur compound containing mining wastewater by acidophilic microorganisms. Res Microbiol 2016; 167(7): 568-75.
[http://dx.doi.org/10.1016/j.resmic.2016.04.010] [PMID: 27155452]

[71] Islam MA, Karim A, Woon CW, *et al.* Augmentation of air cathode microbial fuel cell performance using wild type *Klebsiella variicola*. RSC Advances 2017; 7(8): 4798-805.
[http://dx.doi.org/10.1039/C6RA24835G]

[72] Gurav R, Bhatia SK, Choi TR, *et al.* Chitin biomass powered microbial fuel cell for electricity production using halophilic *Bacillus circulans* BBL03 isolated from sea salt harvesting area. Bioelectrochemistry 2019; 130: 107329.
[http://dx.doi.org/10.1016/j.bioelechem.2019.107329] [PMID: 31325898]

[73] Kumari S, Mangwani N, Das S. Low-voltage producing microbial fuel cell constructs using biofilm-forming marine bacteria. Curr Sci 2015; 925-32.

[74] Bond DR, Holmes DE, Tender LM, Lovley DR. Electrode-reducing microorganisms that harvest energy from marine sediments. Science 2002; 295(5554): 483-5.
[http://dx.doi.org/10.1126/science.1066771] [PMID: 11799240]

[75] Gatti MN, Milocco RH. A biofilm model of microbial fuel cells for engineering applications. Int J Energy Environ Eng 2017; 8(4): 303-15.
[http://dx.doi.org/10.1007/s40095-017-0249-1]

[76] Samrot AV, Senthilkumar P, Pavankumar K, Akilandeswari GC, Rajalakshmi N, Dhathathreyan KS. RETRACTED: Electricity generation by Enterobacter cloacae SU-1 in mediator less microbial fuel cell. Int J Hyd Energy 2010; 35(15): 7723-9.

[77] Kim IS, Chae KJ, Choi MJ, Verstraete W. Microbial fuel cells: recent advances, bacterial communities and application beyond electricity generation. Environ Eng Res 2008; 13(2): 51-65.
[http://dx.doi.org/10.4491/eer.2008.13.2.051]

[78] Aghababaie M, Farhadian M, Jeihanipour A, Biria D. Effective factors on the performance of microbial fuel cells in wastewater treatment – a review. Environ Technol Rev 2015; 4(1): 71-89. [Internet].
[http://dx.doi.org/10.1080/09593330.2015.1077896]

[79] Liu H, Cheng S, Logan BE. Production of electricity from acetate or butyrate using a single-chamber microbial fuel cell. Environ Sci Technol 2005; 39(2): 658-62.
[http://dx.doi.org/10.1021/es048927c] [PMID: 15707069]

[80]   Hong SW, Kim HS, Chung TH. Alteration of sediment organic matter in sediment microbial fuel cells. Environ Pollut 2010; 158(1): 185-91.
[http://dx.doi.org/10.1016/j.envpol.2009.07.022] [PMID: 19665268]

[81]   Jung SP, Pandit S. Important factors influencing microbial fuel cell performance. InMicrobial electrochemical technology 2019 Jan 1 (pp. 377-406). Elsevier.
[http://dx.doi.org/10.1016/B978-0-444-64052-9.00015-7]

[82]   Jiang D, Li B. Granular activated carbon single-chamber microbial fuel cells (GAC-SCMFCs): A design suitable for large-scale wastewater treatment processes. Biochem Eng J 2009; 47(1-3): 31-7.
[http://dx.doi.org/10.1016/j.bej.2009.06.013]

[83]   Bullen RA, Arnot TC, Lakeman JB, Walsh FC. Biofuel cells and their development. Biosens Bioelectron 2006; 21(11): 2015-45.
[http://dx.doi.org/10.1016/j.bios.2006.01.030] [PMID: 16569499]

[84]   Mathuriya AS. Inoculum selection to enhance performance of a microbial fuel cell for electricity generation during wastewater treatment. Environ Technol 2013; 34(13-14): 1957-64.
[http://dx.doi.org/10.1080/09593330.2013.808674] [PMID: 24350449]

[85]   Logan BE. Exoelectrogenic bacteria that power microbial fuel cells. Nat Rev Microbiol 2009; 7(5): 375-81.
[http://dx.doi.org/10.1038/nrmicro2113] [PMID: 19330018]

[86]   Sammes N, Ed. Fuel cell technology: reaching towards commercialization. Springer Science & Business Media 2006.
[http://dx.doi.org/10.1007/1-84628-207-1]

[87]   Schröder U. Anodic electron transfer mechanisms in microbial fuel cells and their energy efficiency. Phys Chem Chem Phys 2007; 9(21): 2619-29.
[http://dx.doi.org/10.1039/B703627M] [PMID: 17627307]

[88]   Davis F, Higson SPJ. Biofuel cells—Recent advances and applications. Biosens Bioelectron 2007; 22(7): 1224-35.
[http://dx.doi.org/10.1016/j.bios.2006.04.029] [PMID: 16781864]

[89]   Shukla AK, Suresh P, Sheela B, Rajendran AJ. Biological fuel cells and their applications. Curr Sci 2004; 87(4): 455-68.

[90]   Fatemi S, Ghoreyshi AA, Najafpour G, Rahimnejad M. Bioelectricity generation in mediator-less microbial fuel cell: application of pure and mixed cultures. Iranian (Iranica). Iran J Energy Environ 2012; 3(2).
[http://dx.doi.org/10.5829/idosi.ijee.2012.03.02.0516]

[91]   Logan BE. Scaling up microbial fuel cells and other bioelectrochemical systems. Appl Microbiol Biotechnol 2010; 85(6): 1665-71.
[http://dx.doi.org/10.1007/s00253-009-2378-9] [PMID: 20013119]

[92]   Mahdi Mardanpour M, Nasr Esfahany M, Behzad T, Sedaqatvand R. Single chamber microbial fuel cell with spiral anode for dairy wastewater treatment. Biosens Bioelectron 2012; 38(1): 264-9.
[http://dx.doi.org/10.1016/j.bios.2012.05.046] [PMID: 22748963]

[93]   Yuan H, He Z. Graphene-modified electrodes for enhancing the performance of microbial fuel cells. Nanoscale 2015; 7(16): 7022-9.
[http://dx.doi.org/10.1039/C4NR05637J] [PMID: 25465393]

[94]   Feng Y, Yang Q, Wang X, Logan BE. Treatment of carbon fiber brush anodes for improving power generation in air–cathode microbial fuel cells. J Power Sources 2010; 195(7): 1841-4.
[http://dx.doi.org/10.1016/j.jpowsour.2009.10.030]

[95]   Zhang X, Cheng S, Liang P, Huang X, Logan BE. Scalable air cathode microbial fuel cells using glass fiber separators, plastic mesh supporters, and graphite fiber brush anodes. Bioresour Technol 2011;

102(1): 372-5.
[http://dx.doi.org/10.1016/j.biortech.2010.05.090] [PMID: 20566288]

[96]   Saito T, Mehanna M, Wang X, *et al.* Effect of nitrogen addition on the performance of microbial fuel cell anodes. Bioresour Technol 2011; 102(1): 395-8.
[http://dx.doi.org/10.1016/j.biortech.2010.05.063] [PMID: 20889061]

[97]   Hou R, Luo C, Zhou S, Wang Y, Yuan Y, Zhou S. Anode potential-dependent protection of electroactive biofilms against metal ion shock *via* regulating extracellular polymeric substances. Water Res 2020; 178: 115845.
[http://dx.doi.org/10.1016/j.watres.2020.115845] [PMID: 32353609]

[98]   Tan CH, Oh HS, Sheraton VM, *et al.* Convection and the extracellular matrix dictate inter-and intra-biofilm quorum sensing communication in environmental systems. Environ Sci Technol 2020; 54(11): 6730-40.
[http://dx.doi.org/10.1021/acs.est.0c00716] [PMID: 32390423]

[99]   Varjani SJ, Gnansounou E, Pandey A. Comprehensive review on toxicity of persistent organic pollutants from petroleum refinery waste and their degradation by microorganisms. Chemosphere 2017; 188: 280-91.
[http://dx.doi.org/10.1016/j.chemosphere.2017.09.005] [PMID: 28888116]

[100] Cheng XL, Xu Q, Sun JD, *et al.* Quorum sensing signals improve the power performance and chlortetracycline degradation efficiency of mixed-culture electroactive biofilms. iScience 2022; 25(5): 104299.
[http://dx.doi.org/10.1016/j.isci.2022.104299] [PMID: 35573194]

[101] Sahreen S, Mukhtar H, Imre K, Morar A, Herman V, Sharif S. Exploring the function of quorum sensing regulated biofilms in biological wastewater treatment: A review. Int J Mol Sci 2022; 23(17): 9751.
[http://dx.doi.org/10.3390/ijms23179751] [PMID: 36077148]

[102] Rodrigo MA, Cañizares P, Lobato J, Paz R, Sáez C, Linares JJ. Production of electricity from the treatment of urban waste water using a microbial fuel cell. J Power Sources 2007; 169(1): 198-204.
[http://dx.doi.org/10.1016/j.jpowsour.2007.01.054]

[103] Angelaalincy M, Senthilkumar N, Karpagam R, Kumar GG, Ashokkumar B, Varalakshmi P. Enhanced extracellular polysaccharide production and self-sustainable electricity generation for PAMFCs by *Scenedesmus* sp. SB1. ACS Omega 2017; 2(7): 3754-65.
[http://dx.doi.org/10.1021/acsomega.7b00326] [PMID: 30023702]

[104] Sun D, Chen J, Huang H, Liu W, Ye Y, Cheng S. The effect of biofilm thickness on electrochemical activity of *Geobacter sulfurreducens.* Int J Hydrogen Energy 2016; 41(37): 16523-8.
[http://dx.doi.org/10.1016/j.ijhydene.2016.04.163]

[105] Logan BE. Microbial fuel cells. John Wiley & Sons; 2008 Feb 8.

[106] Kumar R, Singh L, Zularisam AW, Hai FI. Microbial fuel cell is emerging as a versatile technology: a review on its possible applications, challenges and strategies to improve the performances. Int J Energy Res 2018; 42(2): 369-94.
[http://dx.doi.org/10.1002/er.3780]

[107] Reguera G, Nevin KP, Nicoll JS, Covalla SF, Woodard TL, Lovley DR. Biofilm and nanowire production leads to increased current in *Geobacter sulfurreducens* fuel cells. Appl Environ Microbiol 2006; 72(11): 7345-8.
[http://dx.doi.org/10.1128/AEM.01444-06] [PMID: 16936064]

[108] Angelaalincy MJ, Navanietha Krishnaraj R, Shakambari G, Ashokkumar B, Kathiresan S, Varalakshmi P. Biofilm engineering approaches for improving the performance of microbial fuel cells and bioelectrochemical systems. Front Energy Res 2018; 6: 63.
[http://dx.doi.org/10.3389/fenrg.2018.00063]

[109] Kouzuma A, Meng XY, Kimura N, Hashimoto K, Watanabe K. Disruption of the putative cell surface polysaccharide biosynthesis gene SO3177 in *Shewanella oneidensis* MR-1 enhances adhesion to electrodes and current generation in microbial fuel cells. Appl Environ Microbiol 2010; 76(13): 4151-7.
[http://dx.doi.org/10.1128/AEM.00117-10] [PMID: 20453127]

[110] Yang Y, Wu Y, Hu Y, *et al.* Engineering electrode-attached microbial consortia for high-performance xylose-fed microbial fuel cell. ACS Catal 2015; 5(11): 6937-45.
[http://dx.doi.org/10.1021/acscatal.5b01733]

[111] Santoro C, Arbizzani C, Erable B, Ieropoulos I. Microbial fuel cells: From fundamentals to applications. A review. J Power Sources 2017; 356: 225-44.
[http://dx.doi.org/10.1016/j.jpowsour.2017.03.109] [PMID: 28717261]

[112] Nawaz A, ul Haq I, Qaisar K, *et al.* Microbial fuel cells: Insight into simultaneous wastewater treatment and bioelectricity generation. Process Saf Environ Prot 2022; 161: 357-73.
[http://dx.doi.org/10.1016/j.psep.2022.03.039]

[113] Yaqoob AA, Ibrahim MNM, Rodríguez-Couto S. Development and modification of materials to build cost-effective anodes for microbial fuel cells (MFCs): An overview. Biochem Eng J 2020; 164: 107779.
[http://dx.doi.org/10.1016/j.bej.2020.107779]

[114] kumar GG, Sarathi VGS, Nahm KS. Recent advances and challenges in the anode architecture and their modifications for the applications of microbial fuel cells. Biosens Bioelectron 2013; 43: 461-75.
[http://dx.doi.org/10.1016/j.bios.2012.12.048] [PMID: 23452909]

[115] Naaz T, Kumar A, Vempaty A, *et al.* Recent advances in biological approaches towards anode biofilm engineering for improvement of extracellular electron transfer in microbial fuel cells. Environ Eng Res 2023; 28(5): 220666.
[http://dx.doi.org/10.4491/eer.2022.666]

[116] Rahimnejad M, Adhami A, Darvari S, Zirepour A, Oh SE. Microbial fuel cell as new technology for bioelectricity generation: A review. Alex Eng J 2015; 54(3): 745-56.
[http://dx.doi.org/10.1016/j.aej.2015.03.031]

[117] Song YC, Kim DS, Woo JH, Subha B, Jang SH, Sivakumar S. Effect of surface modification of anode with surfactant on the performance of microbial fuel cell. Int J Energy Res 2015; 39(6): 860-8.
[http://dx.doi.org/10.1002/er.3284]

[118] Md Khudzari J, Gariépy Y, Kurian J, Tartakovsky B, Raghavan GSV. Effects of biochar anodes in rice plant microbial fuel cells on the production of bioelectricity, biomass, and methane. Biochem Eng J 2019; 141: 190-9.
[http://dx.doi.org/10.1016/j.bej.2018.10.012]

[119] Chen S, He G, Liu Q, *et al.* Layered corrugated electrode macrostructures boost microbial bioelectrocatalysis. Energy Environ Sci 2012; 5(12): 9769-72.
[http://dx.doi.org/10.1039/c2ee23344d]

[120] Karthikeyan R, Wang B, Xuan J, Wong JWC, Lee PKH, Leung MKH. Interfacial electron transfer and bioelectrocatalysis of carbonized plant material as effective anode of microbial fuel cell. Electrochim Acta 2015; 157: 314-23.
[http://dx.doi.org/10.1016/j.electacta.2015.01.029]

[121] Chen S, Hou H, Harnisch F, *et al.* Electrospun and solution blown three-dimensional carbon fiber nonwovens for application as electrodes in microbial fuel cells. Energy Environ Sci 2011; 4(4): 1417-21.
[http://dx.doi.org/10.1039/c0ee00446d]

[122] Sekoai PT, Awosusi AA, Yoro KO, *et al.* Microbial cell immobilization in biohydrogen production: a short overview. Crit Rev Biotechnol 2018; 38(2): 157-71.

[http://dx.doi.org/10.1080/07388551.2017.1312274] [PMID: 28391705]

[123] Guo K, Freguia S, Dennis PG, *et al.* Effects of surface charge and hydrophobicity on anodic biofilm formation, community composition, and current generation in bioelectrochemical systems. Environ Sci Technol 2013; 47(13): 7563-70.
[http://dx.doi.org/10.1021/es400901u] [PMID: 23745742]

[124] Lai B, Tang X, Li H, Du Z, Liu X, Zhang Q. Power production enhancement with a polyaniline modified anode in microbial fuel cells. Biosens Bioelectron 2011; 28(1): 373-7.
[http://dx.doi.org/10.1016/j.bios.2011.07.050] [PMID: 21820889]

[125] Marshall KC, Blainey BL. Role of bacterial adhesion in biofilm formation and biocorrosion. Proceedings of the International Workshop on Industrial Biofouling and Biocorrosion. Stuttgart. September 13–14, 1990; 29-46.
[http://dx.doi.org/10.1007/978-3-642-76543-8_3]

[126] Rabaey K, Boon N, Höfte M, Verstraete W. Microbial phenazine production enhances electron transfer in biofuel cells. Environ Sci Technol 2005; 39(9): 3401-8.
[http://dx.doi.org/10.1021/es048563o] [PMID: 15926596]

[127] Xu YS, Zheng T, Yong XY, *et al.* Trace heavy metal ions promoted extracellular electron transfer and power generation by *Shewanella* in microbial fuel cells. Bioresour Technol 2016; 211: 542-7.
[http://dx.doi.org/10.1016/j.biortech.2016.03.144] [PMID: 27038263]

[128] Christwardana M, Kwon Y. Yeast and carbon nanotube based biocatalyst developed by synergetic effects of covalent bonding and hydrophobic interaction for performance enhancement of membraneless microbial fuel cell. Bioresour Technol 2017; 225: 175-82.
[http://dx.doi.org/10.1016/j.biortech.2016.11.051] [PMID: 27889476]

[129] Vaara M. Agents that increase the permeability of the outer membrane. Microbiol Rev 1992; 56(3): 395-411.
[http://dx.doi.org/10.1128/mr.56.3.395-411.1992] [PMID: 1406489]

[130] Liu J, Qiao Y, Lu ZS, Song H, Li CM. Enhance electron transfer and performance of microbial fuel cells by perforating the cell membrane. Electrochem Commun 2012; 15(1): 50-3.
[http://dx.doi.org/10.1016/j.elecom.2011.11.018]

[131] Xu YS, Zheng T, Yong XY, *et al.* Trace heavy metal ions promoted extracellular electron transfer and power generation by *Shewanella* in microbial fuel cells. Bioresour Technol 2016; 211: 542-7.
[http://dx.doi.org/10.1016/j.biortech.2016.03.144] [PMID: 27038263]

[132] Yanuka-Golub K, Reshef L, Rishpon J, Gophna U. Community structure dynamics during startup in microbial fuel cells – The effect of phosphate concentrations. Bioresour Technol 2016; 212: 151-9.
[http://dx.doi.org/10.1016/j.biortech.2016.04.016] [PMID: 27092994]

[133] Xiang K, Qiao Y, Ching CB, Li CM. GldA overexpressing-engineered E. coli as superior electrocatalyst for microbial fuel cells. Electrochem Commun 2009; 11(8): 1593-5.
[http://dx.doi.org/10.1016/j.elecom.2009.06.004]

[134] Yong YC, Yu YY, Yang Y, *et al.* Increasing intracellular releasable electrons dramatically enhances bioelectricity output in microbial fuel cells. Electrochem Commun 2012; 19: 13-6.
[http://dx.doi.org/10.1016/j.elecom.2012.03.002]

[135] Yong YC, Yu YY, Li CM, Zhong JJ, Song H. Bioelectricity enhancement *via* overexpression of quorum sensing system in Pseudomonas aeruginosa-inoculated microbial fuel cells. Biosens Bioelectron 2011; 30(1): 87-92.
[http://dx.doi.org/10.1016/j.bios.2011.08.032] [PMID: 21945141]

[136] Yong XY, Shi DY, Chen YL, *et al.* Enhancement of bioelectricity generation by manipulation of the electron shuttles synthesis pathway in microbial fuel cells. Bioresour Technol 2014; 152: 220-4.
[http://dx.doi.org/10.1016/j.biortech.2013.10.086] [PMID: 24292201]

[137] Costa KC, Glasser NR, Conway SJ, Newman DK. Pyocyanin degradation by a tautomerizing

demethylase inhibits *Pseudomonas aeruginosa* biofilms. Science 2017; 355(6321): 170-3.
[http://dx.doi.org/10.1126/science.aag3180] [PMID: 27940577]

[138] Förster J, Famili I, Fu P, Palsson BØ, Nielsen J. Genome-scale reconstruction of the *Saccharomyces cerevisiae* metabolic network. Genome Res 2003; 13(2): 244-53.
[http://dx.doi.org/10.1101/gr.234503] [PMID: 12566402]

[139] Berríos-Rivera S, San KY, Bennett GN. The effect of NAPRTase overexpression on the total levels of NAD, the NADH/NAD+ ratio, and the distribution of metabolites in *Escherichia coli*. Metab Eng 2002; 4(3): 238-47.
[http://dx.doi.org/10.1006/mben.2002.0229] [PMID: 12616693]

[140] Yong XY, Feng J, Chen YL, *et al.* Enhancement of bioelectricity generation by cofactor manipulation in microbial fuel cell. Biosens Bioelectron 2014; 56: 19-25.
[http://dx.doi.org/10.1016/j.bios.2013.12.058] [PMID: 24445069]

[141] Pandit S, Khilari S, Roy S, Ghangrekar MM, Pradhan D, Das D. Reduction of start-up time through bioaugmentation process in microbial fuel cells using an isolate from dark fermentative spent media fed anode. Water Sci Technol 2015; 72(1): 106-15.
[http://dx.doi.org/10.2166/wst.2015.174] [PMID: 26114278]

[142] Schmitz S, Rosenbaum MA. Boosting mediated electron transfer in bioelectrochemical systems with tailored defined microbial cocultures. Biotechnol Bioeng 2018; 115(9): 2183-93.
[http://dx.doi.org/10.1002/bit.26732] [PMID: 29777590]

[143] Min D, Cheng L, Zhang F, *et al.* Enhancing extracellular electron transfer of *Shewanella oneidensis* MR-1 through coupling improved flavin synthesis and metal-reducing conduit for pollutant degradation. Environ Sci Technol 2017; 51(9): 5082-9.
[http://dx.doi.org/10.1021/acs.est.6b04640] [PMID: 28414427]

[144] McAnulty MJG, Poosarla VG, Kim KY, Jasso-Chávez R, Logan BE, Wood TK. Electricity from methane by reversing methanogenesis. Nat Commun 2017; 8(1): 15419.
[http://dx.doi.org/10.1038/ncomms15419] [PMID: 28513579]

[145] Li F, Li Y, Sun L, *et al.* Engineering *Shewanella oneidensis* enables xylose-fed microbial fuel cell. Biotechnol Biofuels 2017; 10(1): 196.
[http://dx.doi.org/10.1186/s13068-017-0881-2] [PMID: 28804512]

[146] Nguyen DT, Tamura T, Tobe R, Mihara H, Taguchi K. Microbial fuel cell performance improvement based on FliC-deficient E. coli strain. Energy Rep 2020; 6: 763-7.
[http://dx.doi.org/10.1016/j.egyr.2020.11.133]

[147] Ge Z, Li J, Xiao L, Tong Y, He Z. Recovery of electrical energy in microbial fuel cells: brief review. Environ Sci Technol Lett 2014; 1(2): 137-41.
[http://dx.doi.org/10.1021/ez4000324]

[148] Walter XA, You J, Winfield J, Bajarunas U, Greenman J, Ieropoulos IA. From the lab to the field: Self-stratifying microbial fuel cells stacks directly powering lights. Appl Energy 2020; 277: 115514.
[http://dx.doi.org/10.1016/j.apenergy.2020.115514] [PMID: 33144751]

[149] Wang H, Ren ZJ. A comprehensive review of microbial electrochemical systems as a platform technology. Biotechnol Adv 2013; 31(8): 1796-807.
[http://dx.doi.org/10.1016/j.biotechadv.2013.10.001] [PMID: 24113213]

[150] Gude VG. Energy and water autarky of wastewater treatment and power generation systems. Renew Sustain Energy Rev 2015; 45: 52-68.
[http://dx.doi.org/10.1016/j.rser.2015.01.055]

[151] Saba B, Christy AD, Yu Z, Co AC. Sustainable power generation from bacterio-algal microbial fuel cells (MFCs): An overview. Renew Sustain Energy Rev 2017; 73: 75-84.
[http://dx.doi.org/10.1016/j.rser.2017.01.115]

[152] Ye Y, Ngo HH, Guo W, *et al.* Microbial fuel cell for nutrient recovery and electricity generation from

municipal wastewater under different ammonium concentrations. Bioresour Technol 2019; 292: 121992.
[http://dx.doi.org/10.1016/j.biortech.2019.121992] [PMID: 31430674]

[153] He L, Du P, Chen Y, *et al.* Advances in microbial fuel cells for wastewater treatment. Renew Sustain Energy Rev 2017; 71: 388-403.
[http://dx.doi.org/10.1016/j.rser.2016.12.069]

[154] Min B, Kim J, Oh S, Regan JM, Logan BE. Electricity generation from swine wastewater using microbial fuel cells. Water Res 2005; 39(20): 4961-8.
[http://dx.doi.org/10.1016/j.watres.2005.09.039] [PMID: 16293279]

[155] Tharali AD, Sain N, Osborne WJ. Microbial fuel cells in bioelectricity production. Front Life Sci 2016; 9(4): 252-66.
[http://dx.doi.org/10.1080/21553769.2016.1230787]

[156] Kumar G, Bakonyi P, Zhen G, *et al.* Microbial electrochemical systems for sustainable biohydrogen production: Surveying the experiences from a start-up viewpoint. Renew Sustain Energy Rev 2017; 70: 589-97.
[http://dx.doi.org/10.1016/j.rser.2016.11.107]

[157] Feng Y, He W, Liu J, Wang X, Qu Y, Ren N. A horizontal plug flow and stackable pilot microbial fuel cell for municipal wastewater treatment. Bioresour Technol 2014; 156: 132-8.
[http://dx.doi.org/10.1016/j.biortech.2013.12.104] [PMID: 24495538]

[158] Abubackar HN, Biryol İ, Ayol A. Yeast industry wastewater treatment with microbial fuel cells: Effect of electrode materials and reactor configurations. Int J Hydrogen Energy 2023; 48(33): 12424-32.
[http://dx.doi.org/10.1016/j.ijhydene.2022.05.277]

[159] Rashid T, Sher F, Hazafa A, *et al.* Design and feasibility study of novel paraboloid graphite based microbial fuel cell for bioelectrogenesis and pharmaceutical wastewater treatment. J Environ Chem Eng 2021; 9(1): 104502.
[http://dx.doi.org/10.1016/j.jece.2020.104502]

[160] Karuppiah T, Uthirakrishnan U, Sivakumar SV, *et al.* Processing of electroplating industry wastewater through dual chambered microbial fuel cells (MFC) for simultaneous treatment of wastewater and green fuel production. Int J Hydrogen Energy 2022; 47(88): 37569-76.
[http://dx.doi.org/10.1016/j.ijhydene.2021.06.034]

[161] Negassa LW, Mohiuddin M, Tiruye GA. Treatment of brewery industrial wastewater and generation of sustainable bioelectricity by microbial fuel cell inoculated with locally isolated microorganisms. J Water Process Eng 2021; 41: 102018.
[http://dx.doi.org/10.1016/j.jwpe.2021.102018]

[162] Pugazhendi A, Al-Mutairi AE, Jamal MT, Jeyakumar RB, Palanisamy K. Treatment of seafood industrial wastewater coupled with electricity production using air cathode microbial fuel cell under saline condition. Int J Energy Res 2020; 44(15): 12535-45.
[http://dx.doi.org/10.1002/er.5774]

[163] Mittal Y, Dash S, Srivastava P, Mishra PM, Aminabhavi TM, Yadav AK. Azo dye containing wastewater treatment in earthen membrane based unplanted two chambered constructed wetlands-microbial fuel cells: A new design for enhanced performance. Chem Eng J 2022; 427: 131856.
[http://dx.doi.org/10.1016/j.cej.2021.131856]

[164] Ren B, Wang T, Zhao Y. Two-stage hybrid constructed wetland-microbial fuel cells for swine wastewater treatment and bioenergy generation. Chemosphere 2021; 268: 128803.
[http://dx.doi.org/10.1016/j.chemosphere.2020.128803] [PMID: 33143898]

[165] Jamal MT, Pugazhendi A. Treatment of fish market wastewater and energy production using halophiles in air cathode microbial fuel cell. J Environ Manage 2021; 292: 112752.
[http://dx.doi.org/10.1016/j.jenvman.2021.112752] [PMID: 33984645]

[166] Sharma I. Bioremediation techniques for polluted environment: concept, advantages, limitations, and prospects. InTrace metals in the environment-new approaches and recent advances 2020 Dec 7. IntechOpen.

[167] Mandal SK, Das N. Application of microbial fuel cells for bioremediation of environmental pollutants: an overview. J Microbiol Biotechnol Food Sci 2021; 2021: 437-44.

[168] Strycharz SM, Woodard TL, Johnson JP, *et al.* Graphite electrode as a sole electron donor for reductive dechlorination of tetrachlorethene by *Geobacter lovleyi*. Appl Environ Microbiol 2008; 74(19): 5943-7.
[http://dx.doi.org/10.1128/AEM.00961-08] [PMID: 18658278]

[169] Wang J, Song X, Li Q, *et al.* Bioenergy generation and degradation pathway of phenanthrene and anthracene in a constructed wetland-microbial fuel cell with an anode amended with nZVI. Water Res 2019; 150: 340-8.
[http://dx.doi.org/10.1016/j.watres.2018.11.075] [PMID: 30530128]

[170] Wang L, Liu Y, Ma J, Zhao F. Rapid degradation of sulphamethoxazole and the further transformation of 3-amino-5-methylisoxazole in a microbial fuel cell. Water Res 2016; 88: 322-8.
[http://dx.doi.org/10.1016/j.watres.2015.10.030] [PMID: 26512810]

[171] Zhang L, Yin X, Li SFY. Bio-electrochemical degradation of paracetamol in a microbial fuel cell-Fenton system. Chem Eng J 2015; 276: 185-92.
[http://dx.doi.org/10.1016/j.cej.2015.04.065]

[172] Liu H, Hu TJ, Zeng GM, *et al.* Electricity generation using p-nitrophenol as substrate in microbial fuel cell. Int Biodeterior Biodegradation 2013; 76: 108-11.
[http://dx.doi.org/10.1016/j.ibiod.2012.06.015]

[173] Hassan H, Schulte-Illingheim L, Jin B, Dai S. Degradation of 2, 4-dichlorophenol by *Bacillus Subtilis* with concurrent electricity generation in microbial fuel cell. Procedia Eng 2016; 148: 370-7.
[http://dx.doi.org/10.1016/j.proeng.2016.06.473]

[174] Sherafatmand M, Ng HY. Using sediment microbial fuel cells (SMFCs) for bioremediation of polycyclic aromatic hydrocarbons (PAHs). Bioresour Technol 2015; 195: 122-30.
[http://dx.doi.org/10.1016/j.biortech.2015.06.002] [PMID: 26081161]

[175] Adelaja O, Keshavarz T, Kyazze G. Treatment of phenanthrene and benzene using microbial fuel cells operated continuously for possible *in situ* and *ex situ* applications. Int Biodeterior Biodegradation 2017; 116: 91-103.
[http://dx.doi.org/10.1016/j.ibiod.2016.10.021]

[176] Borello D, Gagliardi G, Aimola G, *et al.* Use of microbial fuel cells for soil remediation: A preliminary study on DDE. Int J Hydrogen Energy 2021; 46(16): 10131-42.
[http://dx.doi.org/10.1016/j.ijhydene.2020.07.074]

[177] Zhao H, Kong CH. Elimination of pyraclostrobin by simultaneous microbial degradation coupled with the Fenton process in microbial fuel cells and the microbial community. Bioresour Technol 2018; 258: 227-33.
[http://dx.doi.org/10.1016/j.biortech.2018.03.012] [PMID: 29525598]

[178] Nagel B, Dellweg H, Gierasch LM. Glossary for chemists of terms used in biotechnology (IUPAC Recommendations 1992). Pure Appl Chem 1992; 64(1): 143-68.
[http://dx.doi.org/10.1351/pac199264010143]

[179] Karube I, Matsunaga T, Mttsuda S, Suzuki S. Btotechnol Bioeng. 1997.

[180] Kim M, Sik Hyun M, Gadd GM, Joo Kim H. A novel biomonitoring system using microbial fuel cells. J Environ Monit 2007; 9(12): 1323-8.
[http://dx.doi.org/10.1039/b713114c] [PMID: 18049770]

[181] Cui Y, Lai B, Tang X. Microbial fuel cell-based biosensors. Biosensors (Basel) 2019; 9(3): 92.

[http://dx.doi.org/10.3390/bios9030092] [PMID: 31340591]

[182]   Zheng Q, Xiong L, Mo B, Lu W, Kim S, Wang Z. Temperature and humidity sensor powered by an individual microbial fuel cell in a power management system. Sensors (Basel) 2015; 15(9): 23126-44.
[http://dx.doi.org/10.3390/s150923126] [PMID: 26378546]

[183]   Kim BH, Chang IS, Cheol Gil G, Park HS, Kim HJ. Novel BOD (biological oxygen demand) sensor using mediator-less microbial fuel cell. Biotechnol Lett 2003; 25(7): 541-5.
[http://dx.doi.org/10.1023/A:1022891231369] [PMID: 12882142]

CHAPTER 4

Sustainable Production of Bioenergy through Microbes for Ecosystem Restoration: A Clean and Green Energy Strategy

Omolara Victoria Oyelade[1,*], Jerome O. Ihuma[2], Govindaraj Kamalam Dinesh[3,4,5] and Ravi Raveena[6]

[1] *Department of Physics, Faculty of Science and Technology, Bingham University, Karu Nasarawa State, Nigeria*

[2] *Department of Biological Science, Faculty of Science and Technology, Bingham University, Karu Nasarawa State, Nigeria*

[3] *Division of Environment Science, ICAR-Indian Agricultural Research Institute, New Delhi-110012, India*

[4] *Division of Environmental Sciences, Department of Soil Science and Agricultural Chemistry, SRM College of Agricultural Sciences, SRM Institute of Science and Technology, Baburayanpettai-603201, Chengalpattu, Tamil Nadu, India*

[5] *INTI International University, Persiaran Perdana BBN, Putra Nilai, 71800 Negeri Sembilan, Malaysia*

[6] *Department of Environmental Sciences, Tamil Nadu Agricultural University, Coimbatore, Tamil Nadu, India*

Abstract: Energy crises resulting from the depletion of petroleum resources, hikes in the price of fossil fuel, and unpredictable climate change are some of the recent concerns that have provoked serious research on alternative energy sources that would be sustainable. This book chapter reviews how sustainable bioenergy production through microbes using feedstocks can provide clean and green energy that can consequently facilitate ecosystem restoration. Feedstocks are pivotal to this biotechnological process. Microbes are also equally very vital. Therefore, changing from fossil fuel to bioenergy resource options is essential. Energy transition can, therefore, create emerging opportunities in bioenergy rendering and bioeconomy that will result in the possible use of clean and green energy. In this regard, biofuels are a straightforward substitute for fossil fuels. Renewable feedstocks are suitable ingredients that sustainably produce biofuels using microbial-based bioconversion processes. Microorganisms can massively secrete industrially important enzymes capable of degrading long-chained biopolymers into short-chained monomeric sugars and fermenting them into energy-dense biomolecules. Microbes play a crucial role in

---

* **Corresponding author Omolara Victoria Oyelade:** Department of Physics, Faculty of Science and Technology, Bingham University, Karu Nasarawa State, Nigeria; E-mail: oyeladeov@binghamuni.edu.ng

the sustainable generation of biofuels and bioenergy. Bioenergy research is, therefore, crucial for a nation's economic stability and energy security. Additionally, reducing greenhouse gas emissions while promoting the use of renewable energies and the creation of livelihoods aids in the worldwide effort. Anthropogenic activities are highly reduced, thereby enhancing ecosystem restoration.

Keywords: Biofuels, Bioenergy, Ecosystem, Feedstock, Energy, Microbes, Renewable.

INTRODUCTION

Global climate change and energy security are the two issues that affect all countries [1]. Energy consumption has surged as a result of rapid industrialization and accelerated population expansion globally [2]. The importance of generating renewable fuels as a substitute for conventional fossil fuels has become more crucial because of the significant decline in fossil fuel concentration and the growing global demand for energy. In order to address the escalating levels of greenhouse gases in the Earth's atmosphere, which have consequently led to substantial alterations in global climate patterns, these modifications can have disastrous effects, including temperature and sea level rise. Transportation and energy generation using fossil fuels account for twenty-five and fourteen percent of the total emissions of greenhouse gases, respectively [3]. For instance, because it is both economically viable and ecologically acceptable, bioenergy has the potential to both replace traditional fuels and relieve environmental concerns [4]. Resources from nature, such as vegetation, woody biomass, and other organic wastes, are used to create bioenergy.

It is now the most prevalent renewable energy, providing around twelve percent of the worldwide gross total consumption of energy. Bioenergy comes in many forms; solid biomass, sometimes referred to as solid biofuels, comprises wood (logs, chips, tree bark, and dust particles from wood shavings), crop residue (fruit peels, corn cobs, hay), solid waste materials (trash, rubbish, waste from food processing unit), and gaseous or liquid-biofuels (biogas and bioethanol), which can be used for industrial purposes, transportation fuels, cooking, heating, and electricity generation. One of the most common types of bioenergy is solid biomass. In many nations, particularly developing countries, it has traditionally been used for heating or cooking. Solid biofuels provide more than 80% of the energy needed in Africa, primarily for cooking, and 30% of Austria's overall energy needs for heating. Bioenergy has a promising future because only a fraction of its potential can be used at this point. Although biomass has been utilized for at least 30 years, it is still challenging to use responsibly [5].

Critical features to take into account for its effective use include

1. The raw materials' availability, quality, and cost

2. The technology for conversion

3. Sustainability, which includes land use modification, carbon dioxide exhaustion, and reforestation

The severe threat presented by the increased atmospheric deposition of greenhouse gases, which led to substantial climatic disruptions, can be lessened by using adaptive bacteria to develop sources of sustainable energy from feedstock and organic wastes. In line with Liao *et al.*, due to the range of chemical reactions that different microbes are capable of, which allow for the production of biodiesel form a wide range of substrates, interest in using microorganisms to manufacture various biofuels has been gradually increasing in recent years [6].

Biofuel production is done using a variety of conversion techniques. Several microorganisms produce enzymes that are crucial to the synthesis of biofuels [7]. The microbial enzymes can more effectively break down feedstocks from different biomass materials and create various types of fuel sources like biodiesel and biogas. The organic waste materials from animals and plants are converted to biofuels by the microbial enzymes that use them as substrates. Biofuels are produced from the biomass of animals and feedstocks from crop residues, microbes, fungi, and algae through biological and chemical processes. It is possible to convert biological biomass employing a variety of microbial components, including extracellular enzymes [8]. These microorganisms serve as source materials and produce enzymes appropriate for converting biomass [9, 10]. Examples of biomass-to-biofuel conversion include the bacterial conversion of sugars into ethanol and plant-derived substrates by cellulolytic microorganisms. Methane can be utilized to make methanol, and microalgae and cyanobacteria can photosynthesize atmospheric carbon dioxide into biofuels [6]. When used in bio electrochemical techniques for the synthesis of biohydrogen and bioelectricity, *Geobacter sulfurreducens* and *Shewanella oneidensis* demonstrate particular "molecular machinery" that promotes the movement of ions from bacterial exterior membranes to surfaces that are conductive [11, 12].

BIOENERGY AS AN EMERGING OPPORTUNITY

Sunlight energy is converted to bioenergy in plants, which produce fuels or power that can replace nonrenewable energy sources. With appropriate planning and management, bioenergy can aid in the fight against global warming while providing various economic and environmental benefits to rural areas [13].

Compared to using fossil fuels, bioenergy systems are safer and more environmentally responsible [14, 15]. The cost of bioenergy will decrease due to technological developments, and well-managed bioenergy systems may be able to support rural development while generating jobs and enhancing the environment.

It is possible to produce bioenergy using available marginal and degraded areas [16]. A large variety of climate-friendly bioenergy tree species and low labor costs are further advantages for its production. One of the numerous benefits of bioenergy is that it is from non-edible plants that grow in unsuitable agricultural terrain, like rocky, damaged, and abandoned fields, where it is difficult to cultivate crops. Without harming food crops, fuel is produced by carefully exploiting these damaged and abandoned lands for biomass production. To address significant environmental problems, forestry-based bioenergy production techniques (such as agroforestry, reforestation, and afforestation) can stabilize the land, stop soil erosion and flooding, stop the loss of biodiversity, and provide rural communities with access to economic opportunities.

One of the largest contributors to the global demand for energy is the creation of bioenergy or energy made from biological resources, including plants, woody biomass, and other organic materials and trash. It accounts for around twelve percent of the ultimate energy usage in the globe, making it the major source of today's renewable energy use [17, 18]. The majority of bioenergy is used for conventional building heating and cooking. The Sustainable Development Goal 7 (SDG 7) aims to provide universal access to environmentally friendly gas for cooking by 2030, with Sub-Saharan African families making up over 50 percent of all houses without use to clean cooking in 2019. However, from 2010 to 2019, the yearly growth rate was just 0.2-1.8 percent globally, which is much lower than the needed 3% [17, 19]. Modern applications of bioenergy are in the production of heat and power for buildings and industries using biomass and biogas/biomethane, transportation using biofuels and biomethane and biomass materials used as industrial feedstocks. The energy shift needs modern bioenergy to be successful.

By 2050, it may account for 17% of the final energy demand or a quarter of the world's energy supply, according to the 1.5°C Scenario from the International Renewable Energy Agency (IRENA) [20]. Scaling up bioenergy would be necessary to produce fuel for transportation and heat for buildings and industrial processes. The chemical sector will use it as feedstock to make chemicals and plastics.

Bioenergy may provide the indirect emissions required to reach the net-zero emission objectives when combined with carbon capture and storage (CCS) technology in the electricity sector and some manufacturing sectors. Modern bioenergy uses include the effective use of liquid biofuels and biomethane for transportation and other purposes, as well as solid and gaseous biofuels for heating and production of electricity. Compared to hydropower and other contemporary renewable energy sources, this met 6% of the global overall demand for energy in 2019. The market for contemporary bioenergy, both liquid and solid, was projected to be worth approximately USD 79 billion in 2019. Of this amount, USD 34 billion came from bioethanol, USD 35 billion from biodiesel, and USD 10 billion from wood pellets [21].

In several countries and areas, such as Brazil, China, the European Union, and the United States, the use of bioenergy has significantly increased in recent years [17, 22]. The rate of growth remains much below what is necessary to achieve the decarbonization goals for a future with net zero emissions. Only a small fraction of 3 percent of transportation fuels and 8% of the total energy of homes and industries used for final applications are provided by bioenergy [20]. The development of bioenergy must center around the idea of sustainability. The sustainability of bioenergy has come under scrutiny, particularly regarding land use. These concerns range from socioeconomic problems like food *versus* fuel and land grabs to environmental ones like carbon stocks and biodiversity. Bioenergy can help with the energy transition, but it is crucial to watch out for any unwanted side effects.

BIOENERGY IN ENERGY TRANSITION

In the global energy shift, bioenergy is crucial. Sustainable bioenergy is essential to decreasing worldwide CO_2 emissions to virtually zero by 2050 and maintaining a rise in the world's temperature to 1.5°C, according to IRENA's 1.5°C Forecast. For many industries with few other renewable possibilities, such as the provision of aviation fuels, Bioenergy could offer medium- and long-term answers. Additionally, it might give the industry both feedstock and renewable heat [21]. By 2050, the 1.5°C scenario's ultimate energy use includes 17% bioenergy. 25% of the world's total primary energy source comes from biomass, that is 153 EJ of biomass produced, with a three-fold increase from 2018. Sustainably obtaining the necessary supply can be extremely difficult. Based on increased productivity and sustainable management, there is a significant physical potential to expand the supply of biomass, according to an IRENA study. The following has to be implemented to meet these goals:

1. Development of efficient technology for producing biofuels from algae and lignocellulose feedstocks (such as grasses, wood, and farm and forest waste).

2. Increase access to seed, water, and fertilizer in developing nations while promoting modern farming practices to accelerate the development of crop yields.

3. Integrated logistics methods for harvesting farm and forest leftovers at a reasonable cost would need to be enhanced.

4. Promote tree planting on degraded land and exchange best practices for sustainable forest management to speed up afforestation.

The connection between bioenergy and environmentally friendly development has emerged as a critical scientific subject as demand for bioenergy increases globally. Possibilities for additional land uses, economic expansion, global warming mitigation, increased energy security, and job opportunities are the main positive outcomes highlighted [23, 24]. Concerns include the possibility of food security and rural livelihoods being disrupted, as well as direct and indirect emissions of greenhouse gases (GHG) from changing agricultural practices, a rise in water scarcity, environmental impacts, a rise in rural poverty, and the eviction of local farmers, pastoralists, and timber users [25]. The unique system, the development environment, and the amount of intervention all influence the kind and extent of the consequences of adopting bioenergy [26].

BIOENERGY IN SUSTAINABLE BIOECONOMY

Bioenergy sustainability requires a transition to a bioeconomy. A bioeconomy is one in which the primary resource for materials, chemicals, or energy is renewable biological resources [27]. As a component of the broader idea of the green economy, the bioeconomic perspective lays an emphasis on the utilization of environmentally friendly raw materials and the application of commercial biotechnology research, development, and innovation in sectors like food, paper, cellulose, or energy [28]. Production and use of energy are necessary to support present economic progress [29].

IMPLICATIONS OF BIOENERGY IN THE ECONOMY

A bioeconomy (bio-based economy), in which environmentally friendly production of sustainable bioresources decreases manmade climate impacts and eliminates the need for fossil fuels, will be created as a result of the replacement of fossil fuels with innovative alternatives such as bioenergy. The intrinsic worth of materials derived from biomass will rise with reduced consumption of resources [30]. Bioresources will convert into more products. These will increase investment in research, innovative technologies, and skills [27]. Action plans to bolster the industry and give it special consideration during the regulatory process must be devised to make the transition to the bioeconomy feasible [31]. One of the developments that can aid in achieving a few of the objectives of environmentally friendly development is bioenergy. The majority of energy production and consumption methods used today necessitate that a substantial portion of the energy originates from renewable resources.

Fossil fuels would need to be replaced in a natural-based economy, and biomass consumption would have to increase significantly [31]. Bioenergy is a viable alternative to hydrocarbons in industrialized countries for use in transport, renewable energy, generation of heat, and domestic heating. Bioenergy predominates as a fuel, particularly in remote areas without access to electricity [32]. Both industrial development and economic growth can be facilitated by this industry [33].

BIOENERGY SOURCES AND THEIR PRODUCTION

Renewable energy derived from biomass is bioenergy. Any organic item containing chemical energy from sunlight stored in it is biomass. Biomass currently accounts for about 50 EJ (1 EJ = 1018 Joules) of the world's yearly primary energy consumption [34]. Fig. (**1**) depicts the biomass feedstocks utilized for energy, 80% of which come from wood (trees, branches, and leftovers) and shrubs. The remaining bioenergy feedstocks come from a range of industrial and recycled waste and byproduct streams, such as biomass processing and recycling and the organic biogenic fraction of solid waste from municipalities [34, 35].

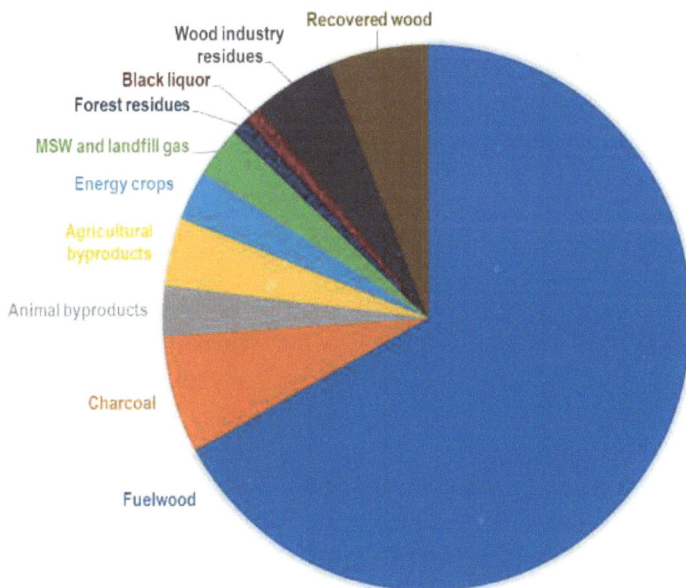

Fig. (1). Shares of global primary biomass sources for energy (modified from [18, 26]).

By utilizing the substantial amounts of underutilized residues and trash, biomass utilization has growth potential. Traditional crop usage for energy may be expanded with a thorough assessment of natural resources and food consumption. In the longer future, lignocellulose crops may be cultivated in marginal, deteriorated, and excess agricultural regions, accounting for the bulk of the biomass resource. As indicated in Fig. (1) above, these crops can be both herbaceous and woody. Aquatic biomass, algae, may also contribute significantly in the long run. Available sources of bioenergy include:

Legumous Plants

These are plant species from the Leguminosae family, including *Vachellia nilotica*, *Dalbergia sissoo*, *Peltophorum pterocarpum*, *Perkia biglobosa*, *Delonix regia*, *etc*. These plants' seed pods can be a source of carbohydrates for the fermentation process, which uses them as a substrate. These plants' pods release a sizable amount of reducing sugars during enzymatic processing.

Algae

Microalgae are prokaryotic or eukaryotic multicellular or unicellular organisms. Microalgae are sometimes used as raw materials for the production of biodiesel. In order to overcome the limitations of first-generation and second-generation biofuels, microalgae can be used.

Monocots

Monocots are plants having a single cotyledon and these are used as bioenergy sources. For example, sorghum, sweet corn or maize corn, wheat, sugarcane, and miscanthus.

Corn

Fermentation is a well-known method used to turn corn into ethanol. In a big production plant, 2.69 kilograms of maize grains may yield one liter of ethanol. As a result of the high cost of growing maize for use as a feedstock, stover from corn is the best replacement. "Corn stover" is the waste that is left behind while cultivating corn grains from the soil. The corn cobs, rice husks, foliage, and stem pieces make up leftovers. There is a lot of it because leftovers from the maize portion lead to very valuable grain production. Corn stover may become a common bioenergy component due to its extensive physical availability. The direct production of energy in biomass cofiring applications and the use of maize stover as a raw material in the manufacture of liquid fuels are potential uses for the material. Corn stover is utilized in bioenergy applications as a second-generation or cellulosic feedstock without impacting the food supply [35, 36].

Maize

In addition to being one of the most widely grown crops in the world, maize can be crucial to the development of biofuels. Maize must be grown in greater amounts if it is to be utilized for the manufacture of biofuels in order to produce both stem biomass and grain. The ability to grow maize as a dual crop is made possible by the availability of agronomic and genetic resources. Because it possesses all of these qualities and is readily available, maize can be considered to be the best example plant for biomass productivity in the field of research [37].

Wheat

A key crop for biofuels might be wheat. Wheat may be converted into ethanol *via* fermentation to create a fuel that can power automobiles. Wheat is a plant known as a C3 species or a plant that uses C3 photosynthesis. These plant species can store enough carbon in their dry mass, which supplies enough biomass for energy conversion [38].

Sugar Cane

Sugarcane is a productive crop for absorbing sunlight and transforming it into chemical energy for use. There is no disputing the promise of sugarcane as a raw material for biomass. A significant amount of bagasse from sugar cane cultivation is created after the delivery of sugarcane for processing. This bagasse is presently burnt in furnaces to generate vapor and electricity. If the technologies used to produce bioethanol are improved, it may be possible to obtain more bagasse. Bagasse can produce electricity, a raw material for the synthesis of bioethanol, and create a variety of other bio-based products [37, 39].

Sorghum

Sorghum is regarded as a different species since it has two distinct grain types, one that is used for biomass and the other for sugar production. The genomic sequence of sorghum, which is currently available, also suggests that it has the potential to be a first- and second-generation biofuel crop. The great potential for biofuel production is in forage sorghums. The method used to produce sorghum for food can apply to the production of sweet sorghum for biofuels. Combining genetic, agronomic, and processing technology can improve sorghum's potential as a bioenergy crop [37].

Edible Vegetable Oils

There is a promise to use edible oils as feedstock for biofuels, such as palm oil, soybean oil, and rapeseed oil. When utilizing edible oils as the starting material for biofuels, it is crucial to take into account the oil's origin (whether it originates from food or non-food sources), composition, and suitability as a feedstock. Despite the enormous potential of biodiesel, the high cost and growing demand for edible oils prevent their utilization as feedstocks.

Non-edible Vegetable Oils

Concerning price and source, the manufacture of biofuels utilizing food oils as feedstock is subject to some restrictions. Therefore, fewer expensive feedstocks are required to overcome this problem. Non-edible oils, notably biodiesel, offer a huge potential as raw materials for the production of biodiesel. Costs associated with producing biofuel might be reduced by using oils that are not edible as raw materials. The numerous sources that are abundantly found in nature and can make excellent feedstocks include *Jatropha curcas*, *Pongamia pinnata*, palm, *Madhuca longifolia, etc.* These plants are abundantly available in developing countries and are far less costly than edible oils. Vegetable oils reduced in viscosity can be used as fuels. Several processes, such as micro-emulsification, pyrolysis, and transesterification, can accomplish this. Due to its considerable yield, rapid reaction time, minimal temperature, and elevated pressure, transesterification is the method that makes biodiesel most commonly used for commercial use [40, 41].

Mahua

Mahua oil found in the seeds of this plant has a concentration of 30%–40%. Orissa, Chhattisgarh, Jharkhand, Bihar, Madhya Pradesh, and Tamil Nadu are the states where it is most prevalent.

Jatropha

It is a semi-evergreen shrub that thrives in marginal or poor soils, is drought-tolerant, and produces seeds with a 37% oil content. Without refinement, oil from the Jatropha plant seed is fuel. Jatropha oil was used as fuel for straightforward diesel engines and produced a clean, smoke-free flame [42].

Karanja

An abundant non-edible oil plant found in tropical Asia is called Karanja. This tree produces dried pods with an oil content of about 27.5% [43].

Neem

Neem belongs to a group of plants called Meliaceae. It is native to India and Burma and thrives in tropical and semitropical climates. This tree grows fast and can reach heights of 15 to 40 meters. Neem tree seed kernels have a healthy amount of fat (between 33% and 45%). Other uses for neem oil include soaps, medications, and insecticides [44, 45].

Animal Fats

Animal fats are suitable as feedstocks for biofuel production. When used as biofuel feedstocks, animal fats accomplish two goals: reduce the demand for waste disposal and help produce biodiesel. Animal fats are readily available sources of feedstock. Biodiesel manufactured from virgin oil-based products, such as soyabean oil, is more resistant to cold temperatures than biodiesel derived from fats from animals [42].

BIOMASS UNIQUENESS AS A RENEWABLE RESOURCE

In many ways, biomass from plant and animal matter is an exceptional source of renewable energy [28, 46]. In opposition to clean energy sources such as solar and wind energy, which create inconsistent electrical energy that requires quick consumption and a connection to the grid, it may be easily stored and transferred. The cost of biomass frequently makes up a sizable amount of the cost of producing bioenergy without waste and residues (usually between 50% and 90%) [47]. As a result, the economics of bioenergy are fundamentally different from those of other renewable energy sources that rely on unpaid resources (*e.g.*, the wind, direct sunlight, geothermal energy, tidal energy, *etc.*). Raw biomass must undergo one or more conversion processes before it can be turned into consumable bioenergy commodities and services. Plant biomass receives solar energy and turns it (through photosynthesis) into chemical energy that is stored in chemical compound bonds of its molecular elements as it develops. The resulting chemical energy can be transformed into a variety of commercially useful intermediate chemicals and energy molecules, or it can be immediately produced as heat *via* burning (and subsequently converted into electricity *via* a combustion engine or turbines). Bioenergy products obtained from feedstock might be gaseous (biogas, synthesis gas, and hydrogen), liquid (biodiesel and bioethanol), or solid (chips, pellets, and charcoal). These products are in energy applications, including transportation fuels. Biomass is a resource that, in contrast to all other forms of sustainable energy, has a strikingly diversified nature (for instance, the spectrum of sunshine is uniform everywhere in the globe). It asks that particular technologies be created for each situation, as will be discussed in the following section.

Biomass Conversion Routes

The various physical characteristics and chemical compositions of feedstocks and the necessary energy service have led to the development of several conversion routes (heat, electricity, transport fuel). Fig. (**2**) presents a summary of the various bioenergy routes. Spontaneous burning of timber to generate energy is one easy method, but other methods necessitate preparation, improving, and transformation steps, which include those required to make liquid fuels for internal combustion engines. There are three primary categories of conversion routes [48]:

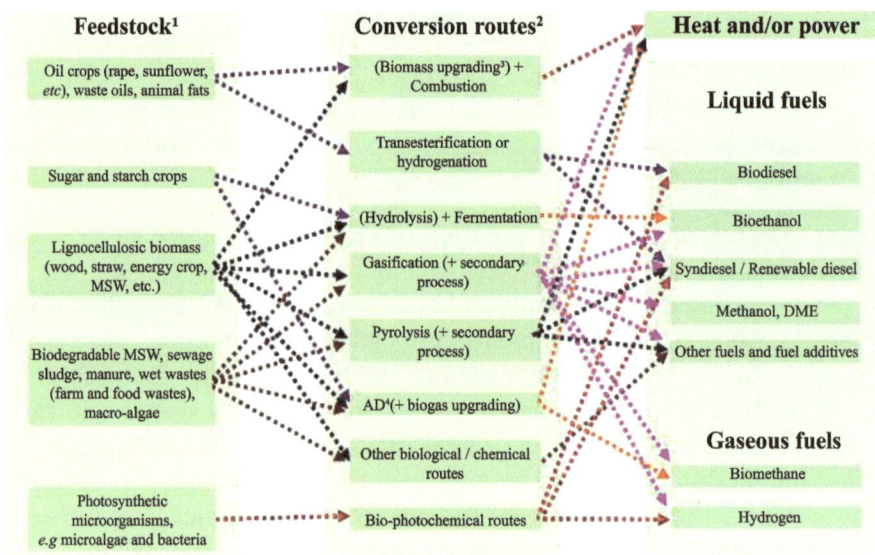

Fig. (2). Bioenergy conversion routes (modified from [49]).

Thermochemical conversion is the process by which biomass is degraded chemically at high temperatures. Combustion, gasification, pyrolysis, and torrefaction are the four thermochemical processes. The differences between these four processes mostly relate to various temperature limits, burning rates, and reaction-time oxygen content.

Oil is extracted from plants, including rapeseed, soybean, and jatropha, to create liquid fuels (biodiesel or vegetable oil) by a process known as physicochemical conversion.

The actions of live microorganisms (enzymes, bacteria) break down the substrate and produce fuels that are both gaseous and liquid in biological conversions. Examples include anaerobic metabolism (often from wet feedstock), the microbial fermentation of substrates including sugar (sugar cane, sugar beet, and sugar), starch (corn/maize, wheat), and lignocelluloses (grass, wood), and the bio-photochemical synthesis of hydrogen using algae that depend on sunlight.

Each biodiesel conversion pathway yields secondary products in addition to the main energy output. The method might benefit greatly from the co-products' economic value. Examples include fertilizers, specialty chemicals, food additives, animal feed, and additions to foods like charcoal.

The most advantageous bioenergy route is determined by the level of technological preparedness, the kind and volume of available feedstocks, and the energy service needed. As a result of having diverse goals, various parties may favor different technologies. While project developers focus on maximizing their financial gain, governments will address additional factors like the potential for carbon reduction, energy security, and overall economic gain [49].

Pre-Treatment and Upgrading Technologies for Biomass

Although biofuel is a good renewable energy source, it differs from conventional energy sources in a number of ways. Because of its inconsistent physical characteristics and a lower density of energy (up to five times lower per unit volume), it is more difficult to transport, handle, and store than fossil fuels. Additionally, pretreatment may be required to fulfill the requirements for quality and homogeneity of many conversion processes because of the considerable variability in chemical composition and water content of feedstocks made from biomass [46]. By converting raw biomass into more flexible, dense, and homogenous (solid or liquid) fuels, biofuel pre-treatment processes (or upgrading) techniques help to reduce supply chain costs and increase the efficiency and reliability of downstream operations. Due to increased transport, it is necessary to keep bioenergy generation and usage separate, while the increasing energy density of biomass may be alluring. The upgrading processes used to increase biomass's energy density include palletization, pyrolysis, torrefaction, and hydrothermal.

Palletization

Pellets are already a common fuel used in industry and houses in industrialized countries (particularly the increasingly ubiquitous pellet boilers). Reduced tiny particles of biomass that are solid are pressed to form pellets. The establishment of quality standards has led to an increase in pellet use and global trade. Particularly for heating applications where more affordable replacements for petroleum-based materials like heating oil and gas are needed, pellets have the potential to provide enormous volumes of standardized solid fuel [50].

However, during storage and transit, pellets typically absorb moisture, lowering their net calorific value and necessitating mitigating measures along the supply chain, such as quality monitoring. Currently, sawdust, a byproduct of sawmills, is used to make most pellets, which may be a constraint on the number of pellets brought into the market.

Hydrothermal Upgrading and Pyrolysis

Pyrolysis is the regulated thermal disintegration of organic matter that produces aqueous bio-oils, a mixture of gases (syngas), and carbon at temperatures of around 500°C (biochar) in an anaerobic atmosphere without oxygen. Rapid and slow pyrolysis are the two main techniques. They can be separated through various pyrolysis reactor retention times, which lead to different gas, solid, and component ratios. Fast pyrolysis is common because it optimizes the synthesis of bio-oil, whereas slow pyrolysis favors the formation of bio-char. Bio-oil should be simpler for handling, storing, and transfer than raw solid biomass. Additionally, because bio-oil has a greater density of energy (per unit volume) than pellets or torrefied biomass, it offers an economic advantage in terms of transportation costs. Evidently, bio-oil could be developed and used as an energy source for transportation, providing a useful way to link to the infrastructure for petroleum. Furthermore, liquefaction under elevated pressures (120–200 atmospheres) and at moderate temperatures (300–400 °C), maybe with the addition of water and other solvents (like methanol), can be used to produce bio-oils. The process is known as hydrothermal upgrading or HTU. The wet feedstock may be used immediately in this process, and the bio-oil produced has the desirable attribute of being harder to dissolve in water than the bio-oil produced by rapid pyrolysis.

These technologies are still at the experimental stage despite the advantages and enormous knowledge obtained, notably for quick pyrolysis. There are still issues to be solved regarding the purity, reliability, and durability of the bio-oil, which tends to deteriorate over time; additionally, only a small number of successful pyrolysis trial units have been established (such as in Finland and Canada).

Torrefaction

The biomass materials (typically wood) are transformed into a dry product that mimics coal using the higher-efficiency combustion method known as torrefaction. It occurs between 200 and 300 °C. Torrefied plant material is hydrophobic and has a substantial amount of energy, allowing it to be transported over extended periods and maintained outside without losing any of its calorific value to water absorption. It is also possible to pelletize torrefied biomass to lower the expense of handling and delivery. Torrefied pellets should offer even better cost efficiency than conventional pellets. Torrefied biomass is an attractive feedstock since it is compatible with a variety of conversion processes due to its homogenous and coal-like characteristics. Torrefaction technology is still in the demonstration phase but might be made commercially available. This would make it easier to obtain resources located far away, including forest residues and far-off forests.

Biomass for Heat Applications

Making heat is how biomass production often uses its energy. Despite the fact that the economics will vary based on the climate and the cost of fossil fuel alternatives, all industrial biomass-to-heat systems are reasonably priced.

Combustion

Burning solid plant material for heat is the most traditional and widely used way to transform it into electricity. Wide varieties of commercial solutions are currently in use and are tailored to the characteristics of the biomass and the breadth of the application since incineration is a relatively simple and widely recognized process.

Domestic Heating

Spontaneous combustion of forest feedstock has been a common practice since the dawn of civilization and is still, by far, the material transformation method that contributes the most to the global energy supply. Even though some more recent devices, such as the growing popular pellet furnaces, achieve an efficiency of up to 90%, almost all domestic biomass equipment in use are classic cooking stoves with inadequate efficiency ratings of between 5 and 30% that are common in less developed countries [19]. Both industrialized and developing countries have plenty of space to expand their usage of biomass as a heating source.

District Heating and Cooling

Despite being a reliable technology, biomass heating's feasibility is dependent on a number of complex techno-economic considerations. Some nations, such as those in northern Europe, already meet a sizeable amount of their heating demands with district heating that is powered by biomass. The primary barriers to wider implementation, notwithstanding the growing business rationale for properly sized district heating networks, are the substantial expense of new energy distribution networks and the difficulty in guaranteeing sufficient overall effectiveness. A strategy for successfully delivering cooling solutions and greater commercial viability of bioenergy projects may be found in effectively employing plants and infrastructure.

Industrial Systems

Heating systems in the 0.5–10 MWth category are more common in industries with high heat demands and abundant biomass residual supplies. Although the industrial sector has the potential to be an important sector for plant material heating, it necessitates specialist solutions that abide by the scientific standards of diverse sectors, such as those concerning flue gas quality and heating temperatures.

Gasification

While wealthy countries utilize gasification for high-value energy outputs (such as power and fuel), many emerging nations use gasifiers for direct heat applications. For instance, numerous smaller biofuel gasifiers (10-500 kWth) are effectively employed for intermittent thermal applications in China, India, and Southeast Asia. There seems to be a difficulty with the upkeep of these units throughout their functioning.

Applications of Biomass for Power and Combined Heat and Power (CHP)

There exist a variety of substrate and transformation technology combinations that have the potential to produce energy and combined heat and power (CHP); however, their development and acceptance may vary throughout different stages. The feasibility of implementing a bioenergy solution for power and combined heat and power (CHP) is contingent upon various factors. These factors include the technological aspects of the bioenergy system, such as capital and operating costs, conversion efficiency, process reliability, and economies of scale. However, the success of such a solution is also heavily influenced by the specific circumstances pertaining to the biomass supply, including its quality, type, availability, and cost. Additionally, the final electricity demand, encompassing factors such as the expenses associated with renewable power production, heat demand and value, grid connectivity, and support policies, plays a crucial role in determining the viability of a bioenergy option.

APPLICATIONS OF BIOFUELS IN TRANSPORTATION

Biofuels are an alternative energy source derived from biomass, such as plants and crops. There are two major biofuels: ethanol from biomass and biofuel. Unlike some other energy sources, plants can be cultivated almost everywhere. Biofuels produce less carbon dioxide than conventional fuels. Biofuels provide several environmental benefits, including a decrease in carbon dioxide emissions. Biofuels do, however, have some unique benefits and drawbacks [51], outlined as follows:

Advantages

1. Biofuels from renewable energy sources are preferable to fossil fuels that are from a finite supply.

2. Biofuel production reduces greenhouse gas emissions

3. Biofuels significantly lower the net production of acid rain.

4. Biofuels have superior environmental performance compared to hydrocarbon-based fuels due to their reduced emission of pollutants during combustion.

5. Biofuels are produced by converting various sources such as agricultural waste, used vegetable oils, non-edible oils, fats from animals, and other economically viable materials.

Disadvantages

1. The burning of biofuels results in significant carbon emissions.

2. In order to attain comparable energy levels to those of petroleum-based fuels, it is essential to use bioenergy on a large scale, given their relatively lower energy output.

3. The crops used for the production of biodiesel are often known as bioenergy crops. The increasing demand for crops may lead to an escalation in food expenses.

4. The use of precious farmland for making biofuel is a matter of significant concern. The cultivation of crops for the purpose of biofuel production has the potential to result in food scarcity.

5. A significant amount of water is necessary for biofuel crop maintenance. It could place a burden on the water resources available.

Biofuel Classifications

Biofuels are classified into four distinct types, namely initial, second, third, and fourth generations. The feedstock serves as the primary basis for categorization. Fig. (**3**) illustrates the process of biofuel generation and the corresponding sources from which they are derived.

Generations of Biofuels

First generation	Second generation	Third generation	Fourth generation
Edible biomass	Non edible biomass	Algal biomass	Breakthrough
• Sugarbeet • Sugarcane • Wheat • Corn • Oil crops	• Wood • Grass • Straw • Waste	• Macroalgae • Microalgae	• Pyrolysis • Solar to fuel • GMOs

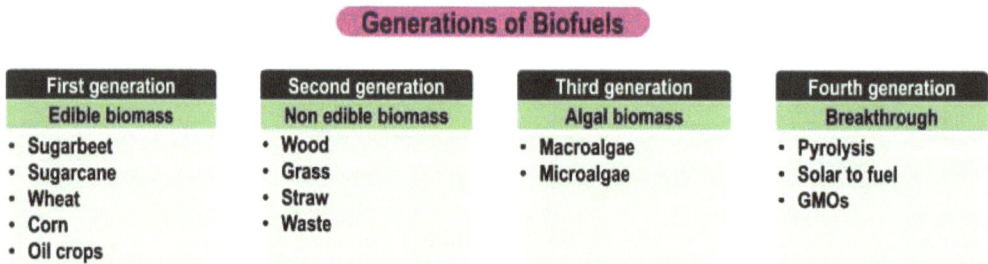

Fig. (3). Generations of Biofuels (modified from [7]).

In the first generation of biofuels, we find the proven techniques for making ethanol from biomass from carbohydrates such as starch and sugar plants, biofuel and green diesel from oil-rich crops and fats from livestock, and biomethane as from the anaerobic breakdown of wet material [52].

The term "second-generation biofuels" encompasses a broad category of novel biodiesel developed from new raw materials. Among these are biofuels derived from lignocellulosic materials (*i.e.*, fibrous biomass such as straw, timber, and grass) and conventional biofuels (including ethanol, butanol, and syndiesel) produced using standard techniques but using unique starch, oil, and sugar-producing plants like Jatropha, cassava, or Miscanthus. These routes rely on biochemistry and thermochemical techniques that are currently in the demonstration of concept stage of development [53].

Biofuels of third-generation are the most economically practical, environmentally benign, and sustainably produced fuels from photosynthetic microalgae. Microalgae are used to create several third-generation biofuels, including biodiesel, biohydrogen, and methane [48, 54]. Due to the photosynthetic nature of microalgal energy generation and the fact that it does not need fertile agricultural land, emissions of greenhouse gases may be reduced, and progress toward carbon neutrality can be accelerated. Microalgal fuel outperforms first- and second-generation biofuels for these reasons [55].

The basic concept of fourth-generation biofuels involves using solar energy to convert carbon dioxide (CO_2) into ethanol. This is achieved by the use of photosynthetic microorganisms, photovoltaics, bioelectrochemical fuel cells (or microbial fuel cells), and artificial cell components specifically designed for the manufacturing of targeted and compatible fuels [56] The use of synthetic biology methodologies is extensively applied for the extraction of lipids. Additionally, this method seeks to produce high-quality biofuels with high octane numbers, a measure of fuel quality [57]. Utilizing bioengineered microalgae also improves carbon dioxide sequestration [56]. The process of converting inorganic CO_2 to organic molecules by photosynthetic organisms is known as carbon dioxide sequestration [58].

SUSTAINABLE PRODUCTION OF BIOENERGY

The demand for bioenergy is predicted to rise as nations explore sustainable low-carbon energy options as part of national climate change mitigation programs. Bioenergy is now the source of renewable energy in the world. As nations implement more comprehensive bioeconomy laws, bioenergy expects to compete with other end users for sustainably sourced biomass. In order to understand how biomass supply chains are implemented to support bioenergy production while contributing to the Sustainable Development Goals (SDG) of the United Nations (UN), a project was initiated by IEA Bioenergy. This initiative involved several tasks to identify and document best practice case studies from around the world [59, 60].

Sustainability in bioenergy is complicated. In theory, using bioenergy can have several advantages, such as reducing GHG emissions by switching to it in place of fossil fuels for transportation, manufacturing, heating, and power generation. Additionally, it can result in socioeconomic and environmental advantages like employment development, increased health from clean cooking, and land restoration. These advantages, however, can only be attained under certain situations. The potential negative impacts on the environment, society, or the economy might arise from inadequate management of biofuel transportation and use, given their extensive linkages with key sectors, including farming, timber production, development of rural areas, and disposal of waste. The effects of bioenergy on changes in land use depend greatly on context, location, and scale [61]. For instance, bioenergy supply can result in land use for energy use rather than the current function (such as food production or ecological service), which could raise worries about problems like food security or biodiversity loss. However, potential gains exist in the use of bioenergy byproducts for soil improvement, phytoremediation plantations for water quality improvement, and agroforestry for biodiversity enhancement (see Fig. **4**). Additionally, there is a significant chance that better land stewardship practices may increase carbon sequestration. Understanding the potential role of biomass in the future requires evaluating the trade-offs, synergies, and several land-use governances-related challenges. As a result, determining the positive and negative effects on sustainability must take into account particular locales as well as other factors that define the management of feedstock production [62].

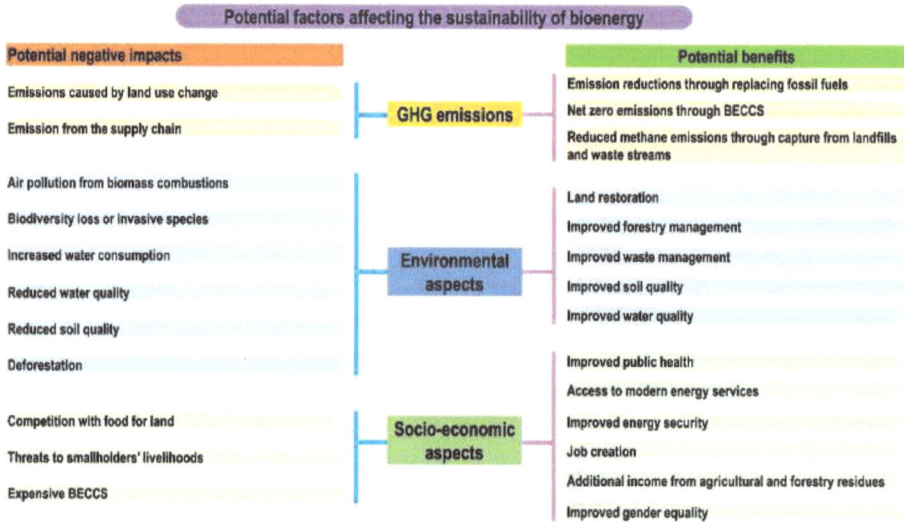

Fig. (4). Potential factors affecting the sustainability of bioenergy (modified from [63]).
Note: BECCS = bioenergy with carbon capture and storage

The viability of a certain bioenergy production plan might vary depending on climatic and biophysical factors, resulting in its feasibility in one location but not in another. The viability of bioenergy deployment in a location influences the socio-economic aspects, population, politics (particularly property ownership), and culture. The carbon balance of bioenergy systems must consider geographical and temporal factors, such as growth and the harvesting rate of biomass. In certain circumstances, the legal framework pertaining to biofuel sources could demonstrate limitations in encompassing all pertinent concerns of the local population in producing nations, contingent upon the efficacy of regional regulation mechanisms within the specific nation. In some scenarios, contingent upon the efficacy of local enforcement mechanisms, the regulatory structure pertaining to biofuel supplies may exhibit limitations in addressing the whole spectrum of concerns relevant to the regional population in nations engaged in manufacturing. Although a sustainable framework in the importing country may have sparked significant change in the supplier country, it may also be failing to govern behaviors in practice, which could reinforce the position of dominant actors and further marginalize vulnerable groups [64]. As a result, various systems link to external influences and activities within a broader network. The comprehension of the sustainability of diverse renewable energy sources and trends in land use in various parts of the world necessitates an extensive expenditure because of the intricate contextual framework involved. It is crucial to have a sustainability framework to support decision-making and guarantee sustainability.

BENEFICIAL MICROBES AND THEIR ROLES IN BIOENERGY PRODUCTION

Microorganisms derived from diverse environments inherently synthesize a wide range of biologically active molecules that serve as sources of energy, pharmaceutical agents, and other vital chemical entities [65]. As previously stated, these microorganisms have shown exceptional proficiency in the production of biofuels *via* the synthesis of enzymes that exhibit activity on many substrates across several processes over an extended period [66]. Alternatives to fossil fuels include biofuels such as ethanol, biobutanol, biohydrogen, and CH_4, among others.

The production of these biofuels from sources that are sustainable may be accomplished in an environmentally friendly manner by using a biological conversion process that is based on microbes. Biofuel produced from lignocellulose feedstock is environmentally friendly; it helps reduce fossil fuel dependency [67], provides an alternate method for delaying the formation of petroleum reserves, and boosts the local economy, particularly in more remote areas. Furthermore, lignocellulosic biomass is an excellent source of substrate for energy from renewable sources since, contrary to popular belief, it does not compete with agriculture for food resources [68]. Poplar, sunflower, and Jatropha are examples of lignocellulose biomass that are predominantly farmed for use as substrates in the manufacture of biofuel. The breakdown of polysaccharides mediated by the activity of particular enzymes is essential to most techniques for converting lignocellulosic biomass to biofuels. They are easily accessible all over the world and plentiful in nature, making them a promising biomass feedstock for the biofuel industry. Since they are not consumed, they provide considerable benefits over the first-generation feedstocks derived from biomass [68].

The global consumption of energy in recent times has been on the increase. At present, energy usage has anticipated a rise in demand [68]. The depletion of reserves of petroleum and coal and the subsequent impact on the environment have long been recognized as precursors to the need for more environmentally friendly energy sources [69]. An important alternative to fossil fuels is the production of sustainable fuels that are both degradable and ecologically beneficial [70]. Other bacterial species can naturally create advanced fuels like butanol from these carbon sources [71].

Microorganisms possess the ability to produce enzymes of industrial significance, such as cellulases, xylanases, chitinases, and amylases. These enzymes facilitate the degradation of biopolymers, including cellulose, hemicelluloses, and chitin, into monomeric sugars. Additionally, microorganisms are capable of fermenting these sugars into biomolecules with high energy density.

In order to achieve commercially viable yields and speeds of biofuel production by bacteria, substantial genetic modifications are necessary. Engineering naturally fuel-producing microorganisms poses significant challenges. One strategy that has been used involves the integration of biofuel processes into genetically manipulable model organisms. *Saccharomyces cerevisiae* and *Zymomonas mobilis*, which are examples of ethanologenic yeasts, have been widely used in the manufacturing industry for the purpose of ethanol generation. *Clostridium acetobutylicum* is used in the process of fermentation to convert feedstock into butanol, another useful organic solvent. Up to 65%-70% of the biomass of certain microorganisms may be made up of lipids. Fatty acid alkyl ester and other lipid-based oleochemicals are produced using oleaginous bacteria such as Yarrowia, Lipomyces, and Rhodosporidium. By employing facultative anaerobic bacteria like *Enterobacter cloacae* in a process called "dark fermentation", biohydrogen is created as a renewable energy source with just water as the outcome. Methane or biogas may be produced by some microorganisms from carbon dioxide. Methanogenic microorganisms thrive under harsh conditions. *Geobacter sulfurreducens* KN400, *Shewanella putrefaciens*, and *Shewanella oneidensis* are all examples of microorganisms. Additionally, several bacteria have been shown to be exoelectrogenic, including *Rhodopseudomonas palustris* DX1, *Candida melibiosica, Saccharomyces cerevisiae*, as well as *Escherichia coli* DH5 [72].

Prokaryote

Because of their widespread usage in other industries and the ease with which they may be genetically modified and examined, microorganisms like *Escherichia coli* (E. coli) and *Saccharomyces cerevisiae* are being investigated for their capacity to create bioenergy [73 - 75].

E. coli has a notable benefit in that it is extensively researched as a model organism for the regulation of genes and expression, and it also offers a wide array of molecular devices for modifying genes. Moreover, *Escherichia coli* strains possess the inherent ability to metabolize diverse carbon sources, encompassing sugars and alcohols, in both oxygen-rich and oxygen-deprived environments. Consequently, they are highly suitable for a wide range of industrial applications, extending beyond biofuel production to encompass the synthesis of hormones, proteins, amino acids, and various high-demand chemicals, notably 1-3 propanediol and Polyhydroxy butyrate [20].

Corynebacterium glutamicum and *Clostridia species* have been effectively used in the production of several biofuels, contingent upon the characteristics of the substrate and the specific biofuel being targeted [69, 76, 77].

Engineered *E. coli* has been seen to create isopropanol at sufficient concentrations of up to 143 g L−1 under optimal circumstances, facilitated by *in situ* product removal using gas-stripping [78]. An important advancement is the substitution of the reversible and flavin-dependent butyryl-CoA dehydrogenase (Bcd) with an irreversible trans-enoyl-CoA reductase (Ter) for the purpose of reducing crotonyl-CoA. This modification effectively shifts the balance onto the production of 1-butanol [72]. The elimination of alternative metabolic pathways in the course of this phenomenon has the potential to enhance the driving forces of NADH and acetyl-CoA, therefore leading to the production of 30 g L−1 of 1-butanol [79]. The present process may be further expanded to generate 1-hexanol with the incorporation of β-keto thiolase (BktB) enzyme, which facilitates the elongation of butyryl-CoA [80]. The chosen process necessitates the establishment of a functioning route for the recycling of NADH [79]. In addition, the inclusion of a long-chain-specific acyl-CoA-thioesterase has the potential to enhance the efficiency of a pivotal enzyme, resulting in the synthesis of 469 mg L−1 1-hexanol and 60 mg L−1 1-octanol [81]. The CoA-dependent mechanism is used in *E. coli* to reverse β-oxidation and generate extended alcohols by means of enzyme deregulation [82]. The use of *Clostridium acetobutylicum* and *Clostridium beijerinckii* has been extensively employed in the industrial sector for the manufacture of biofuels *via* the acetone-butanol-ethanol (ABE) fermentation process [71].

Approaches to Engineering Next-Generation Biofuel Producers:

There are four primary methodologies for developing biofuel generators that are capable of generating next-generation biofuels. The optimal biofuel producers should cultivate inexpensive and sustainable feedstock while also generating substantial quantities of novel fuel.

Low cost: A possible approach is genetically modifying experimental organisms to express both qualities, using their advantageous genetic tractability. Nevertheless, many fuel routes have been identified inside these creatures, yet little advancements have been achieved in modifying them to effectively use environmentally friendly sources of fuel.

Use of native fuel producers: Another strategy is the use of indigenous fuel generators and the modification of substrate requirements. Solventogenic strains of Clostridium have been extensively used in factories for the generation of fuels *via* the acetone-butanol-ethanol (ABE) fermentation technique. Nevertheless, the tolerance for fuel is limited, and the modification of genetics continues to present difficulties.

Renewable feedstock: One alternative strategy involves integrating the manufacture of biofuels within organisms that possess inherent capabilities for utilizing sources of renewable energy, particularly lignocellulosic materials. Certain thermophilic organisms exhibit the ability to thrive on such feedstocks and flourish under elevated temperatures, thereby enhancing the efficiency of biomass deterioration and extraction of fuel processes. Nonetheless, the genetic resources available for these organisms are limited, posing significant challenges in terms of introducing biofuel routes.

Autotrophic organisms have the inherent ability to use CO_2 as a carbon source *via* the process of native fermentation. However, the application of genetic tools remains constrained in terms of facilitating the introduction of biodiesel pathways. The indigenous Escherichia coli bacteria have the ability to synthesize ethanol by an intrinsic mechanism. This process occurs in the absence of oxygen, wherein one mole of glucose is digested to yield two moles of formate and acetate and one mole of ethanol.

Reduction reactions: The last stage of the endogenous ethanol generation process is the conversion of acetyl-CoA to ethanol by the enzymatic action of AdhE [83, 84]. The reduction reaction utilizes two molecules of NADH, but the initial glycolysis process, which converts glucose to pyruvate, only generates one molecule of NADH (one NADH for each glyceraldehyde 3 phosphates to 1,3—Bisphosphoglycerate). This discrepancy results in a disparity in redox reactions. In order to address the redox imbalances, the indigenous Escherichia coli organism achieves equilibrium in ethanol synthesis by oxidizing acetyl-coA into acetate, a process that does not need the use of NADH. Regrettably, the indigenous method of fermentation results in the generation of ethanol at a suboptimal rate, with an estimated output of 0.26 g ethanol per gram of glucose. This falls short of the maximum possible yield of 0.51 g ethanol per gram of glucose [85].

Strategies for Consolidated Bioprocessing

As shown in Fig. (**5**), there are two distinct techniques for integrating the engineering of lignocellulose breakdown and fuel generation within a single strain.

a. The organism with inherent cellulolytic capabilities may be genetically modified to exhibit the expression of fuel pathways.

b. Recombinant expression of cellulolytic enzymes in synthetic organisms is a feasible approach (although expressing them in native energy producers, but more challenging, is not shown in the figure provided)

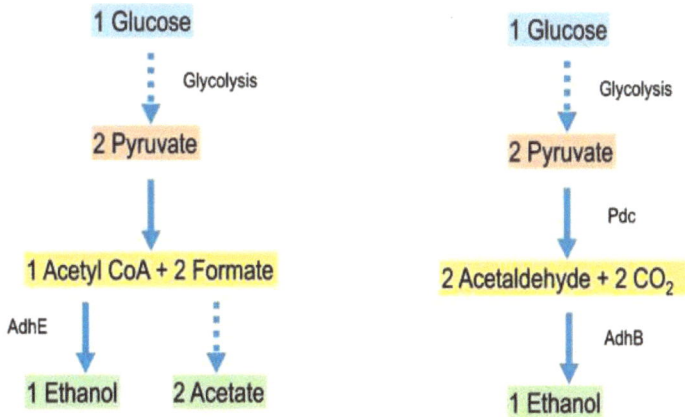

Fig. (5). *Escherichia coli* in biofuel production (modified from [86]).

The methodologies used in the generation of bioethanol from *E. coli* may be categorized into two distinct approaches.

a. The endogenous ethanol production route, also known as the heterofermentative pathway, is a process used by *Escherichia coli* (*E. coli*) for the generation of ethanol.

b. In this study, a metabolic engineering approach was used to enhance the ethanol production route in E. coli. Specifically, the indigenous ethanol production pathway of *E. coli* was modified by introducing and expressing the pdc and adhB genes derived from *Zymomonas mobilis*. The use of broken arrows serves as a visual representation of the intricate routes that include several enzymes and sequential stages.

Eukaryote

Eukaryotic organisms, including photosynthetic algae, both microalgae and macroalgae (commonly known as seaweeds), have garnered significant attention as a potential biofuel source for many decades [87]. There are several species that exhibit higher rates of biomass generation compared to terrestrial plants [87]. A considerable number of eukaryotic microalgae have the ability to accumulate substantial quantities of energy-dense molecules, including triacylglycerol (TAG) and starch. These chemicals serve as valuable precursors for the production of various biofuels, including biodiesel and ethanol [88]. There is a prevailing belief among the scientific community that a significant proportion of crude oil may be attributed to microalgae, particularly diatoms. This belief is based on the examination of their lipid profiles and production levels [89]. Microalgae have notable appeal as a potential fuel source from an environmental perspective due to their capacity to absorb carbon dioxide and their ability to be cultivated on marginal land using sewage or saline water [90].

Despite facing significant challenges, the use of biodiesel and biochemical synthesis by photosynthetic organisms remains very appealing owing to its potential for CO_2 fixation *via* the utilization of sunlight and water. This approach not only offers environmental protection benefits but also contributes to addressing the problem of climate change [89].

A range of host species, including bacteria, fungi, and microalgae, may be potentially used for the production of biodiesel and biochemicals *via* the conversion of CO_2 using solar energy. The generation of biofuels and biochemicals directly from CO_2 and sunshine by photosynthetic organisms necessitates substantial innovation in the creation of large-scale growing, harvesting, and separation of product processes since the generation rate is extremely limited [89].

Microalgae and cyanobacteria are distinct types of organisms with different cellular characteristics. Microalgae are eukaryotic in nature and possess the ability to carry out photosynthesis. They often exhibit a size range of 1 to 100 μm. On the other hand, cyanobacteria are prokaryotic creatures without a nucleus and also exhibit a smaller size range of 1 to 10 μm. Cyanobacteria are believed to have been the progenitors of the chloroplasts seen in eukaryotic algae and terrestrial plants. Consequently, these organisms exhibit several similarities, including their capacity to facilitate photosynthetic water oxidation and photolysis, consequently playing a crucial role in the generation of oxygen from the atmosphere and the synthesis of decreased carbon from organic matter [90].

As indicated in Fig. (**6**), microalgae are single-celled photosynthetic microorganisms capable of converting solar energy into chemical energy with an efficiency ranging from 10 to 50 times higher than that of terrestrial plants [91]. Algae have far greater rates of cellular proliferation compared to plants, resulting in a considerably reduced geographical footprint for energy production purposes [90, 92]. Numerous microalgae have a high oil content, particularly when subjected to nitrogen deprivation, so rendering them suitable for the production of biofuel *via* the use of already available technological processes. The photosynthetic microbes exhibit a higher level of efficiency in the conversion of CO_2 into carbon-rich lipids, which are closely related to biodiesel, compared to oleaginous crops. Importantly, this productivity is achieved without the need to compete for agricultural land [93]. These organisms need aquatic habitats that may range from freshwater to marine conditions. In addition to their ability to perform carbon dioxide fixation, these organisms possess the potential to serve as a viable resource for the cost-effective synthesis of large-scale chemical compounds, owing to the negligible costs associated with the primary inputs to their metabolic processes, namely light and carbon dioxide [94]. Microalgae cells possess a carbon content of roughly 50%, with the production of 1 kilogram of microalgae biomass resulting in the fixation of 1.8 kg of CO_2 [95].

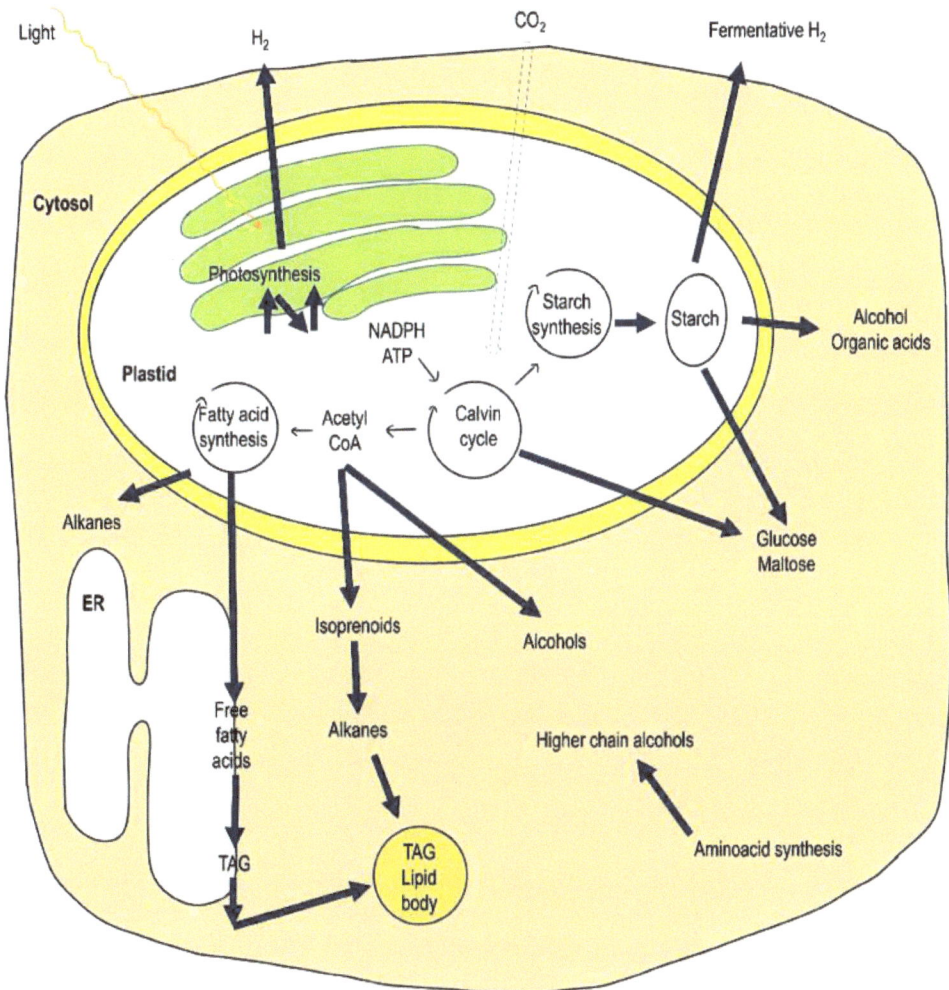

Fig. (6). Metabolic pathway of the microalgae that can be used for biofuel production (modified from [91]).

Algae have been shown to possess substantial oil concentrations, exhibiting varying compositions contingent upon the specific species. Certain species have favorable fatty acid profiles. Similarly, several types of algae have a higher concentration of fatty acid constituents within their dry biomass. Microalgae have the ability to thrive in varying circumstances, including nutrient-deficient environments, and their development is influenced by several environmental variables that may not be universally understood across all regions [96]. The blending technique is a straightforward approach for the extraction of fatty acids and separation of biodiesel, often used on a small or experimental basis (Fig. 7).

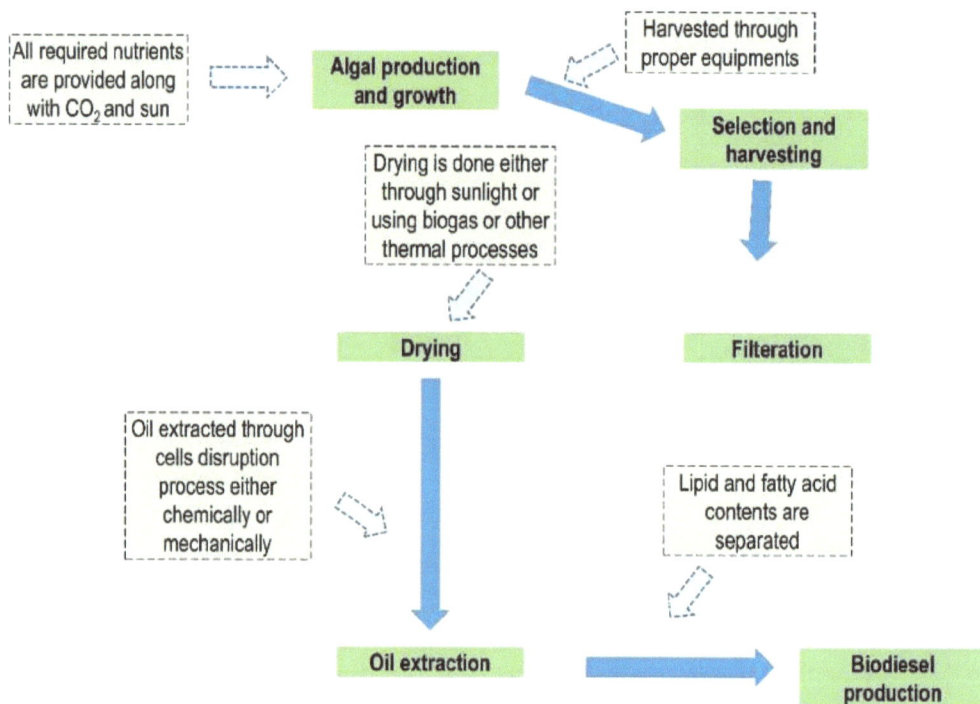

Fig. (7). Process of biofuel production from algae [modified from 100,101].

Algal Mechanism and Metabolism

Understanding the metabolic response to a cultural context is of utmost importance. Systems biology methodologies, such as metabolite profiling and the integration of several levels of information (*e.g.*, metabolites, fluxes, transcript, and protein abundance), prove to be valuable tools in this regard [97, 98]. The study pertains to the metabolic processes of photosynthetic bacteria [99], as shown in Fig. (7). Light energy is assimilated by the cell, whereby pigments absorb light quanta to facilitate photosynthetic electron transport. In this process, ATP is generated at the pathway of respiration using NADPH instead of NADH.

The primary route for carbon dioxide (CO_2) assimilation is the Calvin-Benson-Bassham (CBB) cycle, in which the enzyme RubisCO facilitates the first catalytic reaction. The aforementioned enzyme also exhibits oxygenase activity, enabling it to undergo a reaction with molecular oxygen (O_2), thereby initiating an alternative metabolic route known as photorespiration. Algae possess the photorespiration pathway, which is impeded by elevated levels of oxygen concentration, resulting in the inhibition of photosynthesis. The process of photosynthesis encompasses many key events that take place inside the chloroplasts, including the light reactions, the Calvin-Benson-Bassham (CBB) cycle, and starch synthesis [89]. Subcellular organelles found in algae and plant cells include chloroplasts, mitochondria, and the cytoplasm. Once GAP has been exported from chloroplasts to the cytoplasm, the carbon flow is partitioned to either the sugar synthesis route or the glycolytic pathway, where it is converted to pyruvate. The cytoplasm of plant cells mostly stores sugars like sucrose. PEP carboxylase (Ppc) catalyzes anaplerotic processes that replenish carbon to keep the TCA cycle running in plant cells [89]. The pentose phosphate (PP) pathway is active inside the cytoplasm, whereas the Calvin-Benson-Bassham (CBB) cycle occurs within the chloroplast. Among the many pigments, chlorophyll accounts for a substantial proportion. The chemical δ-aminolevulinic acid (δ-ALA) serves as the primary precursor for chlorophyll synthesis. The traditional succinate-glycine route refers to the enzymatic reaction mediated by δ-ALA synthetase, which involves the association of glycine and succinyl-CoA. Furthermore, many green cells absorb glutamate and -KG into -ALA far more effectively than glycine and succinate. However, the chloroplast is where acetyl-coenzyme is produced, and here is where fatty acid synthesis takes place. (AcCoA) is produced during mitochondrial synthesis [89]. Under autotrophic circumstances, considerable ATP is generated by mitochondrial oxidative phosphorylation, and the fatty acid composition of Chlorella cell lipids changes greatly, especially for the -linolenic acid (C18:3) level. In autotrophic cultures, the CBB cycle is the primary ATP consumer. The following is a list of decreasing ATP yields: Heterotroph> Mixotroph> Autotroph [74], as shown in Fig. (8) below.

POLICIES FOR SUSTAINABLE BIOENERGY PRODUCTION

A renewable energy system may be characterized as an economically viable, dependable, and ecologically sound energy system that efficiently harnesses local resources and infrastructures. The system exhibits a higher degree of flexibility in accommodating new technological, economic, and political solutions, as opposed to a traditional energy system that is characterized by sluggishness and inertia. The active promotion of innovative solutions is also facilitated [100].

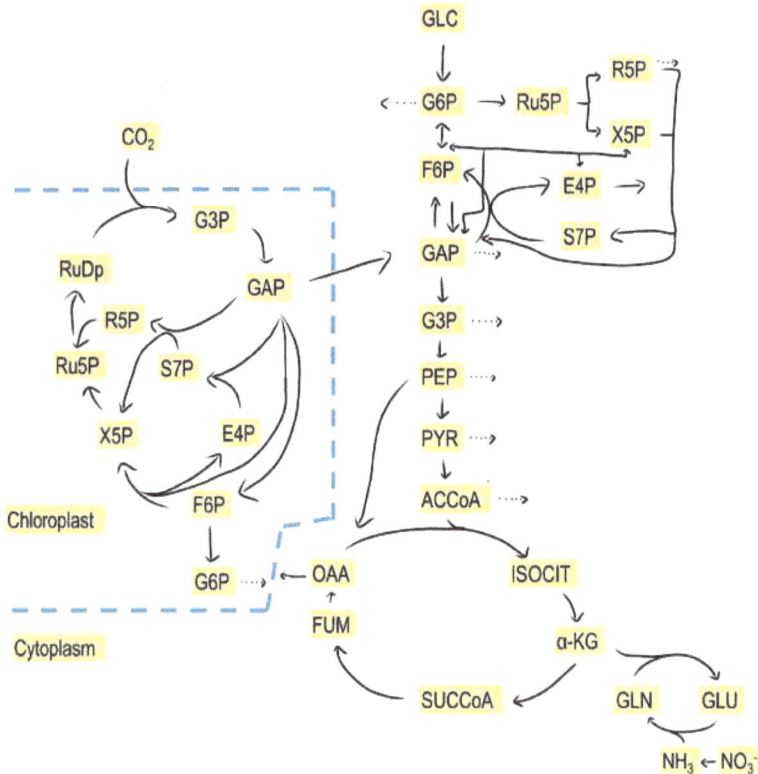

Fig. (8). The mechanism and metabolic path of algae [modified from 100, 101].

The sustainability of bioenergy production and consumption is not inherently guaranteed, and any efforts to promote them must be approached with caution and a balanced perspective. It is essential for governments to actively endorse bioenergy paths that effectively contribute to the reduction of climate change and align with broader sustainability objectives. From a sustainable development standpoint, the bioenergy routes that show the greatest potential are those that use biomass leftovers sourced locally from forestry and agriculture, byproducts generated by associated businesses, and biogenic waste streams. These pathways also employ conversion techniques that are highly efficient [100]. Neither high-carbon stock nor high-biodiversity value land may be used to harvest the raw material. The minimal standards for excellent agricultural and environmental circumstances should be met in the manufacturing of agricultural raw materials farmed within the community. The production and utilization of bioenergy offer various environmental and socio-economic advantages and prospects. These include substantial reductions in greenhouse gas emissions, enhancements in energy security and trade balances, possibilities for economic and social progress, particularly in rural areas, prevention of waste disposal challenges, and improved utilization of renewable energy sources as well as additional resources [101]. Nigeria has emerged as a significant global producer of crude oil and continues to rely on imports for its petroleum product needs in the present day. Due to a scarcity of gasoline stemming from local refineries, the current supply deficit is being addressed by the importation of petroleum. Currently, Nigeria is in the early stages of manufacturing biofuel, with a significant reliance on imported petroleum fuel for blending purposes. According to recent reports, it has been said that the blending of gasoline mostly occurs with imported sources, whereas domestically produced petroleum-based fuel from Nigeria's government-owned refineries in Kaduna and Port-Harcourt is not extensively subjected to blending processes. While the primary operator of petroleum refineries is the government, several private enterprises have been granted authorization to engage in production activities [102].

The formulation of the biofuel strategy in 2007 by the Federal Government exemplified a favorable response toward environmental responsibility [103]. The formulation of the biofuel policy in Nigeria aimed to encourage the implementation, generation, and consumption of biofuels while also attracting investors *via* the provision of incentives such as tax breaks. The biofuel strategy created by the Nigerian National Petroleum Corporation (NNPC) underwent a further review by relevant parties three years later [103]. The initiative aimed at fostering the growth of the biofuel sector in Nigeria was structured into two separate stages. In the first stage, known as the seeding phase, the introduction of bioethanol and biodiesel into the fuel supply is carried out by importing them with Premium Motor Spirit (PMS) and Automotive Gas Oil (AGO) in certain proportions. These proportions consist of up to 10% bioethanol mixed with PMS (referred to as E10) and up to 20% biodiesel mixed with AGO (referred to as B20) [101]. The subsequent stages include the domestic manufacturing of biofuel, feedstock, and biofuel facilities.

The biofuel sector is expected to benefit from its position as a pioneer, with the Ministry of Petroleum Resources serving as the driving force behind the biofuel program [103, 104]. Nevertheless, it is important to acknowledge that biofuel, being a fuel derived from non-petroleum sources and possessing the characteristic of renewability, is often mistakenly regarded as posing a threat to the petroleum sector. Consequently, a potential conflict of interest may arise when attempting to promote the growth of biofuels in the Ministry of Petroleum. The endorsement of the first biofuel refinery in Nigeria by the Nigerian National Petroleum Corporation (NNPC) is anticipated to provide significant job opportunities, potentially benefiting around 58,000 individuals [105]. In 2014, Allied Atlantic Distilleries Limited successfully established the first bioethanol facility in

Nigeria, using cassava as the primary substrate [106] in the Ogun State of Nigeria. The implementation of the biofuel strategy in Nigerian society has not garnered significant attention since there have been reports indicating that the petrol produced in Warri, Kaduna, and Port Harcourt lacks ethanol blending. This is in contrast to the imported petrol entering Nigeria, which undergoes the blending process [102]. Adoption of biofuels had not increased as anticipated by the year 2015, when complete enactment of biofuel policies was predicted. The sociopolitical and economic climates of Nigeria will shape the country's biofuel strategy in the years to come [107]. Due to the present economic slump and the absence of major crude oil importers from the global oil agreement with Nigeria, the government's political will may be weakened toward the enactment of biofuel policies. Another element that may influence the administration's interest in the biofuel business and, hence, the biofuel policy's execution in Nigeria is the recent shift in the country's political environment. The current security concerns posed by Boko Haram insurgencies and conflicts related to herders have the potential to divert the attention of feedstock farmers and the Government away from the effective execution of the biofuel program [107]. Currently, the biofuel strategy, which involves the incorporation of ethanol into petroleum, is in its nascent phase, with the anticipated biofuel supplies for blending being reliant on imports.

To expedite the implementation of bioenergy policies, it is essential to conduct an evaluation of the current operational framework and ascertain the key stakeholders involved. The key stakeholders involved in ensuring the successful execution of the biofuels strategy have been classified as the Government, investors, farmers, traditional land-owners, the young, media researchers, and heads of research institutes. These stakeholders are expected to engage in collaborative interactions to effectively execute the biofuel strategy. The establishment of a robust biofuel industry is a comprehensive undertaking that requires contributions from various sectors. The successful implementation of biofuel policies relies heavily on the Government's demonstration of strong political determination, as well as its coordination efforts with other stakeholders. The Government should actively pursue the realization of biofuel growth in a more assertive manner [108]. The promotion of project continuity needs to be advocated by subsequent administrations. A proposed governmental entity, such as the Biofuel Energy Commission, should be established in order to consistently oversee the execution of the biofuel strategy and conduct periodic evaluations as the implementation proceeds. There is potential to redirect attention from using food as a feedstock for biofuel production towards the conversion of trash into energy within the context of biofuel regulations. In order to enhance biofuel production, it is essential for the Government to provide the necessary and sufficient infrastructural support, including power supply, road network, and water resources. The Federal Government plays a crucial role in facilitating the market demand for biofuel. The 2007 strategy primarily focuses on providing passive assistance rather than actively promoting and enacting measures that are essential for stimulating the market for biofuel. The demand for biofuel has not gathered momentum mainly because of the relatively high price of biodiesel or biofuel compared to fossil fuel, discouraging investors. Farmers need to know more about improving farming practices to enhance quality and yield [104]. Addressing the issue of food security and competition for food resources necessitates the establishment of specialized biofuel plants in currently unused areas in different parts of the nation. There is a need to promote the use of contemporary automated agricultural techniques among farmers. The accommodation of landowners' rights is necessary within the framework of biofuel policy. There is a need to promote private-public partnerships in the realm of investment with regard to the development of biofuels. The unemployed young are educated about the potential career prospects associated with the production of biofuel feedstock. It is vital for end users to possess knowledge about the environmental advantages associated with the use of blended fuel in their automobiles. The media has the capacity to generate awareness among stakeholders and the general public on the environmental advantages and possibilities available to young people. It is important to raise awareness among academics on the need to engage in proactive research aimed at developing cost-effective ways and maximizing the usage of locally sourced feedstock for the manufacture of biofuels. Simultaneously, it is essential for the Government to be prepared to facilitate the allocation of funds towards biofuel research, with the aim of fostering the advancement of domestic technology and promoting the production of biofuels inside the country. It is essential to establish and adequately equip research groups and centers of excellence dedicated to biofuel research, as well as to facilitate research and training endeavors in this field.

CONCLUSION AND FUTURE PERSPECTIVE

The cultivation of agricultural commodities, including maize, sweet potatoes, cassava, palm oil, and sugar cane, has considerable importance as a primary provider of basic food in Nigeria. However, it is worth noting that the production of biofuels in Nigeria may face potential implications due to the use of these agricultural goods [104]. Furthermore, it should be noted that Jatropha, which is considered one of the prominent feedstocks in the biofuel program of Nigeria, has not yet achieved complete establishment. The Ministry of Agriculture has undertaken initiatives to encourage the cultivation of cassava as a means of augmenting food production and mitigating the decline in food reserves. Additionally, there have been attempts to reduce the importation of wheat-based confectioneries by substituting them with cassava-based alternatives, thus conserving food resources. The issue of food *vs* fuel has emerged as a significant topic of concern, particularly in some African nations such as Nigeria, where the principal difficulty is food insecurity. In this context, it is noteworthy that over 70% of the population in Nigeria allocates over 80% of their income towards meeting their food and nourishment needs [105].

One potential obstacle that may provide a future challenge to the production of bioenergy is the integration of biofuel policy development within the agriculture sector. The Nigerian economy is now facing challenges due to its excessive reliance on fossil fuels sourced from the Niger Delta area. This over-dependence has hindered the growth of the non-oil sector, particularly the agricultural sector, which is crucial for the production of feedstock required for the development and consumption of biofuels [104]. The advent of battery-powered vehicles has been widely perceived as a potential detriment to the foreseeable need for biofuels. However, there exist opposing arguments suggesting that despite the advancements in electric vehicle technology in terms of lowering emissions and cost-effectiveness, these improvements may not be sufficient to position electric cars as the optimal choice for buyers, taking into account factors such as time efficiency and conserving the environment [106].

ACKNOWLEDGEMENTS

The authors would like to extend their sincere thanks to their affiliated institutes for their continuous support.

REFERENCES

[1]   González-González LM, Correa DF, Ryan S, Jensen PD, Pratt S, Schenk PM. González-González LM, Correa DF, Ryan S, Jensen PD, Pratt S, Schenk PM. Integrated biodiesel and biogas production from microalgae: Towards a sustainable closed loop through nutrient recycling [Internet]. Vol. 82, Renewable and Sustainable Energy Reviews. Elsevier Ltd; 2018 [cited 2022 Oct 15]. p. 1137–48. Available from: https://eurekamag.com/research/065/111/065111255.php

[2]   Akinsemolu AA. The role of microorganisms in achieving the sustainable development goals. J Clean Prod 2018; 182: 139-55.
[http://dx.doi.org/10.1016/j.jclepro.2018.02.081]

[3]   IPCC. Climate Change 2014: Mitigation of Climate Change - Summary for Policymakers and Technical Summary | GlobalChange.gov [Internet]. [cited 2022 Oct 15]. Available from: https://www.globalchange.gov/browse/reports/ipcc-climate-change-2014-mitigation-cl-mate-change-summary-policymakers-and-technical

[4]   Kumar R, Kumar P. Future microbial applications for bioenergy production: A perspective. Frontiers in Microbiology. Frontiers Research Foundation 2017; Vol. 8: p. 450.

[5]   Cîrstea SD, Cîrstea A, Popa IE, Radu G. The role of bioenergy in transition to a sustainable bioeconomy - Study on EU countries. Amfiteatru Econ 2019; 21(50): 75.
[http://dx.doi.org/10.24818/EA/2019/50/75]

[6]   Liao JC, Mi L, Pontrelli S, Luo S. Fuelling the future: microbial engineering for the production of sustainable biofuels. Nat Rev Microbiol 2016; 14(5): 288-304. Available from: https://pubmed.ncbi.nlm.nih.gov/27026253/10.1038/nrmicro.2016.32

[PMID: 27026253]

[7] Singh V, Tiwari R, Chaturvedi VK, Singh N, Mishra V. Microbiological Aspects of Bioenergy Production: Recent Update and Future Directions. In Springer, Singapore; 2021 [cited 2022 Oct 8]. p. 29–52. Available from: https://link.springer.com/chapter/10.1007/978-981-33-4615-4_2

[8] Parmar A, Singh NK, Pandey A, Gnansounou E, Madamwar D. RETRACTED: Cyanobacteria and microalgae: A positive prospect for biofuels. Bioresour Technol 2011; 102(22): 10163-72. [http://dx.doi.org/10.1016/j.biortech.2011.08.030] [PMID: 21924898]

[9] Okada K, Fujiwara S, Tsuzuki M. Energy conservation in photosynthetic microorganisms. J Gen Appl Microbiol 2020; 66(2): 59-65.https://pubmed.ncbi.nlm.nih.gov/32336724/ [http://dx.doi.org/10.2323/jgam.2020.02.002] [PMID: 32336724]

[10] Elshahed MS. Microbiological aspects of biofuel production: Current status and future directions. Vol. 1, Journal of Advanced Research. Elsevier; 2010. p. 103–11.

[11] Kracke F, Vassilev I, Krömer JO. Microbial electron transport and energy conservation - the foundation for optimizing bioelectrochemical systems. Front Microbiol 2015; 6(JUN): 575. [http://dx.doi.org/10.3389/fmicb.2015.00575] [PMID: 26124754]

[12] Cabrol L, Marone A, Tapia-Venegas E, Steyer JP, Ruiz-Filippi G, Trably E. Microbial ecology of fermentative hydrogen producing bioprocesses: useful insights for driving the ecosystem function. FEMS Microbiol Rev 2017; 41(2): 158-81.https://pubmed.ncbi.nlm.nih.gov/28364728/ [http://dx.doi.org/10.1093/femsre/fuw043] [PMID: 28364728]

[13] Renewables 2021 – Analysis - IEA [Internet]. [cited 2022 Oct 7]. Available from: https://www.iea.org/reports/renewables-2021

[14] Hjersted JL, Henson MA. Steady-state and dynamic flux balance analysis of ethanol production by Saccharomyces cerevisiae. IET Syst Biol 2009; 3(3): 167-79.https://pubmed.ncbi.nlm.nih.gov/19449977/ [http://dx.doi.org/10.1049/iet-syb.2008.0103] [PMID: 19449977]

[15] Lee SY, Kim HU. Systems strategies for developing industrial microbial strains. Nature Biotechnology. Nature Publishing Group 2015; Vol. 33: pp. 1061-72.

[16] Jaung W, Wiraguna E, Okarda B, Artati Y, Goh CS, Syahru R, et al. Spatial assessment of degraded lands for biofuel production in Indonesia. Sustain [Internet]. 2018 Dec 5 [cited 2022 Oct 8];10(12). Available from: www.mdpi.com/journal/sustainability

[17] IRENA. Renewable Capacity Statistics 2020 [Internet]. International Renewable Energy Agency (IRENA). 2020 [cited 2022 Oct 7]. Available from: /publications/2020/Mar/Renewable-Capacit-Statistics-2020

[18] IEA-Bioenergy. Bioenergy–a sustainable and reliable energy source [Internet]. International Energy Agency Bioenergy, Paris, …. 2009 [cited 2022 Oct 7]. Available from: http://www.globalbioenergy.org/uploads/media/0912_IEA_Bioenergy_-_MAIN_REPORT_-_Bioenergy_-_a_sustainable_and_reliable_energy_source._A_review_of_status_and_prospects.pdf

[19] Chambers A, Nakicenovic N. World Energy Outlook 2008 – Analysis - IEA [Internet]. 2008 [cited 2022 Oct 8]. Available from: https://www.iea.org/reports/world-energy-outlook-2008

[20] [IRENA] - International Renewable Energy Agency. World Energy Transitions Outlook: 1.5°C Pathway [Internet]. International Renewable Energy Agency. 2021 [cited 2022 Oct 7]. 1–352 p. Available from: /publications/2022/Mar/World-Energy-Transitions-Outlook-2022

[21] IRENA's World Energy Transitions Outlook outlines pathway to 1.5 C [Internet]. [cited 2022 Oct 7]. Available from: https://www.hydroreview.com/environmental/irenas-world-energy-transitions-outlook-outlines-pathway-to-1-5-c/

[22] IEEFA 2019. IEA: Renewable generation capacity expected to climb by 1,200GW in next five years - Institute for Energy Economics & Financial Analysis [Internet]. [cited 2022 Jan 3]. Available from:

https://ieefa.org/iea-renewable-generation-capacity-expected-to-climb-by-1200gw-in-next-five-years/

[23] Smeets EMW, Faaij APC. The impact of sustainability criteria on the costs and potentials of bioenergy production – Applied for case studies in Brazil and Ukraine. Biomass Bioenergy 2010; 34(3): 319-33. [http://dx.doi.org/10.1016/j.biombioe.2009.11.003]

[24] Souza GM, Victoria R, Joly C, Verdade L, *et al.* Bioenergy Numbers. Bioenergy Sustain Bridg gaps [Internet]. 2016 [cited 2022 Oct 7];29–57. Available from: https://www.researchgate.net/ publication/276274585_Sustainable_Development_and_Innovation_In_Bioenergy_Sustainability_brid ging_the_gaps_Eds_Glaucia_Mendes_Souza_Reynaldo_L_Victoria_Carlos_A_Joly_and_Luciano_M _Verdade_SCOPE_72_ISBN_978-2-9545557-0-6

[25] Delucchi MA. Impacts of biofuels on climate change, water use, and land use. Ann N Y Acad Sci 2010; 1195(1): 28-45.https://pubmed.ncbi.nlm.nih.gov/20536815/ [http://dx.doi.org/10.1111/j.1749-6632.2010.05457.x] [PMID: 20536815]

[26] Edenhofer O. Renewable energy sources and climate change mitigation: special report of the Intergovernmental Panel on Climate Change. Choice Rev Online [Internet]. 2012 [cited 2022 Oct 7]; 49(11): 49-6309-49–6309. Available from: https://books.google.com.ng/books/ about/Renewable_ Energy_Sources_and_Climate_Cha.html?id=AjP9sVg01zoC&redir_esc=y

[27] Staffas L, Gustavsson M, McCormick K. Strategies and policies for the bioeconomy and bio-based economy: An analysis of official national approaches. Sustain [Internet]. 2013 [cited 2022 Oct 7];5(6):2751–69. Available from: https://econpapers.repec.org/RePEc:gam:jsusta:v:5:y:2013:i:6:p: 2751-2769:d:26568

[28] Scarlat N, Dallemand J, Monforti-Ferrario F, Nita V. The role of biomass and bioenergy in a future bioeconomy: Policies and facts. undefined. 2015.

[29] Ozturk I, Aslan A, Kalyoncu H. Energy consumption and economic growth relationship: Evidence from panel data for low and middle income countries. Energy Policy 2010; 38(8): 4422-8.https://econpapers.repec.org/RePEc:eee:enepol:v:38:y:2010:i:8:p:4422-4428 [http://dx.doi.org/10.1016/j.enpol.2010.03.071]

[30] Bioeconomy T, Introduction A. Bioeconomy T, Introduction A. The Bioeconomy: An Introduction to the World of Bioenergy Mccormick, Kes; Willquist, Karin [Internet]. Lund University; 2016 [cited 2022 Oct 7]. Available from: http://lup.lub.lu.se/record/8054628

[31] M'barek R, Philippidis G, Suta C, Vinyes C, Caivano A, Ferrari E, *et al.* Observing and analysing the bioeconomy in the EU – Adapting data and tools to new questions and challenges. Bio-Based Appl Econ 2014; 3(1): 83-91. Available from: http://www.araid.es/en/content/observing-and-analysing-bioeconomy-eu-adapting-data-and-tools-new-questions-and-challenges

[32] Lavidas G. Energy and socio-economic benefits from the development of wave energy in Greece. Renew Energy 2019; 132: 1290-300. [http://dx.doi.org/10.1016/j.renene.2018.09.007]

[33] Bildirici M, Özaksoy F. Woody Biomass Energy Consumption and Economic Growth in Sub-Saharan Africa. Procedia Econ Finance 2016; 38: 287-93. [http://dx.doi.org/10.1016/S2212-5671(16)30202-7]

[34] Chum H, Faaij A, Moreira J, *et al.* Bioenergy. Renew Energy Sources Clim Chang Mitig [Internet]. 2011 Nov 21 [cited 2022 Oct 7];209–332. Available from: https://www.cambridge.org/core/ product/identifier/CBO9781139151153A022/type/book_part

[35] Tiranocyda B. Zych-The Viability Of Corn Cobs As ABioenergy Feedstock [Internet]. [cited 2022 Oct 7]. Available from: https://www.academia.edu/8069712/Zych_The_Viability_Of_Corn_Cobs_As_ ABioenergy_Feedstock

[36] Klingenfeld D. Corn stover as a bioenergy feedstock: Identifying and overcoming barriers for corn stover harvest, storage, and transport [Internet]. Harvard Kennedy School. 2008 [cited 2022 Oct 7]. Available from: http://130.203.136.95/viewdoc/summary?doi=10.1.1.554.402

[37]   van der Weijde T, Alvim Kamei CL, Torres AF, Vermerris W, Dolstra O, Visser RGF, *et al.* The potential of C4 grasses for cellulosic biofuel production. Frontiers in Plant Science. Frontiers Research Foundation 2013; Vol. 4: p. 107.https://research.wur.nl/en/publications/the-potential-of-c4-grasse--for-cellulosic-biofuel-production Internet

[38]   McKendry P. Energy production from biomass (part 1): overview of biomass. Bioresour Technol 2002; 83(1): 37-46.
[http://dx.doi.org/10.1016/S0960-8524(01)00118-3] [PMID: 12058829]

[39]   Kassam A. Bioenergy Development: Issues and Impacts for Poverty and Natural Resource Management. By E. Cushion, A. Whiteman and G. Dieterle. Washington DC: World Bank (2010), pp. 249, £25.95. ISBN 978-0-8213-7629-4. Exp Agric 2010; 46(4): 563-3.https://www.cambridge.org/core/journals/experimental-agriculture/article/abs/bioenergy-development-issues-and-impacts-for-poverty-and-natural-resource-management-by-e-cushion-a-whiteman-and-g-dieterle-washington-dc-world-bank-2010-pp-249-2595-isbn-9780821
[http://dx.doi.org/10.1017/S001447971000044X]

[40]   Shikha K, Rita CY. Biodiesel production from non edible-oils: A review. J Chem Pharm Res 2012; 4(9): 4219-30.

[41]   Liaquat AM, Masjuki HH, Kalam MA, Varman M, Hazrat MA, Shahabuddin M, *et al.* Application of blend fuels in a diesel engine. In: Energy Procedia. 2012. p. 1124–33.
[http://dx.doi.org/10.1016/j.egypro.2011.12.1065]

[42]   Ahmad M, Ajab M, Zafar M, Sult S. Biodiesel from Non Edible Oil Seeds: a Renewable Source of Bioenergy. In: Economic Effects of Biofuel Production [Internet]. InTech; 2011 [cited 2022 Oct 8]. Available from: www.intechopen.com
[http://dx.doi.org/10.5772/24687]

[43]   Ahmad M, Teong LK, Sultana S, Zafar M. Biodiesel Production from Non Food Crops : A Step towards Self Reliance in Energy. undefined. 2013;239–43.

[44]   Mahmood K, Sarwar S, Mehran MT. Current status of electron transport layers in perovskite solar cells: materials and properties. RSC Advances 2017; 7(28): 17044-62. Available from: https://pubs.rsc.org/en/content/articlehtml/2017/ra/c7ra00002b
[http://dx.doi.org/10.1039/C7RA00002B]

[45]   Atabani AE, Silitonga AS, Badruddin IA, Mahlia TMI, Masjuki HH, Mekhilef S. A comprehensive review on biodiesel as an alternative energy resource and its characteristics. Vol. 16, Renewable and Sustainable Energy Reviews. Pergamon; 2012. p. 2070–93.
[http://dx.doi.org/10.1016/j.rser.2012.01.003]

[46]   Yuan JS, Tiller KH, Al-Ahmad H, Stewart NR, Stewart CN Jr. Plants to power: bioenergy to fuel the future. Trends Plant Sci 2008; 13(8): 421-9.
[http://dx.doi.org/10.1016/j.tplants.2008.06.001] [PMID: 18632303]

[47]   Vandevyvere H, Stremke S. Urban planning for a renewable energy future: Methodological challenges and opportunities from a design perspective. Sustainability (Basel) 2012; 4(6): 1309-28.
[http://dx.doi.org/10.3390/su4061309]

[48]   Gavrilescu M, Chisti Y. Biotechnology—a sustainable alternative for chemical industry. Biotechnol Adv 2005; 23(7-8): 471-99. Available from: https://pubmed.ncbi.nlm.nih.gov/15919172/
[http://dx.doi.org/10.1016/j.biotechadv.2005.03.004] [PMID: 15919172]

[49]   Bauen A, Howes J, Bertuccioli L, Chudziak C. 2009. Review of the potential for biofuels in aviation: Final report For CCC. Fuel [Internet]. 2009 [cited 2022 Oct 8];(284):6021–5. Available from: http://www.thegrounds.nl/roundtable/documents/AT_E4tech

[50]   February 2007, Joint T40 & Eubionet workshop, NL [Internet]. [cited 2022 Oct 8]. Available from: http://www.bioenergytrade.org/pastevents/0000009a190b44008/

[51]   Rinkesh. Advantages and Disadvantages of Biofuels. Conserv Energy Futur [Internet]. 2019 [cited

2022 Oct 8];(January 2014). Available from: https://www.biotecharticles.com/Environmental-Biotechnology-Article/Advantages-and-Disadvantages-of-Biofuels-163.html

[52] Lee YC, Lee K, Oh YK. Recent nanoparticle engineering advances in microalgal cultivation and harvesting processes of biodiesel production: A review. Bioresour Technol 2015; 184: 63-72.https://pubmed.ncbi.nlm.nih.gov/25465786/
[http://dx.doi.org/10.1016/j.biortech.2014.10.145] [PMID: 25465786]

[53] Robak K, Balcerek M. Review of second generation bioethanol production from residual biomass [Internet]. Vol. 56, Food Technology and Biotechnology. Food Technol Biotechnol 2018; 56(2): 174-87.https://pubmed.ncbi.nlm.nih.gov/30228792/
[http://dx.doi.org/10.17113/ftb.56.02.18.5428] [PMID: 30228792]

[54] Kapdan IK, Kargi F. Bio-hydrogen production from waste materials. Enzyme Microb Technol 2006; 38(5): 569-82. Available from: https://www.researchgate.net/publication/222705744_Bio-hydrogen_Production_from_Waste_Materials
[http://dx.doi.org/10.1016/j.enzmictec.2005.09.015]

[55] Into P, Pontes A, Sampaio JP, Limtong S. Yeast diversity associated with the phylloplane of corn plants cultivated in Thailand. Microorganisms 2020; 8(1): 80. Available from: https://novaresearch.unl.pt/en/publications/yeast-diversity-associated-with-the-phylloplane-of-corn-plants-cu
[http://dx.doi.org/10.3390/microorganisms8010080] [PMID: 31936155]

[56] Dutta K, Daverey A, Lin JG. Evolution retrospective for alternative fuels: First to fourth generation. Vol. 69, Renewable Energy. Pergamon; 2014. p. 114–22.

[57] Hays SG, Ducat DC. Engineering cyanobacteria as photosynthetic feedstock factories. Photosynth Res [Internet]. 2015 [cited 2022 Oct 8];123(3):285–95. Available from:

[58] Stitt M, Lunn J, Usadel B. Arabidopsis and primary photosynthetic metabolism – more than the icing on the cake. Plant J 2010; 61(6): 1067-91. Available from: https://onlinelibrary.wiley.com/doi/full/10.1111/j.1365-313X.2010.04142.x
[http://dx.doi.org/10.1111/j.1365-313X.2010.04142.x] [PMID: 20409279]

[59] IEA. Energy Technology Perspectives 2017. Int Energy Agency Publ [Internet]. 2017 [cited 2022 Oct 15];371. Available from: https://www.iea.org/reports/energy-technology-perspectives-2017

[60] Vasco Brummer, Carsten Herbes HW. Examples of Positive Bioenergy and Water Relationships. Glob Bioenergy Partnersh [Internet]. 2016 [cited 2022 Oct 15];(March):21–4. Available from: file:///C:/Users/Premkumar/Downloads/AG6_Examples_of_Positive_Bioenergy_and_Water_Relationships_Final.pdf

[61] Global Warming of 1.5 °C. One Earth 2019; 1(3): 374-81. Available from: https://www.ipcc.ch/sr15/

[62] Viaintermedia.com. Biomass - IRENA, FAO and IEA Bioenergy develop brief on Bioenergy for Sustainable Development - Renewable Energy Magazine, at the heart of clean energy journalism.

[63] IRENA IREA. Bioenergy for the Transition: Ensuring Sustainability and Overcoming Barriers [Internet]. Available from: https://www.irena.org/publications/2022/Aug/Bioenergy-for-the-Transition

[64] Tomei J. The sustainability of sugarcane-ethanol systems in Guatemala: Land, labour and law. Biomass Bioenergy 2015; 82: 94-100. Available from: https://www.researchgate.net/publication/277337583_The_sustainability_of_sugarcane-ethanol_systems_in_Guatemala_Land_labour_and_law
[http://dx.doi.org/10.1016/j.biombioe.2015.05.018]

[65] Chubukov V, Mukhopadhyay A, Petzold CJ, Keasling JD, Martín HG. Synthetic and systems biology for microbial production of commodity chemicals. Vol. 2, npj Systems Biology and Applications. Nature Publishing Group; 2018.

[66] Boyle NR, Morgan JA. Flux balance analysis of primary metabolism in Chlamydomonas reinhardtii. BMC Syst Biol 2009; 3(1): 4. Available from: https://pubmed.ncbi.nlm.nih.gov/19128495/
[http://dx.doi.org/10.1186/1752-0509-3-4] [PMID: 19128495]

[67]   Naik SN, Goud VV, Rout PK, Dalai AK. Production of first and second generation biofuels: A comprehensive review. Renew Sustain Energy Rev 2010; 14(2): 578-97.
[http://dx.doi.org/10.1016/j.rser.2009.10.003]

[68]   Adegboye MF, Lobb B, Babalola OO, Doxey AC, Ma K. Draft genome sequences of two novel cellulolytic Streptomyces strains isolated from South African rhizosphere soil. Genome Announc 2018; 6(26): e00632-18. Available from: https://pubmed.ncbi.nlm.nih.gov/29954919/
[http://dx.doi.org/10.1128/genomeA.00632-18] [PMID: 29954919]

[69]   Lan EI, Dekishima Y, Chuang DS, Liao JC. Metabolic engineering of 2-pentanone synthesis in *Escherichia coli*. AIChE J 2013; 59(9): 3167-75. Available from: https://onlinelibrary.wiley.com/doi/full/10.1002/aic.14086
[http://dx.doi.org/10.1002/aic.14086]

[70]   Pandey A, Larroche C, Dussap CG, Gnansounou E, Khanal SK, Ricke S. Biofuels: Alternative feedstocks and conversion processes for the production of liquid and gaseous biofuels. Biomass, Biofuels, Biochemicals: Biofuels: Alternative Feedstocks and Conversion Processes for the Production of Liquid and Gaseous Biofuels. Elsevier; 2019. p.1–886

[71]   Gronenberg LS, Marcheschi RJ, Liao JC. Next generation biofuel engineering in prokaryotes. Curr Opin Chem Biol 2013; 17(3): 462-71.
[http://dx.doi.org/10.1016/j.cbpa.2013.03.037] [PMID: 23623045]

[72]   Renna Eliana Warjoto. This is how microorganisms can produce renewable energy for us [Internet]. The Conversation. 2020 [cited 2022 Dec 5]. Available from: https://theconversation.com/this-is-h-w-microorganisms-can-produce-renewable-energy-for-us-149933

[73]   Alper H, Stephanopoulos G. Engineering for biofuels: exploiting innate microbial capacity or importing biosynthetic potential? Nat Rev Microbiol 2009 710 [Internet]. 2009 [cited 2022 Dec 5];7(10):715–23. Available from: https://www.nature.com/articles/nrmicro2186
[http://dx.doi.org/10.1038/nrmicro2186]

[74]   Liu T, Khosla C. Genetic engineering of Escherichia coli for biofuel production. Annu Rev Genet 2010; 44(1): 53-69. Available from: https://pubmed.ncbi.nlm.nih.gov/20822440/
[http://dx.doi.org/10.1146/annurev-genet-102209-163440] [PMID: 20822440]

[75]   Wen M, Bond-Watts BB, Chang MCY. Production of advanced biofuels in engineered E coli. Current Opinion in Chemical Biology. Elsevier Current Trends 2013; Vol. 17: pp. 472-9.

[76]   Fischer CR, Klein-Marcuschamer D, Stephanopoulos G. Selection and optimization of microbial hosts for biofuels production. Metab Eng 2008; 10(6): 295-304. Available from: https://pubmed.ncbi.nlm.nih.gov/18655844/
[http://dx.doi.org/10.1016/j.ymben.2008.06.009] [PMID: 18655844]

[77]   Yang C, Hua Q, Shimizu K. Energetics and carbon metabolism during growth of microalgal cells under photoautotrophic, mixotrophic and cyclic light-autotrophic/dark-heterotrophic conditions. Biochem Eng J 2000; 6(2): 87-102.
[http://dx.doi.org/10.1016/S1369-703X(00)00080-2] [PMID: 10959082]

[78]   Inokuma K, Liao JC, Okamoto M, Hanai T. Improvement of isopropanol production by metabolically engineered Escherichia coli using gas stripping. J Biosci Bioeng 2010; 110(6): 696-701.
[http://dx.doi.org/10.1016/j.jbiosc.2010.07.010] [PMID: 20696614]

[79]   Shen CR, Lan EI, Dekishima Y, Baez A, Cho KM, Liao JC. Driving forces enable high-titer anaerobic 1-butanol synthesis in Escherichia coli. Appl Environ Microbiol 2011; 77(9): 2905-15. Available from: https://www.ncbi.nlm.nih.gov/pmc/articles/pmid/21398484/?tool=EBI
[http://dx.doi.org/10.1128/AEM.03034-10] [PMID: 21398484]

[80]   Dekishima Y, Lan EI, Shen CR, Cho KM, Liao JC. Extending carbon chain length of 1-butanol pathway for 1-hexanol synthesis from glucose by engineered Escherichia coli. J Am Chem Soc 2011; 133(30): 11399-401. Available from: https://pubmed.ncbi.nlm.nih.gov/21707101/

[http://dx.doi.org/10.1021/ja203814d] [PMID: 21707101]

[81]   Machado HB, Dekishima Y, Luo H, Lan EI, Liao JC. A selection platform for carbon chain elongation using the CoA-dependent pathway to produce linear higher alcohols. Metab Eng 2012; 14(5): 504-11. [http://dx.doi.org/10.1016/j.ymben.2012.07.002] [PMID: 22819734]

[82]   Dellomonaco C, Clomburg JM, Miller EN, Gonzalez R. Engineered reversal of the β-oxidation cycle for the synthesis of fuels and chemicals. Nature 2011; 476(7360): 355-9. Available from: https://pubmed.ncbi.nlm.nih.gov/21832992/ [http://dx.doi.org/10.1038/nature10333] [PMID: 21832992]

[83]   Kessler D, Leibrecht I, Knappe J. Pyruvate-formate-lyase-deactivase and acetyl-CoA reductase activities of *Escherichia coli* reside on a polymeric protein particle encoded by *adhE*. FEBS Lett 1991; 281(1-2): 59-63. Available from: https://pubmed.ncbi.nlm.nih.gov/2015910/ [http://dx.doi.org/10.1016/0014-5793(91)80358-A] [PMID: 2015910]

[84]   Schmitt B. Aldehyde dehydrogenase activity of a complex particle from *E. Coli*. Biochimie 1975; 57(9): 1001-4. [http://dx.doi.org/10.1016/S0300-9084(75)80355-5] [PMID: 769846]

[85]   Jarboe LR, Grabar TB, Yomano LP, Shanmugan KT, Ingram LO. Development of ethanologenic bacteria. Adv Biochem Eng Biotechnol 2007; 108: 237-61. Available from: https://pubmed.ncbi.nlm. nih.gov/17665158/ [http://dx.doi.org/10.1007/10_2007_068] [PMID: 17665158]

[86]   Sheehan J, Dunahay T, Benemann J, Roessler P. A look back at the U.S. Department of Energy's aquatic species program: biodiesel from algae. NREL/TP-580-24190, National Renewable Energy Laboratory, USA; 1998. NREL/TP-580-24190, Natl Renew Energy Lab USA [Internet]. 1998 [cited 2022 Dec 5];(July). Available from: https://www.scirp.org/reference/ReferencesPapers.aspx? ReferenceID=1580546

[87]   Dismukes GC, Carrieri D, Bennette N, Ananyev GM, Posewitz MC. Aquatic phototrophs: efficient alternatives to land-based crops for biofuels. Curr Opin Biotechnol 2008; 19(3): 235-40. Available from: https://pubmed.ncbi.nlm.nih.gov/18539450/ [http://dx.doi.org/10.1016/j.copbio.2008.05.007] [PMID: 18539450]

[88]   Radakovits R, Jinkerson RE, Darzins A, Posewitz MC. Genetic engineering of algae for enhanced biofuel production. Eukaryotic Cell. Eukaryot Cell 2010; Vol. 9: pp. 486-501. Available from: https://pubmed.ncbi.nlm.nih.gov/20139239/

[89]   Ramachandra TV, Mahapatra DM, B K, Gordon R. Milking diatoms for sustainable energy: Biochemical engineering *versus* gasoline-secreting diatom solar panels. Ind Eng Chem Res 2009; 48(19): 8769-88. [http://dx.doi.org/10.1021/ie900044j]

[90]   Chisti Y. Biodiesel from microalgae. Biotechnol Adv 2007; 25(3): 294-306. Available from: https://pubmed.ncbi.nlm.nih.gov/17350212/ [http://dx.doi.org/10.1016/j.biotechadv.2007.02.001] [PMID: 17350212]

[91]   Larkum AWD, Ross IL, Kruse O, Hankamer B. Selection, breeding and engineering of microalgae for bioenergy and biofuel production. Trends Biotechnol 2012; 30(4): 198-205. Available from: https://pubmed.ncbi.nlm.nih.gov/22178650/ [http://dx.doi.org/10.1016/j.tibtech.2011.11.003] [PMID: 22178650]

[92]   Khan SA, Rashmi, Hussain MZ, Prasad S, Banerjee UC. Prospects of biodiesel production from microalgae in India. Vol. 13, Renewable and Sustainable Energy Reviews. Pergamon; 2009. p. 2361–72.

[93]   Wijffels RH, Barbosa MJ. An outlook on microalgal biofuels. Science 2010; 329(5993): 796-9. Available from: https://pubmed.ncbi.nlm.nih.gov/20705853/ [http://dx.doi.org/10.1126/science.1189003] [PMID: 20705853]

[94]   Gielen D, Boshell F, Saygin D, Bazilian MD, Wagner N, Gorini R. The role of renewable energy in the global energy transformation. Energy Strategy Reviews 2019; 24: 38-50.
[http://dx.doi.org/10.1016/j.esr.2019.01.006]

[95]   Georgianna DR, Mayfield SP. Exploiting diversity and synthetic biology for the production of algal biofuels. Nature 2012; 488(7411): 329-35. Available from: https://pubmed.ncbi.nlm.nih.gov/22895338/
[http://dx.doi.org/10.1038/nature11479]

[96]   Richmond A, Hu Q. Handbook of Microalgal Culture: Applied Phycology and Biotechnology: Second Edition [Internet]. Handbook of Microalgal Culture: Applied Phycology and Biotechnology: Second Edition. wiley; 2013 [cited 2022 Dec 18]. 1–719 p. Available from: https://onlinelibrary.wiley.com/doi/book/10.1002/9781118567166

[97]   Veyel D, Erban A, Fehrle I, Kopka J, Schroda M. Rationales and approaches for studying metabolism in eukaryotic microalgae. Metabolites 2014; 4(2): 184-217.
[http://dx.doi.org/10.3390/metabo4020184] [PMID: 24957022]

[98]   Mettler T, Mühlhaus T, Hemme D, *et al.* Systems analysis of the response of photosynthesis, metabolism, and growth to an increase in irradiance in the photosynthetic model organism Chlamydomonas reinhardtii. Plant Cell 2014; 26(6): 2310-50. Available from: https://academic.oup.com/plcell/article/26/6/2310/6098504
[http://dx.doi.org/10.1105/tpc.114.124537] [PMID: 24894045]

[99]   Yan Y, Liao JC. Engineering metabolic systems for production of advanced fuels [Internet]. Vol. 36, Journal of Industrial Microbiology and Biotechnology. Springer; 2009 [cited 2022 Dec 18]. p. 471–9. Available from: https://link.springer.com/article/10.1007/s10295-009-0532-0

[100]   Babalola PO, Olayinka Oyedepo S. Bioenergy technology development in Nigeria-pathway to sustainable energy development Article in International Journal of Environment and Sustainable Development · Viability of Hydroelectricity in Nigeria and the Future Prospect View project Renewable Energy View project. 2019 [cited 2023 Mar 6]; Available from: https://www.researchgate.net/publication/332634332

[101]   Krug M, Rivza P, Rivza S. Policies and measures to promote sustainable bioenergy production and use in the Baltic Sea Region. 2012.

[102]   Odetoye TE, Ajala EO, Titiloye JO. A Review of A Bioenergy Policy Implementation in Sub-Saharan Africa: Opportunities and Challenges- A Case of Nigeria, Ghana and Malawi. FUOYE J Eng Technol. 2019 Mar 31;4(1).

[103]   Krug M. Part-financed by the European Union (European Regional Development Fund and European Neighbourhood and Partnership Instrument) Policies and measures to promote sustainable bioenergy production and use in the Baltic Sea Region. 2014 [cited 2023 Mar 6]; Available from: http://www.bioenergypromotion.org/

[104]   Abila N. Biofuels adoption in Nigeria: Attaining a balance in the food, fuel, feed and fibre objectives. Vol. 35, Renewable and Sustainable Energy Reviews. Pergamon; 2014. p. 347–55.

[105]   Nigeria: NNPC Endorses First Biofuel Refinery in Country - allAfrica.com [Internet]. [cited 2023 Mar 6]. Available from: https://allafrica.com/stories/200802260245.html

[106]   New ethanol plant berths in Ogun - Vanguard News [Internet]. [cited 2023 Mar 6]. Available from: https://www.vanguardngr.com/2014/01/new-ethanol-plant-berths-ogun/

[107]   Data surprise: Biofuels still beating electrics on cost, emissions : Biofuels Digest [Internet]. The Digest. [cited 2023 Mar 6]. Available from: https://www.biofuelsdigest.com/bdigest/2017/10/02/data-surprise-biofuels-still-beating-electrics-on-cost-emissions/

[108]   Oshewolo S. Design to Fail? Nigeria's Quest for Biofuel. Afro Asian J Soc Sci [Internet]. 2012 May 1 [cited 2023 Mar 6];3(33):15. Available from: https://www.academia.edu/65631052/Designed_to_Fail_Nigerias_Quest_for_Biofuel

Microbiome in Mitigating Ghg Emission and Climate Change Impacts

CHAPTER 5

Role of Microbes and Microbiomes in GHG Emissions and Mitigation in Agricultural Ecosystem Restoration

Sethupathi Nedumaran[1,*], Deepasri Mohan[2], Helen Mary Rose[1], Murugesan Kokila[1], Muthusamy Shankar[3], Selvaraj Keerthana[4], Ravi Raveena[4], Kovilpillai Boomiraj[5] and Sudhakaran Mani[6]

[1] *Division of Environment Science, ICAR-IARI, New Delhi, India*

[2] *Division of Environmental Sciences, Sher-e-Kashmir University of Agricultural Sciences & Technology of Jammu, Jammu and Kashmir, India*

[3] *Division of Plant Genetic Resources, ICAR-Indian Agricultural Research Institute, New Delhi, India*

[4] *Department of Environmental Sciences, Tamil Nadu Agricultural University, Coimbatore, Tamil Nadu, India*

[5] *Climate Research Centre, Tamil Nadu Agricultural University, Coimbatore, Tamil Nadu, India*

[6] *Department of Environmental Science, JKK Munirajah College of Agricultural Science, Namakkal, India*

Abstract: Microbes are crucial for the survival of life on Earth as they affect the major biogeochemical cycles that make our planet congenial for life, providing essential elements like carbon and nitrogen in required forms and quantities. Microbes also play a significant role as either generators or consumers of greenhouse gases, such as carbon dioxide (CO_2), methane (CH_4), and nitrous oxide (N_2O), through various processes in our environment. The distribution of these chemicals on the Earth and in the atmosphere is severely reliant on the equilibrium of these microbial progressions. The consumption of GHGs by microbes is facilitated through their use as substrates in processes like photo/chemoautotrophy, methanotrophy, and nitrous oxide reduction. The CO_2 emitted from the organic matter decomposition and terrestrial respiration is subsequently subjected to photosynthetic fixation partially and is mitigated through carbon sequestration into soil and biomass. The biogenic release of methane through the biological anaerobic decomposition of organic materials by methanogens constitutes an important source of atmospheric CH_4, while methanotrophs, through CH_4 oxidation, facilitate methane emission mitigation. The microbial nitrification-denitrification processes are the significant source of N_2O emission, while the N_2O-reducing bacteria are responsible for decreasing N_2O emissions *via* nitrous oxide reduc-

---

[*] **Corresponding author Sethupathi Nedumaran:** Division of Environment Science, ICAR-IARI, New Delhi, India; E-mail: sethumartha@gmail.com

Govindaraj Kamalam Dinesh, Shiv Prasad, Ramesh Poornima, Sangilidurai Karthika, Murugaiyan Sinduja & Velusamy Sathya (Eds.)

tion enzymatic processes. The complexity of the interactions between these microbes with neighboring biotic and bacterial variables in order to regulate Earth's greenhouse gas emissions is a factor that affects their activity. Hence, interdisciplinary approaches, including microbial ecology, environmental genomics, soil and plant sciences, *etc.*, should be concentrated on mitigating greenhouse gases.

Keywords: Climate change, CO_2, GHG, Microbes, Mitigation, Microbiomes.

INTRODUCTION

Climate change and global warming are extensively recognized as serious contemporary global issues for humanity. These global environmental issues are caused by increased anthropogenic emission of greenhouse gases (GHGs) in the atmosphere, which exerts a warming effect due to the enhanced greenhouse effect. The greenhouse effect refers to the traps of solar radiation in Earth's atmosphere, facilitated by the existence of greenhouse gases (GHGs) that include carbon dioxide (CO_2), water vapor, methane (CH_4), and nitrous oxide (N_2O) that let incoming solar radiation to pass through, but then absorb the heat sent from the surface of the Earth, thus increasing temperature across the world. It is evident from the IPCC's sixth assessment report (AR6) that the cumulative anthropogenic CO_2 emissions have warmed Earth's ecosystems unequivocally [1]. Other threats due to GHGs are briefly listed in Table **1**. In various sectors, numerous indirect and direct sources involve the emission of GHGs in the atmospheric environment. The energy sector, which produces electricity and heat, is considered a significant emission source, while agriculture, forests, and other land uses, industrial and transport sectors are other sources of GHG contribution to the environment.

Table 1. Impact of greenhouse gases in the atmosphere.

Brief threats	References
Continuous increase in the Earth's surface temperature	[2]
Melting of glaciers (21% of Himalayan glaciers were lost in the past 40 years)	[3]
Radiation exposure	[4]
Changes in the atmospheric composition	[4]
Sea level rise	[5]
Violation of the agricultural system	[5]
Increased flood risk	[6]

Microorganisms play prominent roles in the carbon and nutrient cycle, agriculture, and the global food web, and their role in climate change needs consideration [7]. Microbial processes and their diversity in different ecosystems have significant roles related to global fluxes of GHGs and climate change. They are involved in both the emission and consumption of greenhouse gases (GHG), *viz.* carbon dioxide (CO_2), methane (CH_4), and nitrous oxide (N_2O) that are responsible for 98% of increased warming conditions [8, 9], which are ultimately the cause for climate change. Microbes play a vital role either as producers or users of these GHGs in the environment, as they can recycle or transform indispensable elements such as carbon and nitrogen [10]. Microbes consume these GHGs as resources for their growth through photo/ chemoautotrophy (algae, cyanobacteria, nitrifiers), methanotrophy (methane oxidizers), and nitrous oxide reduction (denitrifiers). Photosynthetic microbes consume CO_2 from the atmosphere, while the heterotrophs break down organic matter to release GHGs. The net carbon flux in different ecosystems is determined by the balance between microbes that depends on temperature and other climatic factors [11] since they also store and emit carbon into the atmosphere in massive quantities [12].

ROLE OF MICROBES AND MICROBIOMES IN GHG EMISSIONS

The greenhouse gases such as CO_2, CH_4, and N_2O are produced mainly by microbial processes as their essential by-products [13]. In soil, the microbes play a significant role in the C-N cycle by decomposing organic materials and releasing CO_2 back into the atmosphere, accounting for 25% of the CO_2 naturally released into the atmosphere through microbial respiration, a primary channel for carbon efflux from ecosystems. Similarly, methanogenic microbes are known to produce the greenhouse gas CH_4 (Non-fossil) , which has 27.9 as per AR6 WG I report times more global warming potential than CO_2. The primary cause of nitrous oxide emissions in the soil is by the bacteria through nitrification and denitrification under aerobic and anoxic environments. N_2O has a 273 as per AR6 WG I times greater global warming potential than CO_2, contributing to around 19% of the overall global warming effect. Globally, naturally vegetated soils are estimated to generate 6.6 Tg of N_2O per year and 3.8 Tg of nitrous oxide per year in the Earth's atmosphere [14].

Role of Microbes and Microbiomes in CO_2 Emissions

Approximately 4.1 petagrams of carbon are added to the atmosphere as CO_2 per year, which has been predicted to rise dramatically by 2100 [15]. The atmospheric carbon amounts produced by microbial decomposition in the soil are around 7.5-9 folds of the annual anthropogenic emissions worldwide. The terrestrial environment is tightly linked with atmospheric CO_2 concentrations, such as carbon sequestration into the soil as biomass, emissions from respiration, and decomposition of organic matter subjected to partial photosynthetic fixation (Fig.

1). Studies show that microbial respiration is responsible for major CO_2 emissions in the terrestrial atmosphere [16, 17], and the heterotrophic microbial communities in soil are the key drivers for the recycling of soil organic matter, based soil organic matter decomposition, and increasing CO_2 levels [18]. In oceans, phytoplankton perform photosynthesis, while CO_2 is released to in the atmosphere by the heterotrophic and autotrophic respiration processes [19, 20]. Microbial activities are significantly influenced by several environmental factors, like temperature affecting respiration rates and moisture levels influencing decomposition rates, resulting in increased/ decreased carbon concentration levels in the environment.

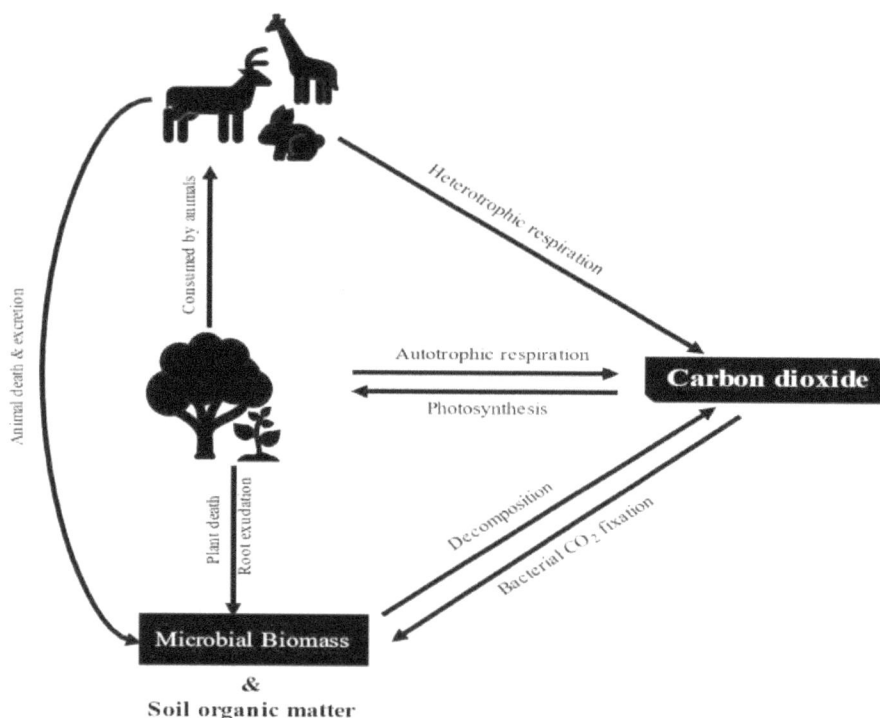

Fig. (1). Terrestrial carbon cycle and CO_2 emissions.

Bacterioplankton, a heterotrophic microbe, plays a vital role in the carbon biogeochemical cycle where microbes transform organic carbon to biomass, and the remaining are transformed into CO_2 and then released into the atmosphere through different processes. The change in vegetation diversity and composition affects the metabolic activities of microbes and the carbon cycle. Studies also show that the microbial communities actively and passively promote the release of root exudates that elevate the CO_2 concentration in the atmosphere [21].

Role of Microbes and Microbiomes in Methane Emissions

The atmospheric concentration of CH_4 is approximately 1,780 ppb, which is higher than pre-industrial levels [22, 23]. Globally, paddy fields or wetlands are regarded as important anthropogenic sources of CH_4 emission, accounting for 6–20% of the total CH_4 emissions [24]. In addition, animal manure contributes 30-45% percent of global CH_4 emissions [25]. Approximately 85 percent of global methane emissions were caused by microorganisms [26]. The biological soil crust has a close relationship with the microbial communities responsible for CH_4 emission into the atmosphere (Fig. **2**). The bog soils are known to possess huge numbers of microbes and enzymes (methyl coenzyme M reductase) to transform CO_2 into methane and release larger amounts of methane. The biogenic release of CH_4 through the biological anaerobic decomposition of organic materials by *methanogenic archaea* is an important source for atmospheric CH_4 [27]. Methanogens are anaerobic microorganisms belonging to the phylum Euryarchaeota, under six orders, such as Methanocellales, Methanococcales, Methanomicrobiales, Methanopyrales, Methanosarcinales, and Methano-bacteriales. Various types of methanogenic archaea including hydrogenotrophic and acetoclastic methanogens influence the CH_4 emission concentrations in anoxic environment [28]. Methanotrophic bacteria and various anaerobic fungi also play an important role in CH_4 emission from the terrestrial environment (Table **2**). For CH_4 oxidation involves different types of bacterial groups, including both methanotrophs responsible for the direct metabolization of CH_4) and non-methanotrophic methylotrophs (involving methanol for CH_4 oxidation) [29].

Table 2. Microbial communities responsible for GHG emissions.

Microorganism	Habitat	References
CO₂		
Bacterioplankton picophytoplankton	Aquatic &Marine ecosystems	[42]
Burkholderiacepacia	Soil	[5]
Brucella suis	Soil	
Ostreococcustauri	Marine	[43]
CH₄		
Methanococcoides	Freshwater, saltwater	[44]
Methanosarcinabarkeri	Lakes	[28]

(Table 2) cont.....

Microorganism	Habitat	References
Methanothrixsoehngenii	Soil, water	[23]
Methanococcus	Marine	
Methanosarcinaacetivorans	Deepa Sea, Oil wells, dump yards	[45]
Methylococcuscapsulatus	Soil	[46]
Nitrification		
(a) Oxidation of ammonia (NH_3) to nitrite (NO_2^-)		
Nitrosospirabriensis	Soil	[47]
Nitrosomonas europaea	Soil, sewage sediments, and water	
Nitrosococcusoceanus	Marine	
Nitrosovibrio tenuis	Soil	
Nitrosococcusmobilis	Soil	
(b) Conversion of nitrite (NO_2^-) to nitrate (NO_3^-)		
Nitrospiragracilis	Marine	[13]
Nitrobacter winogradskyi	Soil	
Denitrification		
Bacillus subtilis	Soil	[48]
Pseudomonas aeruginosa		[49]
Fusarium oxysporium		
Alcaligenes faecalis	Soil and water	[50]
CH_4 & N_2O emissions through biodegradation		
(a) Bacterial genera		
Pseudomonas	Soil, compost, coastal marine	[51]
Enterobacter	Soil, sewage sediments, and water	
Saccharomonospora	Compost, manure, peat	
(b) Fungal genera		
Chaetomium	Air, soil, and plant debris	[51]
Dothideomycetes	Fresh and saltwater	

Due to the presence of hydrocarbons, these methanotrophs' activities support the eukaryotes and heterotrophic bacteria to enhance the methane emissions in the terrestrial environment. The microbe *Methanoflorens stordalenmirensis* plays a crucial role in releasing huge quantities of methane in the Arctic as carbon is trapped in permafrost.

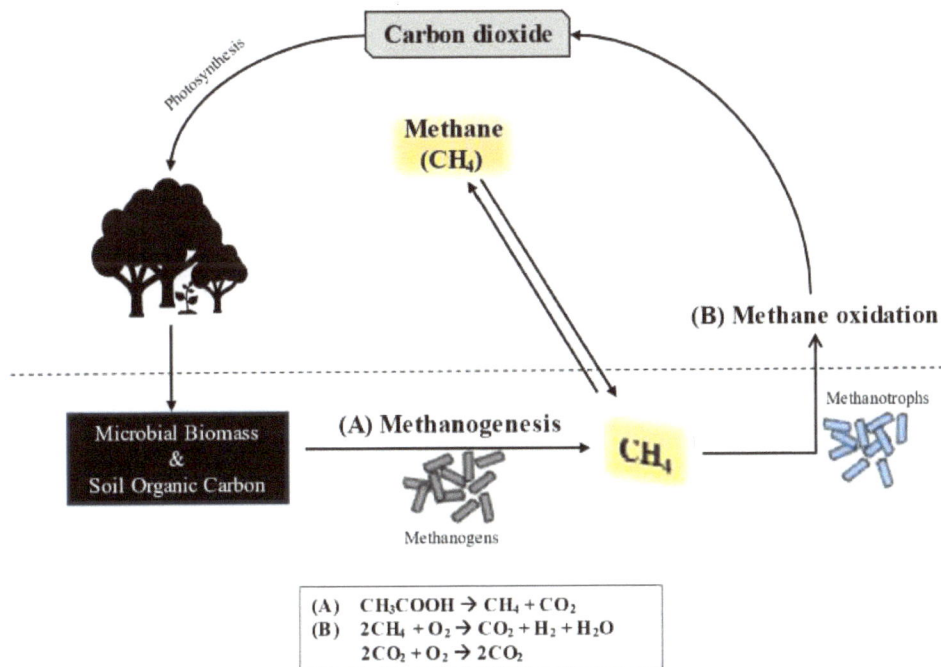

Fig. (2). Biogenic methane emissions.

Role of Microbes and Microbiomes in N$_2$O Emissions

The atmospheric N$_2$O concentration is around 322 ppb in the environment, which is 20 percent higher than in the preindustrial era, and it is still increasing every year with a mean ratio of 0.8 ppb [30]. Various anthropogenic activities such as fossil fuel combustion, industrial emissions, and continuous use of nitrogenous enrichers in agricultural fields are known to alter the nitrogen cycle. Nearly 75% of N$_2$O emissions were from the agriculture sector, pertaining to the significant increase in the use of synthetic nitrogen fertilizers [31].

N$_2$O is mainly produced through nitrification or denitrification processes with the help of soil microbes, depending on the availability of oxygen levels. Further, it is derived from aerobic autotrophic nitrification as a by-product ($NH_4^+ \rightarrow NH_2OH \rightarrow NO_2^- \rightarrow NO_3^-$) and as an intermediate product during the anaerobic denitrification ($NO_3^- \rightarrow NO_2^- \rightarrow NO^- \rightarrow N_2O \rightarrow N_2$), and both pathways are associated with the soil microbial metabolic activities [32].

Theoretically, under very low-oxygen environmental conditions, the microbial nitrification-denitrification processes produce a huge concentration of N$_2$O from the terrestrial environment [33, 34]. Under normal conditions, ammonium (NH_4^+) nitrification and nitrate (NO_3^-) denitrification are considered the significant microbial processes involved in the N_2O emission (Fig. **3**) [35, 36]. The production of N_2O in the ocean is dominated by archaeal nitrification [37]. Furthermore, earthworms play a significant role in N_2O emission (*via* nephridia) by decomposing dead and decaying organic matter [38].

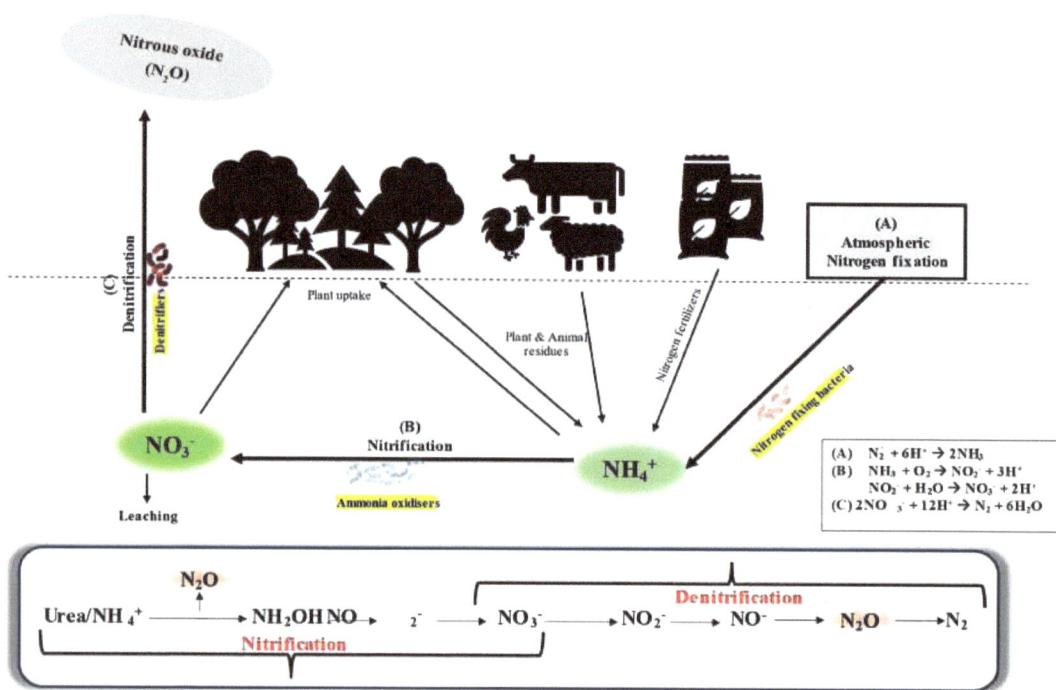

Fig. (3). Biotic N_2O emissions from the Nitrogen cycle.

Role of Microbes and Microbiomes in Ammonia Emissions

Ammonia (NH_3) is a common toxicant emitted from wastes (landfills, compost, slurry, *etc.*), fertilizers (urea) and natural processes (sediments). Though NH_3 is not a greenhouse gas, it is a transboundary pollutant representing a major loss pathway *via* ammoniacal nitrogen (N) volatilization, where approximately 90 percent of the global NH_3 is from farmlands [39]. The decomposition and volatilization of urea produces ammonia. As for NH_3, it can also be redeposited once it is emitted into the atmosphere, leading to water eutrophication and subsidizing the local nitrogen pool. In addition, an increase in nitrogen concentration indirectly contributes to N_2O production through nitrification and denitrification [40, 41].

According to IPCC EP4 (default emission factor), indirect N_2O emission is 1% of the loss of nitrogen deposition and volatilization. Bacterial disintegration of organic matter that accumulates in detritus results in the production of ammonia. In sediments, microorganisms produce ammonia or mineralize organic nitrogen through nitrate reduction (less frequently). In anoxic sediments, where nitrification is inhibited, ammonia is notably abundant. Ammonia produced in sediments can be toxic to benthic organisms [52].

ROLE OF MICROBES AND MICROBIOMES IN GHG MITIGATION

Microbes evolved in the atmosphere about 3.5 billion years ago in the ocean, and later, they moved to land some 2 billion years ago, producing and removing GHGs [12]. In the course of Earth's history, the activities of microorganisms have been critically important in determining GHG concentrations as they are accountable for the GHG fluxes between soil and the atmospheric environment. Studies showed that soil bacteria are classified as copiotrophic sp. (*Actinobacteria*) and oligotrophic sp. (*Acidobacteria*) due to their ability and capacity for carbon sequestration. Various other microbial communities have the ability to reduce GHG emissions through indirect or direct consumption (Table 3). Various practices in mitigating GHG emissions include managing the different microbial communities and several approaches toward agriculture-management strategies. Table **4** gives an overview of the opportunities for microbial mitigation of GHGs under direct and indirect effects categories.

Table 3. Microbial communities that reduce GHG concentrations.

Genus	GHG	Habitat	References
Synechococcus	CO_2	Ocean	[53]
Prochlorococcus	CO_2	Ocean	[53]
Scytonema	CO_2	Terrestrial rocks, wood, soil	[54]
Methylokorus	CH_4	Soil	[53]
Methylobacillus	CO_2 & CH_4	Marine and freshwater	[53]

Table 4. The potential of microbes to mitigate climate change in agroecosystems.

Microbe	Function	Effect	Effectiveness	References
(a) Direct effects				
CO_2 fixing micro-organisms (Engineered)	Reduced CO_2 emissions from soil	Promotes microbial CO_2 sequestration	In the culture medium, comparable fixation rates to the autotrophic microbes' capacity were achieved.	[55]

(Table 4) cont.....

Microbe	Function	Effect	Effectiveness	References
Methanotrophs	Decreased emissions of soil CH_4&atmospheric CH_4removal.	Increased biological CH_4 oxidation.	Diminished 6.9 percent of CH_4 emissions– 12 percent in paddy field.	[56]
N_2O reducing bacteria	Decrease soil N_2O emissions	Increased N_2O reduction	N_2O emissions diminished by 28% - 189% in soil microcosms and field levels.	[57]
(b) Indirect effects:				
PGPR (Plant Growth Promoting Rhizobacteria)	Better plant carbon sequestration	Increased whole-plant biomass production	Plant growth increased in planted soil pots with enhanced CO_2 by an average of 42 percent and a 91 percent increase by plant-derived C inputs to soil.	[58]
PGPR *Azospirillum lipoferum*	Increased N_2O [gross (+113%) + net (+37%)] production with high carbon limitation Decreased N_2O [gross (-15%) + net (-40%)] production with low carbon limitations.	Increase nitrite reducer abundance in sites with high C limitation. Decrease nirS-denitrifier and N_2O reducer abundance in low carbon limitation sites.	Variable outcomes *in situ* from -6% to +25% in implanted soil mesocosms and field environments.	[59 - 61]

Role of Microbes and Microbiomes in CO_2 Mitigation Options

Carbon dioxide does not react with anything and is extremely stable. Therefore, it is a big challenge to increase the value of bio-fixed carbon dioxide. The natural CO_2 fluxes from terrestrial, marine, and atmosphere are huge, and they shadow the annual emissions from fossil fuel combustion in the global carbon cycle. The CO_2 levels greatly depend on the balance in the net carbon exchange between photosynthesis and respiration among various autotrophs and heterotrophs. Researchers have found that using conventional fuel accounts for 56 percent of all anthropogenic carbon dioxide emissions [62 - 64]. Altogether, CO_2 is responsible for 77% of the global warming effects.

In light of this, CO_2 capture is essential, notwithstanding the other greenhouse gases, such as NO_x, hydrocarbons, and SO_2. Capturing carbon dioxide has been the subject of many studies in the past. Technologies for capturing and storing carbon dioxide have become increasingly popular means of reducing the global impact of carbon dioxide emissions. The process of capturing and transporting carbon dioxide from the source to the storage facility occurs in real time (Table 6). Biospecies contributing to CO_2 fixation are primary producers, such as marine phytoplankton and land plants. Marine phytoplankton, like unicellular microalgae, accomplish roughly 50 billion tonnes of carbon per year of biological CO_2 fixation [24]. Microalgal systems offer a potential and sustainable alternative to conventional carbon mitigation strategies through bio-mitigation or biological carbon sequestration.

Microalgae are a collection of microorganisms with a wide range of diversity and growth rates; they can survive through photoautotrophy, heterotrophy, or mixotrophy. They have 10 to 50 times higher CO_2 fixing capacity per unit area than terrestrial vegetation. Therefore, it can be grown on soil that is otherwise unsuitable for agriculture. The advantages of microalgae-based CO_2 fixation can be amplified by using microalgal biomass in producing food, feed, chemicals, and biofuels [65].

Microalgal Fixation Of Carbon Dioxide (CO₂)

Microalgae have the ability to fix carbon dioxide from various sources that can be differentiated as

  i. Atmospheric carbon dioxide
 ii. Carbon dioxide emitted from the industrial outlet (e.g., flaring gas, flue gas, *etc.*)
iii. Fixed carbon dioxide in the form of soluble inorganic carbonates (Example: $NaHCO_3$ and Na_2CO_3).

Microalgae can reduce CO_2 either in their natural environments like oceans and lakes or in microalgal ranches. Different algal species with CO_2 sequestration potential are listed in Table **5**. For photosynthesis and self-reproduction, microalgae may actively take up CO_2 from exhaust fumes [64]. Because algae are resistant to outside changes, pure CO_2 may not be suitable for high-quality production [66], and hence, flue gases are the ideal source for CO_2 from both micro and macroalgae. Up to 99 percent of CO_2 emitted during incineration delays the supply of exhaust gases and speeds up the growth of cultivated organisms. Though flue gases are essential for algal growth, it is crucial to note that flue gases comprise 140 other chemical components that can have an impact on culture and the atmosphere. Sulphur oxide (SO_2), for example, continues to be toxic to algae. While there is not much CO_2 in the atmosphere right now, saturating the environment with released gases is still more effective. Microalgae typically develop at a rate of about one-tenth of the CO_2 provided to them.

According to a study [67], microalgae are 50-fold more capable than plants at biofixing carbon dioxide. If solar energy is provided 9% of the time, algae may produce, on average, 280 tonnes of dry biomass per hectare per year. These microalgae can take up roughly 513 tonnes of carbon dioxide during their growth. The most crucial challenge is choosing the exact algae that can absorb and tolerate CO_2, given the makeup of the flue gases, which have 3–30% of carbon dioxide. The ideal pH level must be continuously maintained if it is impossible to grow CO_2-resistant algae precisely. The algae will be able to grow and successfully absorb CO_2 emissions under these circumstances. Moreover, because NO_X and SO_X produce acids when they mix with water, which is harmful to most crops, one should choose algae species that are resistant to these pollutants. It is imperative to note that studies have shown that a 4000 m^3 volume pond can absorb around 2,200 tonnes of CO_2 per year when cultivating algae under naturally occurring settings (pond and sunlight).

Table 5. CO_2 sequestration capabilities of different algal species [24].

Algal Species	% CO_2 Sequestered
Chlorella vulgaris	25
Scenedesmus obliquus	24
Spirulina sp.	53.29
Phaeodactylum tricornutum	63
Chlorococcum littorale	40
Chlorella kessleri	18
Dunaliella	27
Haematococcuspluvialis	20
Botryococcusbraunii	25 – 30
Scenedesmus obliquus	30

Experimental research [67] on the amount of CO_2 captured, crop growth, and kinetic velocity coefficient of algae like *Chlorella vulgaris* and *Nannochloropsis gaditana* shows that algal species of *Chlorella, Chlorococcum, Spirulina, Scenedesmus,* and *Nannochloropsis*not only exhibit massive and vigorous growth rates but also have high levels of tolerance to CO_2 concentrations and atmospheric factors. According to scientists, dry biomass (1 kg) absorbs 1.88 kg of CO_2. However, this broad hypothesis is unreliable, necessitating a unique experimental measurement for each algae taxa (Table 6).

Table 6. Various biological CO_2 sequestration methods [65].

Method	Mechanisms	Prospects	Limitations	References
Afforestation	Throughout a tree's life, atmospheric CO_2 is incorporated into biomass.	Chemical free	• Limits carbon sequestration • Requires large area of land • Possible risk to food supply and biological diversity	[68, 69]
Oceanic fertilization	Stimulate the emergence of photosynthetic organisms using excess iron sources	Increases CO_2 sequestration	• Higher cost • High Level of ambiguity • Impacts the ocean (alteration in plankton built-up) • Probable initiation/trigger of high methane production	[70]
Microalgae-based sequestration	CO_2 utilization through microalgal photosynthesis	• Greaterphotosynthetic efficiency • Efficient and faster sequestration rate compared to higher plants under low CO_2concentration • No competition for farmlands • Co-production of food, fuel, feed, and fine chemicals	• Sensitivity to other toxic substances in gas exhaust • Less cost-effective for the construction of photobioreactors and harvesting algal biomass	[69, 71 - 74]

Role of Microbes and Microbiomes in N_2O Mitigation

N_2O is a prominent greenhouse gas (GHG), constituting only about 0.03 percent of all GHG emissions, with over 300 times more potential to contribute to global warming than carbon dioxide (CO_2), according to its radiative capacity (IPCC, 2007). It has an atmospheric lifetime of 109 years. When N is applied to agricultural soil, it will be lost as nitrate (NO_3^-) either by gas emission or leaching, primarily nitrous oxide (N_2O), making the agricultural sector greatly influence the global environment. The denitrification and nitrification processes, which influence and also regulate the soil's physical-chemical properties like humidity, temperature, pH, oxygen, and nitrogen availability, are responsible for nitrogen loss in the soil as N_2O [75 - 78]. In general, the respiratory N_2O reductase (N_2OR) found in denitrifying bacteria convert N_2O to N_2, which is a major sink for N_2O.

Denitrification enzymes require numerous metal cofactors, such as Cu, Fe, Mo, and Zn. Due to the essential prerequisite of N_2OR for Cu (and sulfur) activity and in the absence of equivalent paths, it can lower N_2O production and can play a crucial role in achieving this final denitrification step.

Numerous bacteria possess scavenging mechanisms, like siderophores, that are expelled from bacterial cells to chelate Fe in order to facilitate extraction from agricultural soils or sequestration from the ocean and transportation of Fe (II) particles for active uptake of receptors. Additionally, Fe can also be stored between cells within proteins as ferritins during Fe stress for recovery. Apart from a few methanotrophic bacteria that emit Cu-chelating chemicals, no such copper sequestration or storage systems are known to exist in bacteria. Accordingly, the availability of copper to the cells solely depends on both Cu concentrations in the external environment and the soil's chelation status. It was initially shown that under laboratory conditions, denitrifying bacterial cultures grown in Cu-deficient media emit significantly more N_2O than when grown in copper-sufficient media, which leads to the conclusion that N_2OR is a solely copper-dependent enzyme [79]. A study found that during nodule degradation of soybean, both *Bradyrhizobium japonicum* and other soil microbes release N_2O by their denitrification processes from nodule N (N_2O source). In contrast, *nosZ*-competent *B. japonicum* entirely takes up N_2O (N_2O sink).

The balance of the N_2O source and sink determines the net N_2O flux from the soybean rhizosphere. Also, studies revealed that N_2O flux is mainly allied with the dynamics of ammonia-oxidizing bacteria rather than the ammonia-oxidizing archaea in nitrogen-rich grassland soils [80]. The mechanism study [81] found that in *Sheanella oneidensis*, MR-1 is an electron-transfer activity that is promoted by the formation of nanotubes within cells, which leads to an increase in denitrification enzyme activity, cell viability, ATP level and carbon source metabolism. As a result of the increased electron generation, transmission, and consumption by *Sheanella oneidensis* MR-1, the performance of the denitrification process was boosted by a decrease in nitrite accumulation and N_2O emission (Tables **7** and **8**).

Table 7. Microbes involved in N_2O gas mitigation.

Microbes	References
Sheanellaoneidensis MR-1 + Paracoccusdenitrificans	[81]
Bradyrhizobium japonicum	[79]
Ammonia-oxidizing bacteria	[80]

Table 8. Microbial Production and Consumption of N_2O and CH_4 [82].

Microbial Organism	Gene Name(s)	Enzyme	General Reaction
Proteobacterial methanotrophs, NC 10 verrucomicrobial methanotrophs,	*pmo*, pxm, *mmo*	Methane monooxygenase	$CH_4 \rightarrow CH_3OH$
Methanogens ANME	*Mcr*	Methyl Co-M reductase	$CO_2 \rightarrow CH_4$ $CH_4 \rightarrow CO_2$
AOB, AOA, and comammox	*Amo*	Ammonia monooxygenase	$NH_3 \rightarrow NO_2^-$
anammox, AOB, comammox,	*Hao*	Hydroxylamine dehydrogenase	$NH_2OH \rightarrow NO$
AOA, AOB, commamox	-	Nitric oxide oxidase	$NO \rightarrow NO_2^-$
Denitrifiers, denitrifying methanotrophs, and AOB (few)	*Nor*	Nitric oxide reductase	$NO \rightarrow N_2O$
NC10	*norZ* (nod)	Nitric oxide dismutase	$2NO \rightarrow O_2 + N_2$
AOA	-	Hydroxylamine-oxidizing enzyme	$NH_2OH + NO \rightarrow NO_2^-$ or $NH_2OH \rightarrow NO$
Comammox, NOB	*Nxr*	Nitrite oxidoreductase	$NO_2^- \rightarrow NO_3^-$
Denitrifiers, denitrifying methanotrophs, anammox, comammox, AOB, AOA, NOB, NC10.	*nirS*, *nirK*, unknown	Nitrite reductase (dissimilatory)	$NO_2^- \rightarrow NO$
ANME-2d, Denitrifiers, denitrifying methanotrophs.	*Nar*	Nitrate reductase (dissimilatory)	$NO_3^- \rightarrow NO_2^-$
Denitrifiers, nondenitrifyingN_2O reducers	*nosZ*	Nitrous oxide reductase	$N_2O \rightarrow N_2$

LIMITATIONS

Diverse communities of microbes that produce and consume the GHGs live in different habitats. While these habitats highly differ in terms of spatial scales and process times, it is a significant challenge to quantify their changes and contributions in response to environmental conditions. Microorganisms are the drivers of climate change that noticeably respond to the changes. The composite effects of the climate change variables (temperature, moisture, *etc.*) negatively impact the microbial world.

FUTURE PERSPECTIVES AND WAY FORWARD

To improve the understanding of microbial regulation of GHG efflux in the atmosphere, adequate screening, identification, and classification of microorganisms based on physiological and functional roles in greenhouse gas production and emission must be performed. A better understanding of aboveground and subsurface interactions and nutrient cycling is essential for plant interaction with soil groups, and the most significant regulator of carbon in soil and nitrogen dynamics is also lacking. Advanced carbon sequestration concepts and recycling of CO_2 to valuable products like fuels and fertilizers are also needed. The microbial aspect of climate change needs to be focused in detail in the present scenario, and the incorporation of microbial data (growth kinetics, enzymes, *etc.*) in climate change models will pave the way for better prediction of GHG fluxes in the future. Hence, interdisciplinary approaches, including soil and plant science, microbial ecology, environmental genomics, and ecosystem modeling, must be used to overcome the challenges (Fig. **4**).

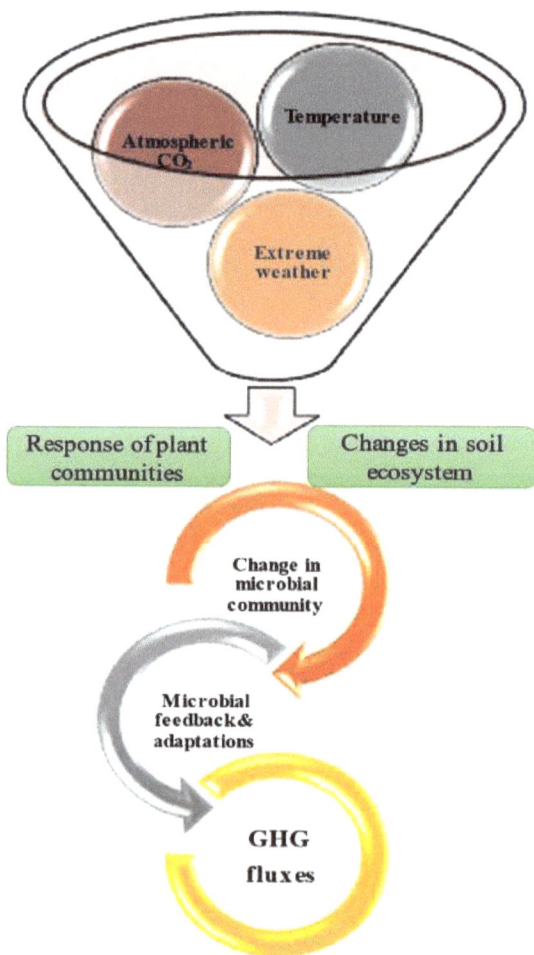

Fig. (4). Comprehensive effect of climate change on microbial community and the GHG emissions.

CONCLUSION

Microbial processes play a significant role in determining the global fluxes as important biogenic greenhouse gases such as carbon dioxide, nitrous oxide, and methane are anticipated to react swiftly to climate alteration. The intricate interactions between microbes and other abiotic and biotic factors should be considered and understood in detail since microbes regulate the GHG flux on Earth. We acquire such an improved knowledge and understanding of microbial responses, and a framework to manage microbial systems in order to reduce greenhouse gas emissions can be developed. The possibility of preventing climate change by reducing greenhouse gas emissions by regulating microbial processes is alluring. It is now high time to consider the aspects, study and understand the acting mechanisms more precisely and appropriately utilize them for developing solutions. It seems that microorganisms can be a crucial natural resource in reducing climate change effects. However, without proper interventions and actions, it can become the most emerging accelerator to the issues that will be faced in the future.

ACKNOWLEDGEMENTS

The authors would like to thank the support provided by ICAR – IARI, Sher---Kashmir University of Agricultural Sciences & Technology of Kashmir, and TNAU.

REFERENCES

[1]   Masson-Delmotte VP, Zhai P, Pirani SL, *et al.* Ipcc, 2021: Summary for policymakers. In: Climate change 2021: The physical science basis. Contribution of working group 1 to the sixth assessment report of the intergovernmental panel on climate change. 2021.
[http://dx.doi.org/http://hdl.handle.net/10204/12710]

[2]   Xu W, Cai YP, Yang ZF, Yin XA, Tan Q. Microbial nitrification, denitrification and respiration in the leached cinnamon soil of the upper basin of Miyun Reservoir. Sci Rep 2017; 7(1): 42032.
[http://dx.doi.org/10.1038/srep42032] [PMID: 28165035]

[3]   Aichele R, Felbermayr G. Kyoto and the carbon footprint of nations. J Environ Econ Manage 2012; 63(3): 336-54.
[http://dx.doi.org/10.1016/j.jeem.2011.10.005]

[4]   Ali QA, Khayyam U, Nazar U. Energy production and CO_2 emissions: The case of coal fired power plants under China Pakistan economic corridor. J Clean Prod 2021; 281: 124974.
[http://dx.doi.org/10.1016/j.jclepro.2020.124974]

[5]   Ma S, Xiong J, Cui R, *et al.* Effects of intermittent aeration on greenhouse gas emissions and bacterial community succession during large-scale membrane-covered aerobic composting. J Clean Prod 2020; 266: 121551.
[http://dx.doi.org/10.1016/j.jclepro.2020.121551]

[6]   Zheng J, Fan J, Zhang F, *et al.* Interactive effects of mulching practice and nitrogen rate on grain yield, water productivity, fertilizer use efficiency and greenhouse gas emissions of rainfed summer maize in northwest China. Agric Water Manage 2021; 248: 106778.

[http://dx.doi.org/10.1016/j.agwat.2021.106778]

[7]   Walsh DA. Consequences of climate change on microbial life in the ocean. Microbiol Today (Nov 2015 issue) Microbiol Soc Engl. 2015; https://microbiologysociety.org/publication/past-issues/water/article/consequences-of-climate-change-on-microbial-life-in-the-ocean-water.html

[8]   Tiedje JM, Bruns MA, Casadevall A, *et al.* Microbes and climate change: a research prospectus for the future. MBio 2022; 13(3): e00800-22.
[http://dx.doi.org/10.1128/mbio.00800-22] [PMID: 35438534]

[9]   Mangodo C, Adeyemi TOA, Bakpolor VR, Adegboyega DA. Impact of microorganisms on climate change: a review. World News Nat Sci 2020; 31: 36-47.

[10]   Joshi PA, Shekhawat DB. Microbial contributions to global climate changes in soil environments: impact on carbon cycle (short review). Ann Appl Biosci 2014; 1(1): R7-9.

[11]   Weiman S. Microbes help to drive global carbon cycling and climate change. Microbe Wash DC 2015; 10(6): 233-8.
[http://dx.doi.org/10.1128/microbe.10.233.1]

[12]   Zimmer C. The microbe factor and its role in our climate future. Yale Environ Yale Environ 2010; p. 360.

[13]   Singh BK, Bardgett RD, Smith P, Reay DS. Microorganisms and climate change: terrestrial feedbacks and mitigation options. Nat Rev Microbiol 2010; 8(11): 779-90.
[http://dx.doi.org/10.1038/nrmicro2439] [PMID: 20948551]

[14]   Laishram B, Devi OR, Ngairangbam H. Insight into microbes for climate smart agriculture. Vigyan Varta 2023; 4(4): 53-6.

[15]   Stocker TF, Qin D, Plattner GK, Tignor MMB, Allen SK, Boschung J, *et al.* Climate Change 2013: The physical science basis. contribution of working group I to the fifth assessment report of IPCC the intergovernmental panel on climate change. 2014; pp. 1-14.

[16]   Schwen A, Jeitler E, Böttcher J. Spatial and temporal variability of soil gas diffusivity, its scaling and relevance for soil respiration under different tillage. Geoderma 2015; 259-260: 323-36.
[http://dx.doi.org/10.1016/j.geoderma.2015.04.020]

[17]   Xu G, Li Y, Hou W, Wang S, Kong F. Effects of substrate type on enhancing pollutant removal performance and reducing greenhouse gas emission in vertical subsurface flow constructed wetland. J Environ Manage 2021; 280: 111674.
[http://dx.doi.org/10.1016/j.jenvman.2020.111674] [PMID: 33218830]

[18]   Apostel C, Halicki S, Kuzyakov Y, Dippold MA. Soil microorganisms can overcome respiration inhibition by coupling intra-and extracellular metabolism: 13C metabolic tracing reveals the mechanisms. ISME J 2017; 11(6): 1423-33.
[http://dx.doi.org/10.1038/ismej.2017.3] [PMID: 28157187]

[19]   Giovannoni SJ, Stingl U. Molecular diversity and ecology of microbial plankton. Nature 2005; 437(7057): 343-8.
[http://dx.doi.org/10.1038/nature04158] [PMID: 16163344]

[20]   Bardgett RD, Freeman C, Ostle NJ. Microbial contributions to climate change through carbon cycle feedbacks. ISME J 2008; 2(8): 805-14.
[http://dx.doi.org/10.1038/ismej.2008.58] [PMID: 18615117]

[21]   Shakoor A, Ashraf F, Shakoor S, Mustafa A, Rehman A, Altaf MM. Biogeochemical transformation of greenhouse gas emissions from terrestrial to atmospheric environment and potential feedback to climate forcing. Environ Sci Pollut Res Int 2020; 27(31): 38513-36.
[http://dx.doi.org/10.1007/s11356-020-10151-1] [PMID: 32770337]

[22]   Mitchell LE, Brook EJ, Sowers T, McConnell JR, Taylor K. Multidecadal variability of atmospheric methane, 1000–1800 CE. J Geophys Res Biogeosci 2011; 116(G2).

[23] Miller SM, Wofsy SC, Michalak AM, *et al.* Anthropogenic emissions of methane in the United States. Proc Natl Acad Sci USA 2013; 110(50): 20018-22.
[http://dx.doi.org/10.1073/pnas.1314392110] [PMID: 24277804]

[24] Wang C, Lai DYF, Sardans J, Wang W, Zeng C, Peñuelas J. Factors related with CH4 and N₂O emissions from a paddy field: clues for management implications. PLoS One 2017; 12(1): e0169254.
[http://dx.doi.org/10.1371/journal.pone.0169254] [PMID: 28081161]

[25] Oenema O, Wrage N, Velthof GL, van Groenigen JW, Dolfing J, Kuikman PJ. Trends in global nitrous oxide emissions from animal production systems. Nutr Cycl agroecosystems. 2005; 72:51–65.
[http://dx.doi.org/10.1007/s10705-004-7354-2]

[26] Zimmerman L, Labonte B. Climate change and the microbial methane banquet. Clim Alert 2015; 27(1): 1-6.

[27] Shakoor A, Abdullah M, Yousaf B, Amina , Ma Y. Atmospheric emission of nitric oxide and processes involved in its biogeochemical transformation in terrestrial environment. Environ Sci Pollut Res Int 2016; •••: 1-20.
[http://dx.doi.org/10.1007/s11356-016-7823-6] [PMID: 27771880]

[28] Angel R, Claus P, Conrad R. Methanogenic archaea are globally ubiquitous in aerated soils and become active under wet anoxic conditions. ISME J 2012; 6(4): 847-62.
[http://dx.doi.org/10.1038/ismej.2011.141] [PMID: 22071343]

[29] Paul BG, Ding H, Bagby SC, *et al.* Methane-oxidizing bacteria shunt carbon to microbial mats at a marine hydrocarbon seep. Front Microbiol 2017; 8: 186.
[http://dx.doi.org/10.3389/fmicb.2017.00186] [PMID: 28289403]

[30] Field CB, Barros VR, Mastrandrea MD, Mach KJ, Abdrabo MAK, Adger WN, *et al.* Summary for policymakers.Climate change 2014: impacts, adaptation, and vulnerability Part A: global and sectoral aspects Contribution of Working Group II to the Fifth Assessment Report of the Intergovernmental Panel on Climate Change. Cambridge University Press 2015; pp. 1-32.

[31] Field CB, Barros VR. Climate change 2014–Impacts, adaptation and vulnerability: Regional aspects. Cambridge University Press 2014.
[http://dx.doi.org/10.1017/CBO9781107415379]

[32] Butterbach-Bahl K, Baggs EM, Dannenmann M, Kiese R, Zechmeister-Boltenstern S. Nitrous oxide emissions from soils: how well do we understand the processes and their controls? Philos Trans R Soc Lond B Biol Sci 2013; 368(1621): 20130122.
[http://dx.doi.org/10.1098/rstb.2013.0122] [PMID: 23713120]

[33] Jia W, Liang S, Zhang J, *et al.* Nitrous oxide emission in low-oxygen simultaneous nitrification and denitrification process: Sources and mechanisms. Bioresour Technol 2013; 136: 444-51.
[http://dx.doi.org/10.1016/j.biortech.2013.02.117] [PMID: 23567715]

[34] Li YK, Li B, Guo WZ, Wu XP. Effects of nitrogen application on soil nitrification and denitrification rates and N₂O emissions in greenhouse. J Agric Sci Technol 2015; 17(2): 519-30.

[35] Pengthamkeerati P, Modtad A. Nitrification inhibitor, fertilizer rate, and temperature effects on nitrous oxide emission and nitrogen transformation in loamy sand soil. Commun Soil Sci Plant Anal 2016; 47(4): 1-8.
[http://dx.doi.org/10.1080/00103624.2015.1122795]

[36] Shakoor A, Shahzad SM, Chatterjee N, *et al.* Nitrous oxide emission from agricultural soils: Application of animal manure or biochar? A global meta-analysis. J Environ Manage 2021; 285: 112170.
[http://dx.doi.org/10.1016/j.jenvman.2021.112170] [PMID: 33607561]

[37] Santoro AE, Buchwald C, McIlvin MR, Casciotti KL. Isotopic signature of N₂O produced by marine ammonia-oxidizing archaea. Science (80). 2011; 333(6047):1282–5.

[38]   Lubbers IM, van Groenigen KJ, Fonte SJ, Six J, Brussaard L, van Groenigen JW. Greenhouse-gas emissions from soils increased by earthworms. Nat Clim Chang 2013; 3(3): 187-94.
[http://dx.doi.org/10.1038/nclimate1692]

[39]   Galloway JN, Townsend AR, Erisman JW, Bekunda M, Cai Z, Freney JR, *et al.* Transformation of the nitrogen cycle: recent trends, questions, and potential solutions. Science (80-). 2008; 320(5878):889–92.
[http://dx.doi.org/10.1126/science.1136674]

[40]   Kavanagh I, Fenton O, Healy MG, Burchill W, Lanigan GJ, Krol DJ. Mitigating ammonia and greenhouse gas emissions from stored cattle slurry using agricultural waste, commercially available products and a chemical acidifier. J Clean Prod 2021; 294: 126251.
[http://dx.doi.org/10.1016/j.jclepro.2021.126251]

[41]   Pan B, Lam SK, Mosier A, Luo Y, Chen D. Ammonia volatilization from synthetic fertilizers and its mitigation strategies: A global synthesis. Agric Ecosyst Environ 2016; 232: 283-9.
[http://dx.doi.org/10.1016/j.agee.2016.08.019]

[42]   James AK, Passow U, Brzezinski MA, Parsons RJ, Trapani JN, Carlson CA. Elevated pCO_2 enhances bacterioplankton removal of organic carbon. PLoS One 2017; 12(3): e0173145.
[http://dx.doi.org/10.1371/journal.pone.0173145] [PMID: 28257422]

[43]   SULLIVAN MB, CAVICCHIOLI R, TIMMIS KN, BAKKEN LR, BAYLIS M, BOETIUS A, et al. Scientists' warning to humanity: microorganisms and climate change. 2019; 17(9):569-586.

[44]   Beckmann S, Krüger M, Engelen B, Gorbushina AA, Cypionka H. Role of bacteria, archaea and fungi involved in methane release in abandoned coal mines. Geomicrobiol J 2011; 28(4): 347-58.
[http://dx.doi.org/10.1080/01490451.2010.503258]

[45]   Liu J, Chen H, Zhu Q, *et al.* A novel pathway of direct methane production and emission by eukaryotes including plants, animals and fungi: An overview. Atmos Environ 2015; 115: 26-35.
[http://dx.doi.org/10.1016/j.atmosenv.2015.05.019]

[46]   Hanson RS, Hanson TE. Methanotrophic bacteria. Microbiol Rev 1996; 60(2): 439-71.
[http://dx.doi.org/10.1128/mr.60.2.439-471.1996] [PMID: 8801441]

[47]   Liu R, Hayden HL, Suter H, *et al.* The effect of temperature and moisture on the source of N_2O and contributions from ammonia oxidizers in an agricultural soil. Biol Fertil Soils 2017; 53(1): 141-52.
[http://dx.doi.org/10.1007/s00374-016-1167-8]

[48]   Lagomarsino A, Agnelli AE, Pastorelli R, Pallara G, Rasse DP, Silvennoinen H. Past water management affected GHG production and microbial community pattern in Italian rice paddy soils. Soil Biol Biochem 2016; 93: 17-27.
[http://dx.doi.org/10.1016/j.soilbio.2015.10.016]

[49]   Arat S, Bullerjahn GS, Laubenbacher R. A network biology approach to denitrification in Pseudomonas aeruginosa. PLoS One 2015; 10(2): e0118235.
[http://dx.doi.org/10.1371/journal.pone.0118235] [PMID: 25706405]

[50]   Wang X, Yu P, Zeng C, *et al.* Enhanced Alcaligenes faecalis denitrification rate with electrodes as the electron donor. Appl Environ Microbiol 2015; 81(16): 5387-94.
[http://dx.doi.org/10.1128/AEM.00683-15] [PMID: 26048940]

[51]   Xu S, Reuter T, Gilroyed BH, *et al.* Microbial communities and greenhouse gas emissions associated with the biodegradation of specified risk material in compost. Waste Manag 2013; 33(6): 1372-80.
[http://dx.doi.org/10.1016/j.wasman.2013.01.036] [PMID: 23490363]

[52]   Lapota D, Duckworth D, Ward J. Confounding Factors in Sediment Toxicology-Issue Papers 1-19. San Diego: Sp Nav Warf Syst Center 2000; pp. 1-19.

[53]   Gupta C, Prakash DG, Gupta S. Role of microbes in combating global warming. Int J Pharm Sci Lett 2014; 4: 359-63.

[54]   Sprent JI. Nitrogen in soils| symbiotic fixation.Encyclopedia of Soils in the Environment. Amsterdam, The Netherlands: Elsevier 2005; pp. 231-45.
[http://dx.doi.org/10.1016/B0-12-348530-4/00457-4]

[55]   Gong F, Liu G, Zhai X, Zhou J, Cai Z, Li Y. Quantitative analysis of an engineered CO2-fixing Escherichia coli reveals great potential of heterotrophic CO_2 fixation. Biotechnol Biofuels 2015; 8(1): 86.
[http://dx.doi.org/10.1186/s13068-015-0268-1]

[56]   Rani V, Bhatia A, Kaushik R. Inoculation of plant growth promoting-methane utilizing bacteria in different N-fertilizer regime influences methane emission and crop growth of flooded paddy. Sci Total Environ 2021; 775: 145826.
[http://dx.doi.org/10.1016/j.scitotenv.2021.145826] [PMID: 33631576]

[57]   Akiyama H, Hoshino YT, Itakura M, *et al.* Mitigation of soil N_2O emission by inoculation with a mixed culture of indigenous Bradyrhizobium diazoefficiens. Sci Rep 2016; 6(1): 32869.
[http://dx.doi.org/10.1038/srep32869] [PMID: 27633524]

[58]   Nie M, Bell C, Wallenstein MD, Pendall E. Increased plant productivity and decreased microbial respiratory C loss by plant growth-promoting rhizobacteria under elevated CO2. Sci Rep 2015; 5(1): 9212.
[http://dx.doi.org/10.1038/srep09212] [PMID: 25784647]

[59]   Bounaffaa M, Florio A, Le Roux X, Jayet PA. Economic and environmental analysis of maize inoculation by plant growth promoting rhizobacteria in the French Rhône-Alpes region. Ecol Econ 2018; 146: 334-46.
[http://dx.doi.org/10.1016/j.ecolecon.2017.11.009]

[60]   Dalmon A, Desbiez C, Coulon M, *et al.* Evidence for positive selection and recombination hotspots in Deformed wing virus (DWV). Sci Rep 2017; 7(1): 41045.
[http://dx.doi.org/10.1038/srep41045] [PMID: 28120868]

[61]   Florio A, Bréfort C, Gervaix J, Bérard A, Le Roux X. The responses of NO2−- and N_2O-reducing bacteria to maize inoculation by the PGPR Azospirillum lipoferum CRT1 depend on carbon availability and determine soil gross and net N_2O production. Soil Biol Biochem 2019; 136: 107524.
[http://dx.doi.org/10.1016/j.soilbio.2019.107524]

[62]   Aziz MBA, Kassim KA, Bakar WAWA, Marto A. Fossil free fuels: trends in renewable energy. CRC Press 2019; pp. 55-70.
[http://dx.doi.org/10.1201/9780429327773]

[63]   Thiyagarajan S, Varuvel EG, Martin LJ, Beddhannan N. Mitigation of carbon footprints through a blend of biofuels and oxygenates, combined with post-combustion capture system in a single cylinder CI engine. Renew Energy 2019; 130: 1067-81.
[http://dx.doi.org/10.1016/j.renene.2018.07.010]

[64]   Rahman FA, Aziz MMA, Saidur R, *et al.* Pollution to solution: Capture and sequestration of carbon dioxide (CO_2) and its utilization as a renewable energy source for a sustainable future. Renew Sustain Energy Rev 2017; 71: 112-26.
[http://dx.doi.org/10.1016/j.rser.2017.01.011]

[65]   Zhou W, Wang J, Chen P, *et al.* Bio-mitigation of carbon dioxide using microalgal systems: Advances and perspectives. Renew Sustain Energy Rev 2017; 76: 1163-75.
[http://dx.doi.org/10.1016/j.rser.2017.03.065]

[66]   Alami AH, Alasad S, Ali M, Alshamsi M. Investigating algae for CO_2 capture and accumulation and simultaneous production of biomass for biodiesel production. Sci Total Environ 2021; 759: 143529.
[http://dx.doi.org/10.1016/j.scitotenv.2020.143529] [PMID: 33229076]

[67]   Bhola V, Swalaha F, Ranjith Kumar R, Singh M, Bux F. Overview of the potential of microalgae for CO_2 sequestration. Int J Environ Sci Technol 2014; 11(7): 2103-18.

[http://dx.doi.org/10.1007/s13762-013-0487-6]

[68]   Farrelly DJ, Everard CD, Fagan CC, McDonnell KP. Carbon sequestration and the role of biological carbon mitigation: A review. Renew Sustain Energy Rev 2013; 21: 712-27.
[http://dx.doi.org/10.1016/j.rser.2012.12.038]

[69]   Draper B, Yee WL, Pedrana A, *et al.* Reducing liver disease-related deaths in the Asia-Pacific: the important role of decentralised and non-specialist led hepatitis C treatment for cirrhotic patients. Lancet Reg Health West Pac 2022; 20: 100359.
[http://dx.doi.org/10.1016/j.lanwpc.2021.100359] [PMID: 35024676]

[70]   Williamson P, Wallace DWR, Law CS, *et al.* Ocean fertilization for geoengineering: A review of effectiveness, environmental impacts and emerging governance. Process Saf Environ Prot 2012; 90(6): 475-88.
[http://dx.doi.org/10.1016/j.psep.2012.10.007]

[71]   Singh J, Tripathi R, Thakur IS. Characterization of endolithic cyanobacterial strain, Leptolyngbya sp. ISTCY101, for prospective recycling of CO_2 and biodiesel production. Bioresour Technol 2014; 166: 345-52.
[http://dx.doi.org/10.1016/j.biortech.2014.05.055] [PMID: 24926608]

[72]   Varshney P, Mikulic P, Vonshak A, Beardall J, Wangikar PP. Extremophilic micro-algae and their potential contribution in biotechnology. Bioresour Technol 2015; 184: 363-72.
[http://dx.doi.org/10.1016/j.biortech.2014.11.040] [PMID: 25443670]

[73]   Harun R, Singh M, Forde GM, Danquah MK. Bioprocess engineering of microalgae to produce a variety of consumer products. Renew Sustain Energy Rev 2010; 14(3): 1037-47.
[http://dx.doi.org/10.1016/j.rser.2009.11.004]

[74]   Ryan C, Hartley A, Browning B, Garvin C, Greene N, Steger C. Cultivating clean energy. The promise of algae biofuels Springer, Singapore. 2009;1–65.

[75]   Forte A, Fierro A. Denitrification rate and its potential to predict biogenic N_2O field emissions in a Mediterranean maize-cropped soil in Southern Italy. Land (Basel) 2019; 8(6): 97.
[http://dx.doi.org/10.3390/land8060097]

[76]   Norton J, Ouyang Y. Controls and adaptive management of nitrification in agricultural soils. Front Microbiol 2019; 10: 1931.
[http://dx.doi.org/10.3389/fmicb.2019.01931] [PMID: 31543867]

[77]   Vitale L, Tedeschi A, Polimeno F, *et al.* Water regime affects soil N_2O emission and tomato yield grown under different types of fertilisers. Ital J Agron 2018; 13(1): 74-9.

[78]   Tedeschi A, Volpe MG, Polimeno F, *et al.* Soil fertilization with urea has little effect on seed quality but reduces soil N_2O emissions from a hemp cultivation. Agriculture 2020; 10(6): 240.
[http://dx.doi.org/10.3390/agriculture10060240]

[79]   Inaba S, Ikenishi F, Itakura M, *et al.* N(2)O emission from degraded soybean nodules depends on denitrification by Bradyrhizobium japonicum and other microbes in the rhizosphere. Microbes Environ 2012; 27(4): 470-6.
[http://dx.doi.org/10.1264/jsme2.ME12100] [PMID: 23047151]

[80]   Di HJ, Cameron KC, Sherlock RR, Shen JP, He JZ, Winefield CS. Nitrous oxide emissions from grazed grassland as affected by a nitrification inhibitor, dicyandiamide, and relationships with ammonia-oxidizing bacteria and archaea. J Soils Sediments 2010; 10(5): 943-54.
[http://dx.doi.org/10.1007/s11368-009-0174-x]

[81]   Jiang M, Zheng X, Chen Y. Enhancement of denitrification performance with reduction of nitrite accumulation and N_2O emission by Shewanella oneidensis MR-1 in microbial denitrifying process. Water Res 2020; 169: 115242.

[http://dx.doi.org/10.1016/j.watres.2019.115242] [PMID: 31706124]

[82]   Stein LY. The long-term relationship between microbial metabolism and greenhouse gases. Trends Microbiol 2020; 28(6): 500-11.
[http://dx.doi.org/10.1016/j.tim.2020.01.006] [PMID: 32396828]

CHAPTER 6

Role of Carbon in Microbiomes for Ecosystem Restoration

Ihsan Flayyih Hasan Al-Jawhari[1,*]

[1] *Department of Biology, College of Education for Pure Sciences, University of Thi-Qar, Iraq*

Abstract: The most significant threat to civilization is climate change. Carbon dioxide (CO_2), methane (CH_4), and nitrous oxide (N_2O) are the three predominant greenhouse gases generated and utilized by microbes. Certain bacteria can induce diseases in humans, animals, and plants, exacerbating climate change. When conditions allow, microbes that utilize light- or chemoautotrophic activities (such as cyanobacteria and algae) and methanotrophic processes (which oxidize CH_4) and those that reduce N_2O can also metabolize these three gases (denitrifies). The production or consumption of these gases by bacteria is contingent upon their environment and interactions, which humans frequently modify. At times, we can manipulate environmental variables to enhance the microbial degradation of these gasses. According to a recent Intergovernmental Panel on Climate Change (IPCC) study, 3.3 billion individuals globally are subjected to environmental change. At the same time, unsustainable growth patterns exacerbate ecological and human vulnerability to environmental hazards. As individuals, societal change agents, and microbiologists with expertise, we may assist in identifying methods to reverse the prevailing tendency. This chapter argues that understanding both the direct and indirect effects of climate change on microorganisms is essential to evaluate their potential positive and negative impacts on land-atmosphere carbon exchange and global warming. Furthermore, we suggest that this encompasses examining the complex interactions and feedback mechanisms that emerge during communication among microorganisms, plants, and their physical environment within the climate change framework. Furthermore, the influence of further global changes may exacerbate the effects of the environment on soil bacteria.

Keywords: Algae, climate, cyanobacteria, global warming, habitat, microorganisms.

INTRODUCTION

Environmental change is a significant therapeutic and political challenge of the twenty-first century. The release of the IPCC's Fifth Assessment Report (AR5)

* **Corresponding author Ihsan Flayyih Hasan Al-Jawhari:** Department of Biology, College of Education for Pure Sciences, University of Thi-Qar, Iraq; E-mail: dr.ihsan_2012@yahoo.com

**Govindaraj Kamalam Dinesh, Shiv Prasad, Ramesh Poornima, Sangilidurai Karthika, Murugaiyan Sinduja &
Velusamy Sathya (Eds.)**

and the Special Report on Global Warming of 1.5 °C (SR1.5), encompassing data on over 12,000 species globally, has led to new studies revealing changes consistent with climate change. Two-thirds of springtime phenological events have advanced due to regional temperature changes, and approximately fifty percent of species have shifted their ranges to higher latitudes or altitudes, based on an analysis of over 4,000 species globally (with very high confidence). The distribution of species is changing, and international varieties, particularly those in northern latitudes, are more adept at adapting to environmental changes than indigenous species, potentially leading to the emergence of new invasive species due to anthropogenic increases in greenhouse gases [1]. The most significant challenge is understanding the biological mechanisms governing carbon exchanges among terrestrial, marine, and atmospheric systems and their responses to climate change through climate-ecosystem interactions that may amplify or mitigate local and global environmental adjustments [2]. Earthbound ecological communities are crucial in climatic circumstances as they emit and absorb greenhouse gases like carbon dioxide, methane, and nitrous oxide while sequestering significant amounts of carbon in live plants and soils [3]. The sink activity of terrestrial ecosystems is influenced by various interrelated factors, including anthropogenic and natural disturbances [4], agricultural land use [5], nitrogen enrichment [6], sulfur deposition [7], and fluctuations in atmospheric ozone concentration [8].

The potential for increased temperatures to enhance the release of carbon dioxide from soil to the atmosphere due to improved microbial decomposition of organic matter renders the effect of climate change on the soil carbon sink a significant area of uncertainty. If projected climate change scenarios are accurate, this increase in soil carbon loss could significantly exacerbate the responses of the soil carbon cycle [9]. The balance between photosynthesis and respiration fundamentally determines the carbon content of ecosystem carbon budgets due to climate change, encompassing both autotrophic respiration and heterotrophic soil microbial respiration. Although our comprehension of the assimilatory component of the carbon cycle, specifically photosynthesis and its response to environmental changes, is well established, significant gaps remain in our knowledge regarding soil respiration reactions [10, 11]. Numerous factors influence soil respiration, including complex interactions and feedbacks among climate, plants, symbionts, and free-living heterotrophic soil microorganisms. This results in a lack of understanding regarding soil respiration and its sensitivity to climate change [12]. All organisms rely on the Earth's provision of essential materials. Reutilizing these elements is essential to prevent depletion, as the Earth is a closed system with a finite supply of crucial components, including hydrogen (H), oxygen (O), carbon (C), nitrogen (N), sulfur (S), and phosphorus (P). Decomposing and transforming deceased organic matter into forms that other creatures can utilize mostly rely on bacteria. The principal microbial enzyme systems are believed to drive the Earth's biogeochemical cycles [13]. The combination of photosynthesis and respiration governs the terrestrial carbon cycle in equilibrium [14]. Photosynthesizing plants and chemoautotrophic microbes, which convert atmospheric CO_2 into organic matter, are the primary "carbon-fixing" autotrophic organisms that transfer carbon from the atmosphere to the soil (Fig. **1**). Subsequently, various unique processes responsible for the respiration of both autotrophic and heterotrophic organisms release fixed carbon back into the environment [10].

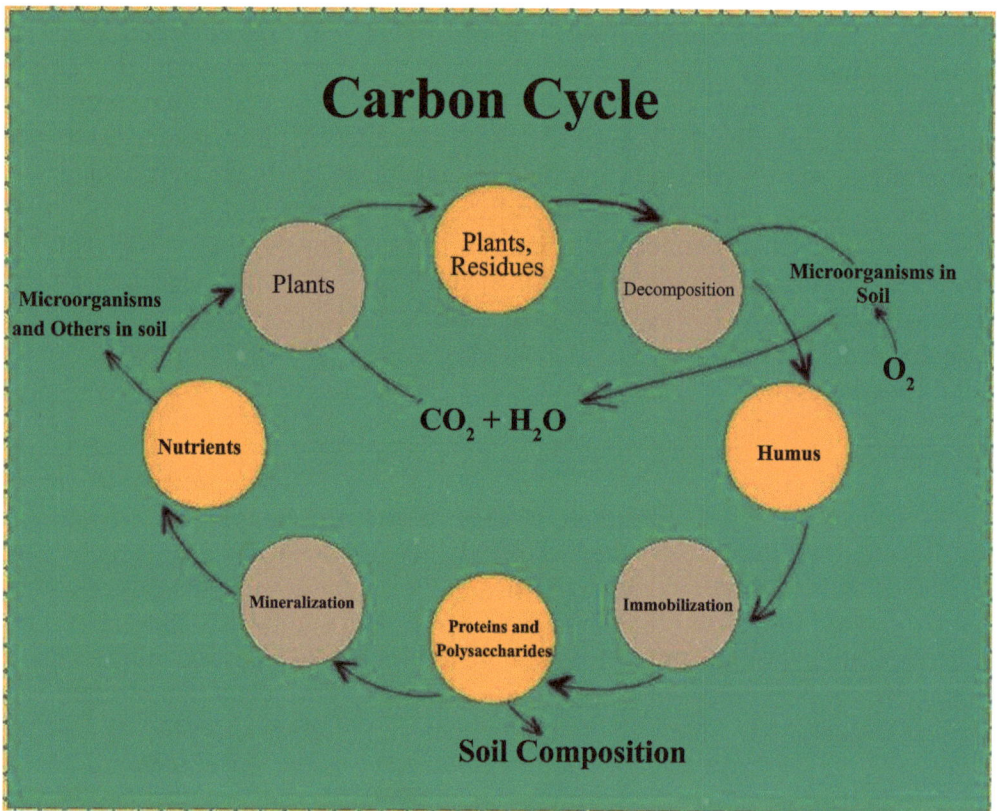

Fig. (1). Carbon cycle in the terrestrial ecosystem.

The "natural carbon-consuming" heterotrophic microorganisms utilize carbon derived from plant, animal, or microbial sources as a substrate for metabolism, sequestering a portion of the carbon in their biomass while releasing the remaining carbon as metabolites or CO_2 back into the environment, which is encompassed in the reverse process [15]. Since numerous soils globally are oxic and unsaturated, carbon dioxide is the primary source of respiration. A study [16] indicates that hydrogenotrophic archaea in peatlands and rice fields reduce CO_2

through methanogenesis. The amount of methane generated is contingent upon the relative activity of methane-oxidizing bacteria and methanogens, including those that ferment acetate [17, 18]. In the oxic soil layers of wetland systems, potentially microbial anaerobic methane oxidation occurs in anoxic conditions in surface-dwelling methanotrophs [19]. Microbes transfer carbon within the environment to fulfill their principal objective: survival through reproduction. Consequently, microorganisms utilize numerous organic and inorganic forms of carbon as carbon and energy sources. The C cycle does not function independently; it acts in constrained synchrony with other essential components of microbial metabolism.

The remaining components serve either as electron donors and acceptors in energy transduction (for instance, N types from the most reduced, $NH4^+$, to the most oxidized, NO_3), or they are rendered inactive and mineralized as part of various biomolecules containing essential elements (*e.g.*, proteins, DNA). The availability of essential nutrients, such as nitrogen and phosphorus, together with environmental factors like pH, soil texture and mineralogy, temperature, and soil moisture content, influences the rate at which microorganisms assimilate and release carbon [20]. The disparate distribution of organic matter in global soils is predominantly governed by the interactions between environmental conditions and biological processes, particularly primary production [21], with the lowest carbon content typically occurring in desert biomes [22], where low mean annual rainfall limits primary production and fosters the prevalence of arid soil conditions. Conversely, wet and cool regions in the Northern Hemisphere host the highest global concentration of carbon [23].

This chapter aims to (i) elucidate the importance of soil microbial communities in soil resource assessment and climate change through carbon cycle feedbacks, (ii) briefly outline the primary methodologies for attributing below-ground utilization of plant-derived carbon to specific microbial groups, and (iii) evaluate whether the implementation of these techniques can provide the necessary information for monitoring agro-ecosystems regarding carbon sequestration and enhanced agricultural sustainability. Understanding the feasible results of soil microorganisms on international warming, both favorable and adverse, requires explicit factors to consider both the straight and indirect effects of climate adjustment on soil microbes, as well as the ability for feedback to greenhouse gas production. In addition, we intend to comprehend exactly how dirt microbial ecology influences environment modification and how soil microbial feedback manages dirt land-atmosphere carbon exchange.

SOIL CO_2 BALANCE

Carbon sequestration is the technique of reducing net CO_2 emissions in agricultural soils to enhance soil carbon storage. Removing carbon-containing compounds from the environment and their retention in soil carbon reservoirs is termed soil carbon sequestration (SCS). The activity of the soil microbial community has been associated with variations in soil's capacity to sequester carbon (SMC). The structure and function of the SMC, essential for the preservation of soil ecosystem services, influence both the availability and turnover of nutrients and the rate of soil organic matter (SOM) decomposition. The sustainability assessment of any soil management method must include a quantification of the impact of agricultural activities on Soil Moisture Content (SMC) and Soil Carbon Stock (SCS) [24]. The volume of gross carbon dioxide exchange between agricultural soils and the atmosphere is significant, as a considerable portion of the biomass produced in agricultural systems is processed through the soil decomposer community. Conversely, there exists a somewhat smaller disparity between carbon dioxide released during decomposition and CO_2 assimilated by photosynthetic activities, which enters the soil as plant detritus. This distinction determines whether the community acts as a generator or sink of CO_2 for its internet carbon balance.

Carbon in Soil Because of Climate Change

The numerous direct and indirect factors involved make the effects of climate change on soil characteristics, particularly soil carbon, a complex issue. Ambient temperature levels may affect the rate of SOM decomposition, which can lead to the emission of greenhouse gases that affect environmental change [25]. Since soil moisture and temperature are two of the most critical factors affecting microbial activity and soil organic carbon (SOC), the impacts of moisture and temperature due to climate change will be emphasized as significant elements [26]. One of the most reliable theories for climate change is that it is warming due to elevated temperatures. Climate warming has been predominantly associated with SOC decomposition due to the influence of temperature on soil microbial communities and their enzymatic and metabolic activity [27]. Nonetheless, the relationship between temperature and soil carbon is less distinct and more constrained than the effects of moisture. The increased climatic temperature may also elevate soil temperature, hence augmenting microbial activity and the rate of SOC decomposition [28].

This is not always the instance as a result of the various temperature levels of sensitivity of biota, particularly the microbial neighborhood, where higher temperatures, such as in colder regions, exhibit more boosted soil respiration, potentially bringing about a efflux of C towards increased climatic CO_2 in contrast with those residing in soils in hotter areas [29]. On the other hand, a greater effect of microbial OM breakdown is seen in hotter locations, indicating that topography, soil structure, and pH are added environmental aspects that might influence the SOC. Ultimately, climate change triggers a decrease in SOC input and an increase in SOC results [28].

Impact of Agricultural Practices on Soil CO2 Balance and Microbiota

As stated in a study [30], agronomic monitoring involves synthesizing soil and crop management techniques that, when implemented effectively, will enhance soil productivity and nutrient availability, fostering improved growth and increased crop yield. These administration strategies can be categorized into targeted and untargeted techniques based on conventional agricultural practices and plant-soil interactions. Targeted techniques typically incorporate biotechnological instruments such as biostimulants and biofertilizers. The soil microbiota will undoubtedly be influenced directly or indirectly, regardless of the approach employed. Appropriate monitoring strategies are crucial to achieving food security for the expanding global population, given that the soil microbiome is regarded as a significant microorganism that may affect fundamental plant health due to its close association with plant roots. Due to its crucial role in the soil carbon pool, the soil microbiome must be noticed in agronomic monitoring procedures, particularly when soil organic matter is involved. For example, additional organic matter applications may lead to increased decomposition and reduced carbon storage due to diminished microbial carbon use efficiency, a positive priming effect from enhanced mineralization of soil organic matter, and increased carbon skimming resulting from the accumulation of microbial byproducts and necromass over time [31].

The stabilization and dissolution of soil organic matter (SOM) directly influence soil organic carbon (SOC). Agronomic practices, such as fertilization, conservation tillage, cover cropping, and crop rotation that enhance soil organic matter (SOM) also influence soil organic carbon (SOC) [32]. The persistence of soil organic matter (SOM) and carbon pools is influenced by biotic and abiotic factors, including soil structure, moisture, carbon-to-nitrogen ratio, soil organic carbon concentration, pH, climate, vegetation, and land use [33]. An ecological community typically does not transition from a net carbon resource to a carbon sink within a relatively short timeframe due to the intricate interactions among its numerous components [34]. Therefore, agronomic management strategies must adopt the most effective methods to mitigate their impact on climate change [32]. While plant food enhances plant yield, soil fertility, and quality, it also compacts the soil, elevates nitrous oxide emissions, and contributes to nitrate leaching into groundwater and surface waters [35]. The soil carbon-to-nitrogen ratio is significantly influenced by the application of fertilizer. An experiment has shown that chemical fertilizers (NPK) reduce soil pH. The effect is significantly enhanced when paired with organic fertilizer. Furthermore, modifications in ammonium nitrogen (NH^4+-N) and nitrate nitrogen (NO^3+-N) result in fluctuations in the populations of microbiome family members following the treatment of organic waste (straw).

The pH affects microbial activity. Soil emissions are influenced by management practices such as liming, as more carbonates can be released as CO_2. Acidic soil diminishes emissions from the ground. Methanogenesis (the production of CH^4) flourishes between pH 4 and 7. Carbon dioxide emissions are highest at neutral pH values. N_2O emissions in acidic soil are decreased. At elevated pH levels, nitrification increases when the equilibrium between NH_3 and NO_3 shifts towards ammonia. Nevertheless, no evidence has established a correlation between pH and NO/N_2O emissions. In acidic soil, denitrification results in nitric oxide (NO) emissions, while nitrification produces NO emissions in alkaline soil.

ENVIRONMENTAL EFFECTS AND THE SIGNIFICANCE OF THE SOIL CARBON CYCLE AND MICROBIAL DECOMPOSERS

The Connection Between Soil and the Atmospheric Carbon Pool

Studies indicate that the carbon stored in soil globally is at least three times more than that in the atmosphere. Terrestrial populations and the environment exchange cause around 8% of the total atmospheric carbon pool annually through net primary production and terrestrial heterotrophic respiration, predominantly by microorganisms. If microbial respiration in soil is ceased, it would require approximately years of current manufacturing rates to deplete atmospheric CO_2 reserves, assuming other components of the carbon cycle, such as marine carbon dioxide exchange, are disregarded [36].

Presently, terrestrial biological groups eliminate around 25% of annual global emissions from fossil fuels and sequester more atmospheric CO_2 through photosynthesis than they release *via* respiration. Nonetheless, land management greatly impacts carbon sequestration and varies considerably between regions. Human activity in both pre-industrial and post-industrial eras has resulted in the loss of 42 to 78 gigatons of carbon from the planet's terrestrial and agricultural soils [37]. Consequently, land restoration to reclaim some of this lost carbon might significantly mitigate emissions from nonrenewable fuel sources [38].

Soil Organic Matter Persists; Microbes Break Down Plant-Derived Carbon

Organic carbon from plants enters the soil system primarily through two pathways: (i) above-ground plant detritus and its leachates, which are dissolved organic carbon transported into the soil by infiltrating rainfall, and (ii) below-ground root litter and exudates, also known as rhizodeposition. The type of plant present and the soil utilized for agriculture and plant care will determine the relative quantity of above- and below-ground inputs. Carbon-containing molecules perpetually circulate from their origins to the soil through rhizodeposition. The chemical constituents of origin exudates comprise simple compounds such as sugars, amino acids, sugar alcohols, and natural acids, together with more structurally complex metabolites [39], which can be swiftly respired in soil (within hours to days) [40]. However, before they can be assimilated into the microbial cell and metabolized, polymers such as lignin, cellulose, and hemicellulose, which are typically the structural components of plant cells, must be depolymerized by extracellular enzymes. The role of mycorrhizal fungi in the sequestration of soil carbon is substantial [41].

This encompasses facultative symbionts capable of mineralizing organic carbon, such as ectomycorrhizal fungi, and obligate symbionts that only derive carbon from the host plant, exemplified by arbuscular mycorrhizal fungus (AMF). As per a study [42], around 85% of all plant households exhibit AMF synergy. Speculative data indicates that up to 20% [43] or possibly 30% [44] of the total carbon adaptation by plants may be conveyed to the fungal partner, with the synergy possessing significant [45]. The soil carbon cycle is disrupted when a portion of the plant carbon transferred to the mycelia is rapidly respired into the environment. The term "soil raw material" (SOM) denotes a continuum of plant waste, exudates, and particles derived from microbes and animals, encompassing the microbial biomass primarily accountable for the degradation of exudates and detrital inputs [46]. This continuum has conventionally been divided into several categories with varying decay speeds, ranging from "energetic" pools that alter within months to "passive" pools that evolve over millennia. The "passive" swimming pool is recognized for containing "black carbon" from combustion and components that are decay-resistant due to their humified characteristics. Humified chemicals are generated through spontaneous condensation reactions between reactive microbial substances and biochemically altered structural macromolecules [47].

Recent research, however, challenges the notion of "recalcitrant" dirt humic materials that underpin predictions of carbon turnover. It indicates that environmental and biological factors may exert a significantly greater influence on the enduring characteristics of soil organic matter than the molecular structure of plant debris inputs and the subsequent formation of humus. Furthermore, while humic macromolecules in soil have not been directly detected *in situ*, extracting humic compounds from soil may be an artifact of the extraction process [48]. The soil organic matter (SOM) consists of partially decomposed detritus and a substantial portion of microbial necromass (dead biomass residues), which remains "passive" due to its physical isolation from or inaccessibility to extracellular enzymes, microorganisms, and essential ecological factors (such as electron acceptors, water, and inorganic nutrients) necessary for decomposition. Soil bacteria facilitate the continuous creation of stable organic matter through their necromass and their role in the decomposition and release of CO_2 from organic material.

CLIMATE CHANGE, MICROBIAL DECOMPOSERS, AND THE SOIL CARBON CYCLE

Since the onset of the commercial era, humans have significantly disrupted the carbon cycle by augmenting atmospheric carbon dioxide, mostly through fossil fuel combustion and converting natural habitats into agricultural land [14, 49]. Nonetheless, our understanding of soil respiration and its representation in Earth system models remains incomplete, contributing to our limited insight into the impacts of human activities on global climate [49, 50]. Moreover, microorganisms have challenges in contributing to environmental changes in the carbon cycle due to their interactions with numerous components and their direct and indirect effects (Fig. **2**) [51].

The microbial activity, and thus natural carbon breakdown and CO_2 emissions from respiration, may be enhanced in response to elevated temperature levels. This is an instance of concise, positive feedback on global warming [20, 51]. Analysis of global field measurements indicates a correlation between elevated temperature levels and enhanced respiration rates from terrestrial sources [52]. The carbon fertilization of essential photosynthetic output, wherein increased atmospheric carbon dioxide enhances photosynthesis, represents an indirect positive response to rising carbon dioxide levels [53]. Additionally, the exudation of root exudates contributes to an increase in the amount of labile carbon available for microbial respiration and decomposition [9]. Due to the significant spatial diversification of terrestrial ecosystems in terms of climate, plant diversity and composition, soil physics and chemistry, microbial community structure, and evolutionary history, the effects of climate change on soil microorganisms vary considerably across different ecosystems. Given the complexity of natural environments, the nonlinear nature of their dynamics, and the time-dependent characteristics, it is prudent to assume that the effects of environmental change on soil microbes will not manifest as a linear increase or decrease over extended ecological timeframes [54].

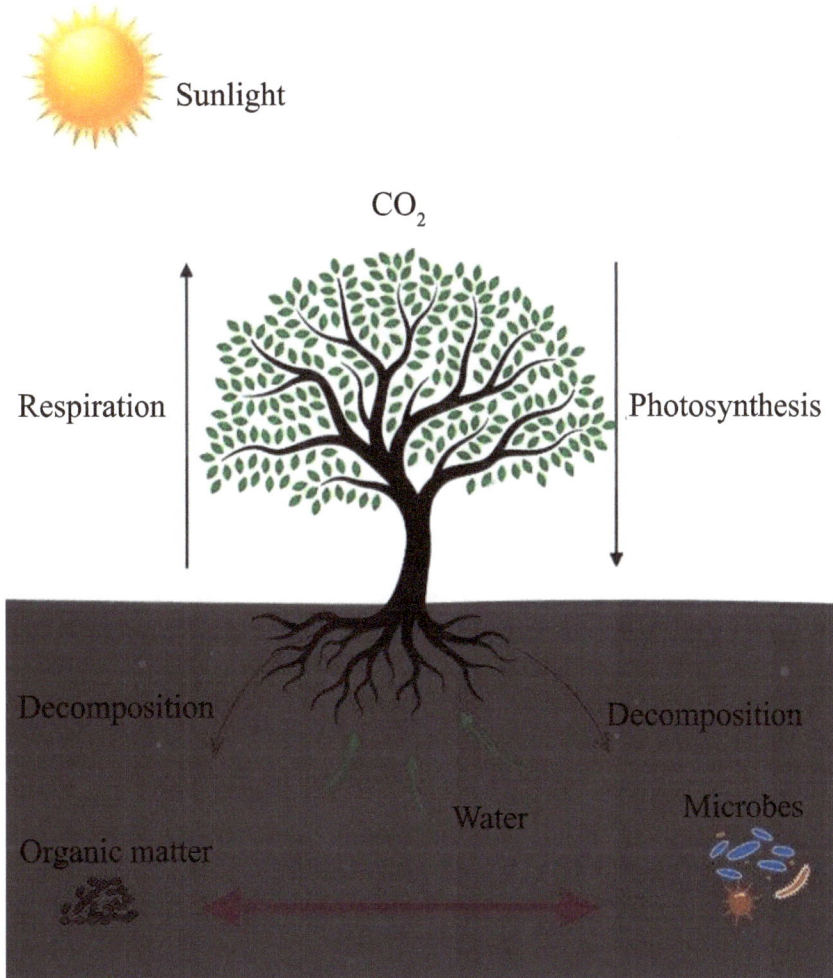

Fig. (2). Role of soil microorganisms in climate change.

It is uncommon to acquire information on the temporal dynamics of microorganisms in response to climate change. Microbe-centered experiments would allow us to examine the responses of various soil microorganisms, including bacteria, archaea, fungi, protists, viruses, and microfauna, to multiple climate change factors in both topsoil and subsoil (*e.g.*, warming, increased CO_2, drought, heightened precipitation, nutrient addition, and their interactions).

OCEAN ENVIRONMENTS

Seventy percent of the Earth's surface is enveloped by the ocean, which possesses an average depth of 4,000 meters. Over 50 distinct biomes, ranging from tropical regions to polar areas and from the illuminated surface layer to the profound abyss, constitute ocean habitats' physical and chemical diversity [55]. Each biome sustains a unique microbe-driven environment that operates as a complex adaptive system with evolving processes and solutions intricately connected to environmental unpredictability [56]. Nautical microorganisms have evolved over 4 billion years to adapt to a constantly changing Earth, acquiring physical plasticity and resilience that may provide some resistance against anthropogenic environmental alterations. Nonetheless, the ocean's microbial life is significantly jeopardized by the current costs of environmental adaptation resulting from heat-trapping greenhouse gasses, which are at unprecedented levels in the Earth's history.

The seas are crucial to the attributes of the global environment. They absorb 90% of the heat that accumulates in the atmosphere, and since the onset of commercial transformation, they have also absorbed 25% of the additional CO_2, leading to ocean acidification. Deoxygenation also occurs due to a warmer, more stratified ocean, and all these threats are attributable to excessive carbon dioxide emissions from human activities [57]. Enhanced stratification will undoubtedly accelerate the rate of future warming. Marine heat waves have been seen to occur with increased frequency and duration in the world's major ocean basins during the past century, a trend expected to persist owing to anthropogenic climate change [58, 59].

Mass extinctions of aquatic life, consisting of photosynthetic microbes, have been triggered by these extended (month-long) periods of anomalously high sea surface area temperature. Fast habitat alterations, such as those induced by marine heat waves, can threaten biodiversity worldwide and also change maritime ecological communities into alternating, less preferable eco-friendly states with reduced general adaptability for future change. A thorough understanding of microorganisms and climate modification is important for scientific research, given that the ecological results of sea acidification and warming are basically irreversible over centuries' worth of time [60]. Due to the ocean's enormity and resulting inaccessibility, only a handful of lasting ocean observatories have actually been able to directly gauge the effects of environment modification on microbial processes [61]. This will certainly permit a more extensive understanding of the microbial oceanography of environment modification.

ECOSYSTEM RESTORATION UNDER CLIMATE CHANGE PERSPECTIVE

Environmental alteration, which is already underway, is anticipated to substantially impact all ecological communities worldwide. In addition to growing temperature levels, a changing environment encompasses varied precipitation patterns, rising CO_2 concentrations, and unforeseen interactions among these processes. These modifications will yield various outcomes on species and ecosystems, potentially leading to the loss of multiple species, changed interspecies interactions, the emergence of new communities, and the destruction of existing ones. The distributions of types originally shifted due to an evolving environment. These can consist of modifications in the position and wealth of people within a type of range, the transfer of varieties into areas where they were not formerly located, or the loss of species from a region where they previously were [62]. The indirect effects of interactions are included in these direct impacts of environment on types of ranges. Adjustments in the distributions of predators, targets, pollinators, and competitors might have an influence on species that are not directly affected by climate [63].

Land supervisors and decision-makers will inevitably face a complex array of decisions due to the diverse and unpredictable reorganization of plant and animal populations resulting from climate change. They can resolve to maintain a commitment to hands-off management that minimizes human interference while acknowledging these changes as a natural series of processes, embracing the likelihood of extinctions and significant alterations in biota. In numerous instances, augmenting an ecosystem's resilience to alterations and safeguarding present or historical flora and fauna in their native forms are the initial stages in adjusting community management to climate change [64]. This strategy is most effective as a temporary solution or to protect a single variety when only minor alterations are expected [65].

CONCLUSION

Microbes remain the primary contributors to climate change across all scales, including terrestrial, aquatic, and urban environments. Microbial metabolism generates and absorbs gases that can impact climate, ranging from the heat of cow rumen to the thawing permafrost in polar regions, the symbiotic coral reef ecosystems in oceans, and the carbon emissions from urban areas. Consequently, all environmental adjustment models must consider microbial contributions to carbon fluxes into and out of the atmosphere. The microbial realm can serve as a vital ally in mitigating the impacts of human greenhouse gas (GHG) emissions, as it is essential to stimulate modifications in microbial activities across various environments to enhance consumption and reduce the production of gases that contribute to atmospheric warming. The soil microbiome significantly influences the dynamics of the soil environment, typically by improving soil fertility and plant productivity. In terrestrial ecosystems, soil serves as a reservoir for substantial amounts of carbon, either as soil organic matter (SOM) or soil organic carbon (SOC). Cooperative interactions between plants and microbes in soils facilitate carbon storage through dynamic biological processes such as photosynthesis, decomposition, and soil respiration. Nonetheless, accurately assessing the relationship and carbon sequestration remains difficult. Recent studies have demonstrated that environmental alteration and human activities have profoundly affected soil ecology, necessitating the use of effective carbon-balancing measures.

Increased research and interdisciplinary collaboration are necessary to address complex issues related to the connections among microbes, environmental changes, and human health. Transformed patterns of host-microbe interactions, altered microbial biogeography, and changes in terrestrial, marine, and urban microbiology result in bacteria's adaptation to a warmer environment, which directly impacts human health and well-being. The field requires significantly more statistically robust, hypothesis-driven, mechanistic research to enhance our comprehension of the roles of microorganisms in climate adaptation and their responses to environmental stimuli, whether natural or anthropogenic.

CONSENT FOR PUBLICATION

The author consented to the release of identifiable information, including photographs, videos, case history, and textual elements in the book. The author certifies that the material is original and not under consideration by any other publication. The author asserts that there are no recognized conflicts of interest related to this publication.

REFERENCES

[1] Parmesan C, Morecroft MD, Trisurat Y *et al.* Climate Change 2022: Impacts, Adaptation and Vulnerability. Contribution of Working Group II to the Sixth Assessment Report of the Intergovernmental Panel on Climate Change [H.-O. Pörtner, D.C. Roberts, M. Tignor, E.S. Poloczanska, K. Mintenbeck, A. Alegría, M. Craig, S. Langsdorf, S. Löschke, V. Möller, A. Okem, B. Rama (eds.)]. Cambridge University Press, Cambridge, UK and New York, NY, USA, 2022, pp. 197–377.
[http://dx.doi.org/10.1017/9781009325844.004]

[2] Heimann M, Reichstein M. Terrestrial ecosystem carbon dynamics and climate feedbacks. Nature 2008; 451(7176): 289-92.
[http://dx.doi.org/10.1038/nature06591] [PMID: 18202646]

[3] Schimel DS, Braswell BH, Holland EA, *et al.* Climatic, edaphic, and biotic controls over storage and turnover of carbon in soils. Global Biogeochem Cycles 1994; 8(3): 279-93.
[http://dx.doi.org/10.1029/94GB00993]

[4]   Magnani F, Mencuccini M, Borghetti M, *et al.* The human footprint in the carbon cycle of temperate and boreal forests. Nature 2007; 447(7146): 849-51.
        [http://dx.doi.org/10.1038/nature05847] [PMID: 17568744]

[5]   Smith P, Martino D, Cai Z, *et al.* Greenhouse gas mitigation in agriculture. Philos Trans R Soc Lond B Biol Sci 2008; 363(1492): 789-813.
        [http://dx.doi.org/10.1098/rstb.2007.2184] [PMID: 17827109]

[6]   Beedlow PA, Tingey DT, Phillips DL, Hogsett WE, Olszyk DM. Rising atmospheric CO$_2$ and carbon sequestration in forests. Front Ecol Environ 2004; 2(6): 315-22.
        [http://dx.doi.org/10.1890/1540-9295(2004)002[0315:RACACS]2.0.CO;2]

[7]   Monteith DT, Stoddard JL, Evans CD, *et al.* Dissolved organic carbon trends resulting from changes in atmospheric deposition chemistry. Nature 2007; 450(7169): 537-40.
        [http://dx.doi.org/10.1038/nature06316] [PMID: 18033294]

[8]   Sitch S, Cox PM, Collins WJ, Huntingford C. Indirect radiative forcing of climate change through ozone effects on the land-carbon sink. Nature 2007; 448(7155): 791-4.
        [http://dx.doi.org/10.1038/nature06059] [PMID: 17653194]

[9]   Friedlingstein P, Cox P, Betts R, *et al.* Climate–carbon cycle feedback analysis: results from the C4MIP model intercomparison. J Clim 2006; 19(14): 3337-53.
        [http://dx.doi.org/10.1175/JCLI3800.1]

[10]   Trumbore S. Trumbore S (2006) Carbon respired by terrestrial ecosystems – recent progress and challenges. Glob Change Biol 12(2):141–153.

[11]   Bahn M, Kutsch WL, Heinemeyer A. 2010. Synthesis: emerging issues and challenges for an integrated understanding of soil carbon fluxes. Chapter 14 in Soil carbon dynamic An Integrated Methodology. Cambridge University Press: 257-271.
        [http://dx.doi.org/10.1017/CBO9780511711794.015]

[12]   De Deyn GB, Cornelissen JHC, Bardgett RD. Plant functional traits and soil carbon sequestration in contrasting biomes. Ecol Lett 2008; 11(5): 516-31.
        [http://dx.doi.org/10.1111/j.1461-0248.2008.01164.x] [PMID: 18279352]

[13]   Falkowski PG, Fenchel T, Delong EF. The microbial engines that drive Earth's biogeochemical cycles. Science 2008; 320(5879): 1034-9.
        [http://dx.doi.org/10.1126/science.1153213] [PMID: 18497287]

[14]   Prentice IC, Farquhar GD, Fasham MJR, *et al.* The carbon cycle and atmospheric carbon dioxide.Climate Change 2001 The Scientific Basis Contribution of Working Group I to the Third Assessment Report of the Intergovernmental Panel on Climate Change. Cambridge: Cambridge University Press 2001; pp. 183-238.

[15]   Liang C, Balser TC. Microbial production of recalcitrant organic matter in global soils: implications for productivity and climate policy. Nat Rev Microbiol 2011; 9(1): 75.
        [http://dx.doi.org/10.1038/nrmicro2386-c1] [PMID: 21113179]

[16]   Lu Y, Lueders T, Friedrich MW, Conrad R. Detecting active methanogenic populations on rice roots using stable isotope probing. Environ Microbiol 2005; 7(3): 326-36.
        [http://dx.doi.org/10.1111/j.1462-2920.2005.00697.x] [PMID: 15683393]

[17]   Chen Y, Dumont MG, Neufeld JD, *et al.* Revealing the uncultivated majority: combining DNA stable☐isotope probing, multiple displacement amplification and metagenomic analyses of uncultivated *Methylocystis* in acidic peatlands. Environ Microbiol 2008; 10(10): 2609-22.
        [http://dx.doi.org/10.1111/j.1462-2920.2008.01683.x] [PMID: 18631364]

[18]   Qiu Q, Noll M, Abraham WR, Lu Y, Conrad R. Applying stable isotope probing of phospholipid fatty acids and rRNA in a Chinese rice field to study activity and composition of the methanotrophic bacterial communities *in situ*. ISME J 2008; 2(6): 602-14.
        [http://dx.doi.org/10.1038/ismej.2008.34] [PMID: 18385771]

[19]   Gupta V, Smemo KA, Yavitt JB, Fowle D, Branfireun B, Basiliko N. Stable isotopes reveal widespread anaerobic methane oxidation across latitude and peatland type. Environ Sci Technol 2013; 47(15).
        [http://dx.doi.org/10.1021/es400484t] [PMID: 23822884]

[20]   Davidson EA, Janssens IA. Temperature sensitivity of soil carbon decomposition and feedbacks to climate change. Nature 2006; 440(7081): 165-73.
        [http://dx.doi.org/10.1038/nature04514] [PMID: 16525463]

[21]   Schmidt MWI, Torn MS, Abiven S, *et al.* Persistence of soil organic matter as an ecosystem property. Nature 2011; 478(7367): 49-56.
        [http://dx.doi.org/10.1038/nature10386] [PMID: 21979045]

[22]   Jobbágy EG, Jackson RB. The vertical distribution of soil organic carbon and its relation to climate and vegetation. Ecol Appl 2000; 10(2): 423-36.
        [http://dx.doi.org/10.1890/1051-0761(2000)010[0423:TVDOSO]2.0.CO;2]

[23]   Gavito ME, Olsson PÃA. Allocation of plant carbon to foraging and storage in arbuscular mycorrhizal fungi. FEMS Microbiol Ecol 2003; 45(2): 181-7.
        [http://dx.doi.org/10.1016/S0168-6496(03)00150-8] [PMID: 19719629]

[24]   Sarjuni MNH, Dolit SAM, Khamis AK, *et al.* Regenerating Soil Microbiome: Balancing Microbial CO_2 Sequestration and Emission. In: Sarvajayakesavalu S, Karthikeyan K, (Eds.) Carbon Sequestration. IntechOpen; 2022.
        [http://dx.doi.org/10.5772/intechopen.104740]

[25]   Karmakar R, Das I, Dutta D, Rakshit A. Potential effects of climate change on soil properties: A review. Sci Int (Lahore) 2016; 4(2): 51-73.
        [http://dx.doi.org/10.17311/sciintl.2016.51.73]

[26]   Huang J, Zhang C, Cheng D, *et al.* Soil organic carbon mineralization in relation to microbial dynamics in subtropical red soils dominated by differently sized aggregates. Open Chem 2019; 17(1): 381-91.
        [http://dx.doi.org/10.1515/chem-2019-0051]

[27]   Huang J, Li Y, Fu C, *et al.* Dryland climate change: Recent progress and challenges. Rev Geophys 2017; 55(3): 719-78.
        [http://dx.doi.org/10.1002/2016RG000550]

[28]   Zhao Y, Ding Y, Hou X, Li FY, Han W, Yun X. Effects of temperature and grazing on soil organic carbon storage in grasslands along the Eurasian steppe eastern transect. PLoS One 2017; 12(10): e0186980.
        [http://dx.doi.org/10.1371/journal.pone.0186980] [PMID: 29084243]

[29]   Johnston ASA, Sibly RM. The influence of soil communities on the temperature sensitivity of soil respiration. Nat Ecol Evol 2018; 2(10): 1597-602.
        [http://dx.doi.org/10.1038/s41559-018-0648-6] [PMID: 30150743]

[30]   Manik SMN, Pengilley G, Dean G, Field B, Shabala S, Zhou M. Soil, crop management practices to minimize the impact of waterlogging on crop productivity. Front Plant Sci 2019; 10: 140.
        [http://dx.doi.org/10.3389/fpls.2019.00140] [PMID: 30809241]

[31]   Bertola M, Ferrarini A, Visioli G. Improvement of soil microbial diversity through sustainable agricultural practices and its evaluation by -omics approaches: A perspective for the environment, food quality and human safety. Microorganisms 2021; 9(7): 1400.
        [http://dx.doi.org/10.3390/microorganisms9071400] [PMID: 34203506]

[32]   Tiefenbacher A, Sandén T, Haslmayr HP, Miloczki J, Wenzel W, Spiegel H. Optimizing carbon sequestration in croplands: A synthesis. Agronomy (Basel) 2021; 11(5): 882.
        [http://dx.doi.org/10.3390/agronomy11050882]

[33]   Zhang K, Maltais-Landry G, Liao HL. How soil biota regulate C cycling and soil C pools in

diversified crop rotations. Soil Biol Biochem 2021; 156: 108219.
[http://dx.doi.org/10.1016/j.soilbio.2021.108219]

[34]   Ray R, Baum A, Rixen T, Gleixner G, Jana TK. Exportation of dissolved (inorganic and organic) and particulate carbon from mangroves and its implication to the carbon budget in the Indian Sundarbans. Sci Total Environ 2018; 621: 535-47.
[http://dx.doi.org/10.1016/j.scitotenv.2017.11.225] [PMID: 29195202]

[35]   Liu Q, Xu H, Yi H. Impact of fertilizer on crop yield and C: N:P stoichiometry in arid and semi-arid soil. Int J Environ Res Public Health 2021; 18(8): 4341.
[http://dx.doi.org/10.3390/ijerph18084341] [PMID: 33923871]

[36]   Sylvia DM, Fuhrmann JJ, Hartel PG, *et al.* Principles and Applications of Soil Microbiology. Upper Saddle River, NJ: Pearson Education 2005.

[37]   Le Quéré C, Raupach MR, Canadell JG, *et al.* Trends in the sources and sinks of carbon dioxide. Nat Geosci 2009; 2(12): 831-6.
[http://dx.doi.org/10.1038/ngeo689]

[38]   Lal R. Soil carbon sequestration impacts on global climate change and food security. Science 2004; 304(5677): 1623-7.
[http://dx.doi.org/10.1126/science.1097396] [PMID: 15192216]

[39]   Bais HP, Weir TL, Perry LG, Gilroy S, Vivanco JM. The role of root exudates in rhizosphere interactions with plants and other organisms. Annu Rev Plant Biol 2006; 57(1): 233-66.
[http://dx.doi.org/10.1146/annurev.arplant.57.032905.105159] [PMID: 16669762]

[40]   Müller S, Van Der Merwe A, Schildknecht H, Visser JH. An automated-system for large-scale recovery of germination stimulants and other root exudates. Weed Sci 1993; 41(1): 138-43.
[http://dx.doi.org/10.1017/S0043174500057714]

[41]   Wallenstein MD, Weintraub MN. Emerging tools for measuring and modeling the *in situ* activity of soil extracellular enzymes. Soil Biol Biochem 2008; 40(9): 2098-106.
[http://dx.doi.org/10.1016/j.soilbio.2008.01.024]

[42]   Smith SE, Read DJ. The Mycorrhizal Symbiosis. San Diego, CA: Academic Press 2008.

[43]   Nakano-Hylander A, Olsson PA. Carbon allocation in mycelia of arbuscular mycorrhizal fungi during colonisation of plant seedlings. Soil Biol Biochem 2007; 39(7): 1450-8.
[http://dx.doi.org/10.1016/j.soilbio.2006.12.031]

[44]   Drigo B, Pijl AS, Duyts H, *et al.* Shifting carbon flow from roots into associated microbial communities in response to elevated atmospheric CO_2. Proc Natl Acad Sci USA 2010; 107(24): 10938-42.
[http://dx.doi.org/10.1073/pnas.0912421107] [PMID: 20534474]

[45]   Jones DL, Hodge A, Kuzyakov Y. (2004) Plant and mycorrhizal regulation of rhizodeposition. New Phytol 163(3):459–480.
[http://dx.doi.org/10.1111/j.1469-8137.2004.01130.x]

[46]   Dungait JAJ, Hopkins DW, Gregory AS, *et al.* (2012) Soil organic matter turnover is governed by accessibility not recalcitrance. Glob Chan Biol 18(6):1781–1796.
[http://dx.doi.org/10.1111/j.1365-2486.2012.02665.x]

[47]   Wolf DC, Wagner GH. Carbon transformations and soil organic matter formation.Principles and Applications of Soil Microbiology. Upper Saddle River, NJ: Pearson Education 2005; pp. 285-332.

[48]   Kleber M, Johnson MG. Advances in understanding the molecular structure of soil organic matter: implications for interactions in the environment. Adv Agron 2010; 106: 77-142.
[http://dx.doi.org/10.1016/S0065-2113(10)06003-7]

[49]   Canadell JG, Kirschbaum MUF, Kurz WA, Sanz M-J, Schlamadinger B, Yamagata Y. Factoring out natural and indirect human effects on terrestrial carbon sources and sinks. Environ Sci Policy 2007;

10(4): 370-84.
[http://dx.doi.org/10.1016/j.envsci.2007.01.009]

[50]   Smith P, Fang C. A warm response by soils. Nature 2010; 464(7288): 499-500.
[http://dx.doi.org/10.1038/464499a] [PMID: 20336128]

[51]   Jin VL, Evans RD. (2007) Elevated CO_2 increases microbial carbon substrate use and nitrogen cycling in Mojave Desert soils. Glob Chan Biol 13(2):452–465.

[52]   Bond-Lamberty B, Thomson A. Temperature-associated increases in the global soil respiration record. Nature 2010; 464(7288): 579-82.
[http://dx.doi.org/10.1038/nature08930] [PMID: 20336143]

[53]   Hungate BA, Johnson DW, Dijkstra P, *et al.* Nitrogen cycling during seven years of atmospheric CO_2 enrichment in a scrub oak woodland. Ecology 2006; 87(1): 26-40.
[http://dx.doi.org/10.1890/04-1732] [PMID: 16634294]

[54]   Tiedje JM, Bruns MA, Casadevall A, *et al.* Microbes and Climate Change: a Research Prospectus for the Future. MBio 2022; 13(3): e00800-22.
[http://dx.doi.org/10.1128/mbio.00800-22] [PMID: 35438534]

[55]   Longhurst A. Ecological geography of the sea. 2nd ed., New York, NY: Academic Press 2006.

[56]   Hagstrom GI, Levin SA. Marine ecosystems as complex adaptive systems: emergent patterns, critical transitions, and public goods. Ecosystems (N Y) 2017; 20(3): 458-76.
[http://dx.doi.org/10.1007/s10021-017-0114-3]

[57]   Gruber N. Warming up, turning sour, losing breath: ocean biogeochemistry under global change. Philos Trans- Royal Soc, Math Phys Eng Sci 2011; 369(1943): 1980-96.
[http://dx.doi.org/10.1098/rsta.2011.0003] [PMID: 21502171]

[58]   Oliver ECJ, Donat MG, Burrows MT, *et al.* Longer and more frequent marine heatwaves over the past century. Nat Commun 2018; 9(1): 1324.
[http://dx.doi.org/10.1038/s41467-018-03732-9] [PMID: 29636482]

[59]   Frölicher TL, Fischer EM, Gruber N. Marine heatwaves under global warming. Nature 2018; 560(7718): 360-4.
[http://dx.doi.org/10.1038/s41586-018-0383-9] [PMID: 30111788]

[60]   Solomon S, Plattner GK, Knutti R, Friedlingstein P. Irreversible climate change due to carbon dioxide emissions. Proc Natl Acad Sci USA 2009; 106(6): 1704-9.
[http://dx.doi.org/10.1073/pnas.0812721106] [PMID: 19179281]

[61]   Karl DM, Bates NR. (2003) Temporal studies of biogeochemical processes determined from ocean time - series observations during the during the JGOFS Era. In: Fasham, M.J.R. (eds) Ocean Biogeochemistry. Global Change — The IGBP Series (closed). Springer, Berlin, Heidelberg.
[http://dx.doi.org/10.1007/978-3-642-55844-3_11]

[62]   Parmesan C. Ecological and evolutionary responses to recent climate change. Annu Rev Ecol Evol Syst 2006; 37(1): 637-69.
[http://dx.doi.org/10.1146/annurev.ecolsys.37.091305.110100]

[63]   Thomas CD. Climate, climate change and range boundaries. Divers Distrib 2010; 16(3): 488-95.
[http://dx.doi.org/10.1111/j.1472-4642.2010.00642.x]

[64]   Heller NE, Zavaleta ES. Biodiversity management in the face of climate change: A review of 22 years of recommendations. Biol Conserv 2009; 142(1): 14-32.
[http://dx.doi.org/10.1016/j.biocon.2008.10.006]

[65]   Galatowitsch S, Frelich L, Phillips-Mao L. Regional climate change adaptation strategies for biodiversity conservation in a midcontinental region of North America. Biol Conserv 2009; 142(10): 2012-22.
[http://dx.doi.org/10.1016/j.biocon.2009.03.030]

Importance of Microbiome in Ecosystem Sustainability

Marine Microbes and Microbiomes: Role and Importance in Ecosystem Sustainability

C. Poornachandhra[1,*]**, M. Sinduja**[2]**, S. Akila**[2]**, A. Manikandan**[3]**, J. Sampath**[1]**, R. Kaveena**[5]**, T. Gokul Kannan**[1] **and Muthusamy Shankar**[4]

[1] *Department of Environmental Sciences, Tamil Nadu Agricultural University, Coimbatore, Tamil Nadu, India*

[2] *National Agro-foundation Research & Development Centre, Chennai, India*

[3] *Institute of Ecology and Earth Sciences, University of Tartu, Tartu, Estonia*

[4] *Division of Plant Genetic Resources, ICAR-Indian Agricultural Research Institute, New Delhi, India*

[5] *Swamy Vivekananda College of Pharmacy, Tiruchengode, India*

Abstract: Marine environments are among the most unfavorable due to salinity, pH, sea surface temperature, wind patterns, ocean currents, and precipitation regimes. Due to the frequent changes in environmental conditions, the microorganisms that live there are better suited to adjusting to unfavorable conditions, which is why they have complex characteristic qualities of adaptation. Consequently, by forming biofilms and producing extracellular polymeric substances, the microorganisms isolated from marine habitats are intended to be better exploited in the bioremediation of soils and water bodies contaminated with toxic pollutants. Many marine bacteria have also been reported to produce bioactive compounds, which found their use in many biotechnological applications. This chapter explores marine microbial diversity, its utilization in bioremediation, and understanding their role in ecosystem sustainability.

Keywords: Ecosystem sustainability, Microbial diversity, Microbiomes, Nutrient cycling, Pharmaceuticals, Remediation.

INTRODUCTION

Marine planktonic microbes dominate ocean biogeochemical processes and biomass. Although several environmental factors have been demonstrated to impact microbial communities, there is disagreement regarding how these factors affect microbial communities [1]. Quantifying the relative contributions of enviro-

* **Corresponding author C. Poornachandhra:** Department of Environmental Sciences, Tamil Nadu Agricultural University,Coimbatore, Tamil Nadu, India; E-mail: poorna155c@gmail.com

Govindaraj Kamalam Dinesh, Shiv Prasad, Ramesh Poornima, Sangilidurai Karthika, Murugaiyan Sinduja & Velusamy Sathya (Eds.)

nmental factors in creating microbial community structure is necessary for predicting how ecosystems react to environmental changes, such as climate change [2]. In this chapter, we concentrate on how environmental selection affects oceanic microbiomes. Stochastic effects make it difficult to find key variables through observational sampling [3], geographical variations in populations, behaviors, and ecological divergence among closely related microbes [4]. Despite these difficulties, it is generally accepted that bacterio-plankton reacts to environmental factors such as temperature and salinity, as well as the abundance of resources such as nutrients and interactions with other organisms [5]. The term "biological diversity" refers to the diversity of all living things, including those belonging to species and ecological complexes [6]. The tropical Indo-Western Pacific region, which encompasses waters along the coasts of Asia, Southeast Africa, Northern Australia, and the Pacific Islands, has the highest overall marine diversity [7].

However, the rate of extinction for biodiversity on Earth is worrying. Therefore, mapping and quantifying marine biodiversity at all structural levels should be done using a method based on ecological and evolutionary processes [8]. The paradigms pertaining to biodiversity patterns in terrestrial systems may not be applicable to marine conditions since marine systems differ from terrestrial systems in many ways [9]. The ability of terrestrial ecosystems to connect their three-dimensional space to either permanent or semi-permanent physical structures fundamentally sets them apart from marine ecosystems. The world Ocean has a 312,000 km long coastline and a volume of 1.46×10^9 km^3 with an average depth of 4000 m [10] and is the planet's largest ecosystem. Despite the fact that humans have exploited it for a variety of reasons for millennia, most studies on biological diversity focus on terrestrial systems, and our understanding of marine biodiversity is much less advanced than that of land [11]. Microorganisms are pervasive and genuinely make up the "unseen majority" in the marine environment. Although marine isolates have been the subject of laboratory-based culture methods for more than ten years, we still do not completely understand the ecology of marine microorganisms. Marine microbes have been studied for a few decades, and new discoveries of previously undiscovered groups like SAR11 and pico autotrophs like *Prochlorococcus* have significantly increased the diversity of marine microbes.

However, the significance of microbial taxonomy, an experimentally complex and labor-intensive process, becomes apparent from sparse and dispersed knowledge about the number of species. When the variety of biological life is measured by the number of species known for each group, the diversity of microorganisms is vastly understated. In comparison to plants and animals, the idea of bacterial species is not only more typological and less evolutionary, but it is also much broader and inclusive. To comprehend the phylogenetic perspective, the mechanism of degradation, and the creation of novel treatment strategies, it is thus essential to study diversity at the genetic level. In terms of microbial genomics, the first two decoded microbial genomes were *Mycoplasma genitalium* and *Haemophilus influenzae*. However, a decade ago, microbial genome sequencing could not gain traction. Today, however, each microbial DNA is sequenced on an individual basis. The entire collection of genes an organism can access is contained in its genome. Exploring microbial diversity is undoubtedly an important and fascinating topic. Additionally, knowledge of the diversity of marine microbes aids in the isolation and identification of novel and promising microbes with high selectivity for resistant substances.

PRESENT STATUS OF MICROBIAL BIODIVERSITY

Approximately 3.5 billion years have passed since the beginning of modern microorganisms' evolutionary history, which has primarily taken place in marine environments. The first division of living things into two very different categories are eukaryotes, which have a nuclear membrane, and prokaryotes, which do not include bacteria, the 'first and simplest division of living beings. The "Five Kingdoms" of life animals, plants, fungi, protists (Protozoa), and monera (Bacteria) were, however, highlighted by taxonomists of the 20th century. The "urkingdoms" or "domains" consist of Bacteria (eubacteria), Archaea (archaebacteria), and Eucarya (eukaryotes) based on the 16S or 18S rRNA composition. These three domains overlap in the water regarding size spectra, physiological traits, metabolic patterns, and ecological roles. Prokaryotes typically have a loosely arranged DNA called the nucleoid and rigid cell wall, including archaea and bacteria. Although there may be considered over 40 divisions of bacteria, the 16S rRNA gene sequences used to infer the division-level diversity of the bacterial domain revealed 36 divisions. Cultivated strains of several described divisions, which were the first to be defined phylogenetically, provide good representations of these divisions.

Marine Microbial Diversity

The world Ocean has a 312,000 km long coastline and a volume of 1.46×10^9 km^3 with an average depth of 4000 m [10] and is the planet's largest ecosystem. The ocean substantially impacts global climate due to its enormous size and volume. Microorganisms play a significant role in our conceptions of life and can be found anywhere in nature. Most of the biomass found in the oceans is made up of microorganisms. The various evolution in the marine microbial ecology is shown in Fig. (**1**). Microorganisms are so numerous that they are thought to account for between 55 and 86% of the planet's prokaryotic biomass or 3.55×10^{29} cells. In just 1 milliliter of ocean water, there are 10^6 bacterial and archaeal cells and 10^7 virus particles [12]. Dissolved oxygen, phosphate, solar radiation, certain metals, poisonous chemicals, and water pH all play essential roles in determining the geographic distribution of marine microbes [13]. There is a possibility that microorganisms adapted to freshwater environments can be identified and thrive in saltwater environments and *vice versa*. Numerous microorganisms can be found in and possibly thrive in various water conditions. Water runoff and airborne transport can carry even microscopic species from land to sea [14]. According to estimates, there are between 2.9 and 5.4 billion cells worth of bacteria in the oceanic sediments, making up 3.6% of all the living things on Earth. Despite these circumstances, researchers have been drawn to the diversity of deep-sea benthic communities, resulting in the identification of novel ecosystems and a wide range of microbes, leading to a paradigm change.

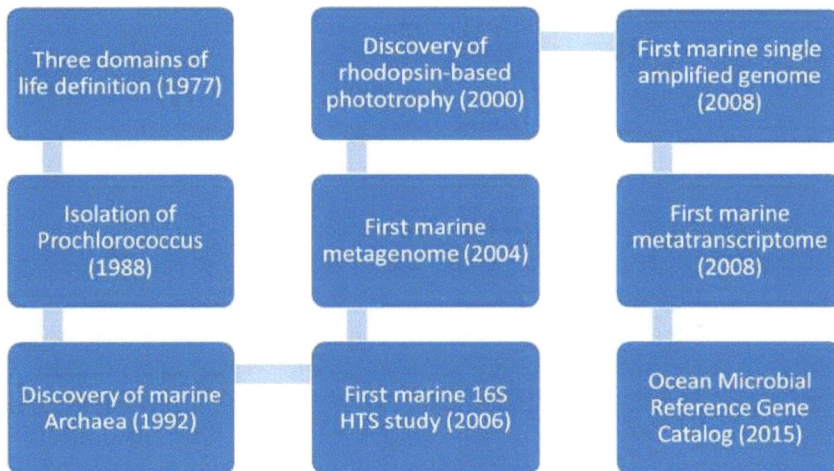

Fig. (1). Key scientific advances in marine microbial ecology.

Numerous marine bacteria require a high Na^+ concentration for transferring substrates into their cells or retaining internal solutes, making them highly salinity tolerant. Microorganisms in the marine environment include all three domains of life, bacteria, archaea, and eukarya, in addition to viruses, as essential biological components. They are the foundation of the marine food web and the engines and engineers of marine ecosystems [15]. Seawater species range from 1500 [16] to one million [17], indicating sequence diversity [18]. The role of these enormous amounts of marine organisms in ecosystem sustainability is still unknown. Microbiology still debates species. It is mainly based on 16S rRNA gene sequence identity and behavioral traits. Marine microorganisms have been cultured and described as 10,000 "species", yet most of them are unculturable under known media and growth conditions.

Prochlorococcus and *Pelagibacter*, the smallest organisms, are the most numerous cyanobacteria in the ocean. Marine bacteria are usually non-motile because Brownian diffusion makes swimming to a nutrient hotspot wasteful [19]. Eight samples from South Indian Ocean microbes analyzed using next-generation sequencing contained 21 bacterial phyla and 541 OTUs, with five samples containing mostly the Proteobacteria phyla. The other three samples had dominant phyla of Firmicutes and Chloroflexi. The most dominant fungal phyla identified belonged to Ascomycota and Basidiomycota. A paucity of database representation can be seen in the eight samples, where 10-58% and 19-26% of the archaeal and fungal OTUs, respectively, were mapped to unclassified taxa. Co-occurrence network research showed that bacterial communities are more dynamic than archaeal and fungal communities [20]. Another study by [21] revealed that the bacterial and archaeal taxa common in organic matter-rich subsurface sediments (like *Dehalococcoida*, *Atribacter*, and *Woesearchaeota*) and chemosynthetic environments (*Helicobacteraceae*) were enriched in the microbial communities in the SWIR axial valley of the Southwest Indian Ridge. In deep marine waters along the Ninetyeast Ridge in the Indian Ocean, *Nitrososphaeria* (*Thaumarchaeota*) dominated with relative abundances ranging from 52.68 to 97.2%, followed by *Thermoplasmata* species [22].

ROLE IN MARINE C, N, S, AND Fe CYCLING

The sustained evolution of microbes and microbiomes are the critical components of aquatic ecosystems (~70% by mass) on 'the Earth's composition, underpinning the role of global element cycles in life and death. Metagenomic approaches in aquatic microbiomes act as a nexus in centralizing the global biogeochemical cycles towards environmental disturbances or resilience influenced by anthropogenic effects [23]. By emphasizing the metabolic activities of microbial ecosystems through detailed metagenomic analyses, several intermediate physicochemical elemental transformations can be widely explored to understand the significance of global biogeochemical cycles (Fig. **2**).

However, complex microbial activity in the subsurface environments is governed by their intrinsic chemical dynamics, such as redox-transport models (or acid-base elements), kinetically influenced biodegradation pathways and secondary metabolism driven by oxygen (O_2), phosphorous (P), carbon (C), hydrogen (H_2), nitrogen (N_2), sulfur (S), manganese (Mn), iron (Fe) and calcium (Ca) cycles. Predominantly, bacteria and archaea-based microbial communities strongly influence C, N, Fe, and S cycles, eventually modulating the global ocean productivity and climate conditions [24]. Herein, we have briefly discussed the distinctive and co-existing relationships of majorly influential biogeochemical cycles such as C, N, Fe, and S, underlining more emphasis on autotrophic and heterotrophic conditions existing on surface and subsurface environments.

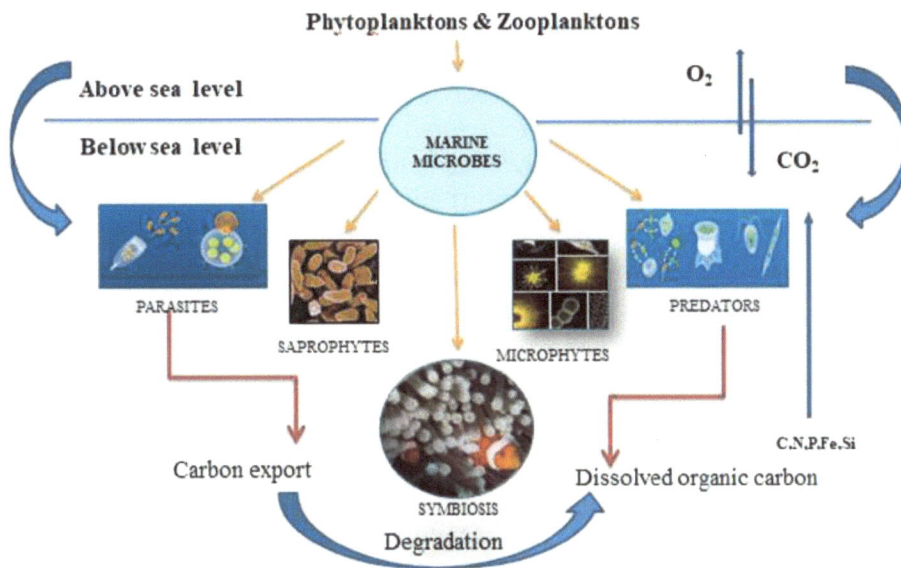

Fig. (2). Functioning of microbes in the marine environment.

Carbon Cycle

The total C metabolism is altered based on temperature fluctuations, nutrient availability, microbial responses towards latitudinal gradients, and spatial and temporal patterns prevailing in aquatic ecosystems [25]. Microbial-mediated complex organic matter decomposition, persistent CO_2 super-saturation, biomass production, and respiration are the crucial factors determining the fluxes associated with C metabolism, especially in the microbial loop and microbial food web components. Intriguingly, the direct effects of CO_2 concentration have been assessed in aquatic ecosystems influenced by natural as well as anthropogenic activities, wherein externally bolstering biological processes such as eutrophication in coastal lagoons generates surplus N and P concentrations, which indirectly affect the rate of C cycle [26]. Such imbalance in low latitude humic coastal lagoons can cause potential risk in tremendously warming temperatures and intensifying metabolic responses between photosynthesis and respiration, eventually increasing organic matter production. Subsequently, high levels of dissolved organic carbon (DOC) and other inorganic C contributions can be accumulated due to the positive effects of warm temperature conditions leading to excess organic decomposition in marine ecosystems.

In addition, net CO_2 and methane (CH_4) gas emissions in eutrophic conditions and their spatio-temporal variations in the sediments of subtropical wetland ecosystems can immensely contribute to the N_2O emissions globally [27]. Such intensified alterations in the biological processes of subterranean ecosystems like cave sediments promote continuous CH_4 consumption and CO_2 emissions, influencing the C and N cycling. Hence, the syntrophic relationship between methanotrophic bacteria (*e.g.*, *Methylomonaceae, Methylomirabilaceae,* and *Methylacidiphilaceae*) and *Crossiella* consume excessive N from heterotrophic ammonification process under the surface of the sediments fixing CO_2 and accelerating CH_4 oxidation [28]. Intriguingly, these subterranean ecosystems act as reservoirs in the ocean margins (with submerged connections to marine waters) for nutrient cycling and C sequestration, thereby establishing interlinked pathways between several biogeochemical cycles of marine ecosystems driven by microbial metabolic activities.

Nitrogen Cycle

Generally, marine microbial communities infiltrate dynamic alterations in N and S cycles based on the external environmental circumstances prevailing in the aquatic ecosystem. However, these two cycles are often interrelated due to the occurrence of autotrophic and denitrification microbial genes that are more relatively available in the ecosystem, thereby expanding the view of N cycling (Fig. **3**). Some related studies integrated denitrification processes with primary carbon oxidation reactions and other CH_4, Fe, and S-based oxidation reactions to observe the proximity between O and N consumption zones in the low-level subsurface ecosystems [29, 30].

N is the key element of life in marine ecosystems. It usually exists in various oxidation states, mostly as nitrites (NO_2^-), nitrates (NO_3^-), organic N, and ammonium (NH_4^+). Concomitantly, microbiomes can contribute to multiple biogeochemical transformations in the N cycle, which can be categorized into two major processes: retention/fixation and sinks/loss. Primary sources of biologically fixed N pool involve nitrification, assimilation, and dissimilatory NO_3^- reduction to ammonia. Marine N_2 fixers (Diazotrophs) mainly include filamentous cyanobacteria such as *Oscillatoria, Trichodesmium, Nodularia, Aphanizomenon,* and *Calothrix*. Other bacteria include *Chlorobium, Klebsiella, Desulfovibrio, Clostridium,* and *Thiobacillus*. On the other hand, loss in fixed N cycles is predominantly caused by denitrification, anaerobic NH_4^+ oxidation (anammox), and nitrate-dependent anaerobic CH_4 oxidation [31]. Microbial transformation of N compounds and N-converting enzymes possess a significant role in balancing the biochemical and ecological diversity of global marine culture, where their conversion processes can be categorized into biological N fixation and N retention pool (processes contributing to the conversion of N species from one form to the other without external loss from the ecosystem); and other fixed N loss through several transboundary marine ecosystems.

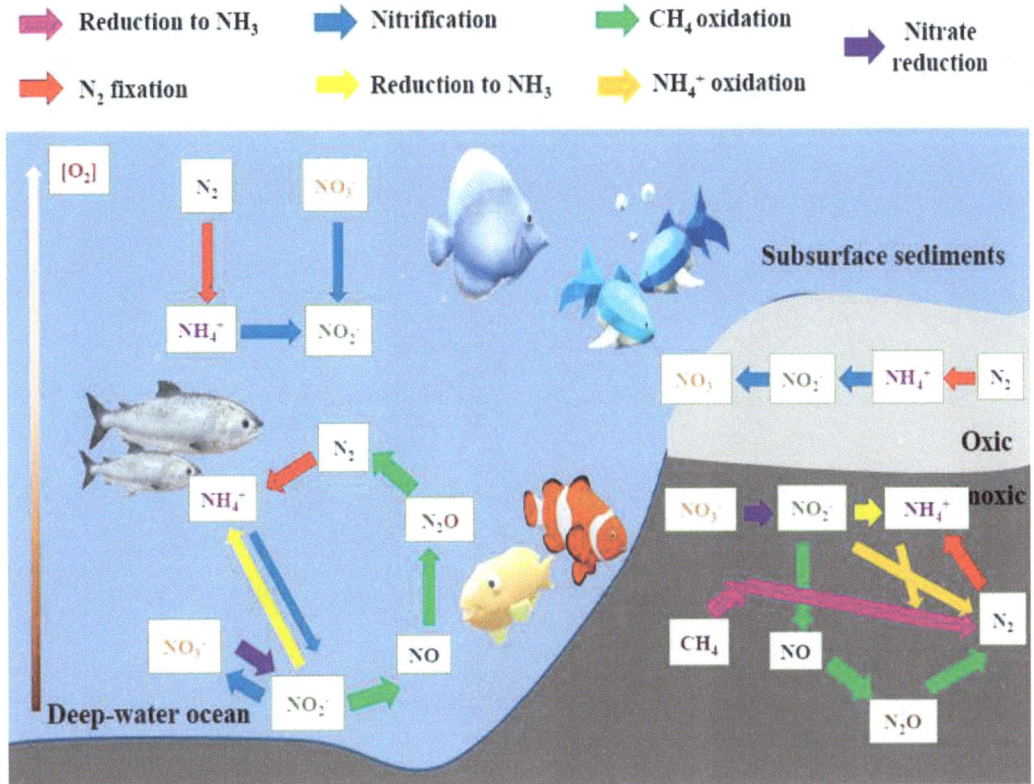

Fig. (3). The nitrogen cycle in the marine ecosystem.

Sulfur (S) Cycle

In a continuum to the previous C and N cycles, the S cycle is sequential and primarily driven by the microbial dissimilatory sulfate reduction to sulfide biochemical oxidation state as a significant terminal pathway towards organic matter biodegradation under anoxic conditions especially by anaerobic microbes (Fig. **4**). Marine food web community is highly benefitted from such organic matter decomposition, which ultimately forms several intermediates during geochemical transformations of elemental sulfur, polysulfides, thiosulfate, and sulfite involving diverse reactions with O, NO_3^-, Mn(IV), Fe(III) and other trace elements. These microbial-catalyzed chemical pathways are primarily driven by Mn and Fe reduction, resulting from the accumulation of buried organic carbon supported by anaerobically oxidized CH_4. Therefore, the majority of the methane produced in sediments from the continental shelf and slope diffuses upward along this gradient until it meets sulfate in the subsurface sulfate-methane transition, where anaerobic methanotrophic archaea (ANME) quantitatively oxidize it. According to the following net equation, sulfate acts as an electron acceptor.

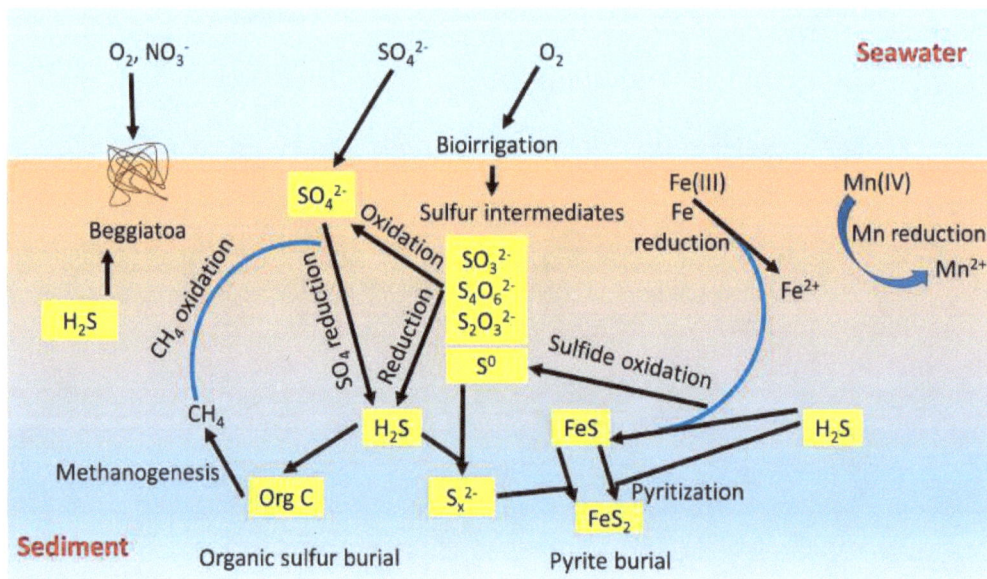

Fig. (4). The sulfur cycle in the marine ecosystem.

$$SO_4^{2-} + CH_4 \rightarrow HCO_3^- + HS^- + H_2O$$

Many sulfate-reducing bacteria (SRB), notably those in the *Desulfovibrionales* and *Desulfobacterales* orders, are members of the Deltaproteobacteria. Other taxa, such as those from the phylum Firmicutes, Chloroflexi, and Atribacteria, grow more prevalent as sediments become deeper. SRB of the *Desulfosarcina*, *Desulfococcus*, and *Desulfobulbus* branches of the Deltaproteobacteria are typically linked to ANME [32]. Recent studies on the dynamics of the microbial population have highlighted the importance of genes that can progressively exhibit substrate promiscuity with one another more sequentially, thus pairing sulfur oxidation with denitrification. Consequently, sulfur oxidizers are paired with oxygen even in deep-water zones where the oxygen depletion is complimented by NO_3^- sources [33]. Marine microbes, particularly sulfur-containing bacteria (*e.g.*, *Beggiatoa* spp.), play a critical role in these catalytic pathways, wherein continuous oxygen flux conditions are regulated with other sulfide oxidizers.

Iron (Fe) Cycle

Proteins essential in fundamental biological activities, such as photosynthesis and respiration, need the micronutrient Fe. Despite being the fourth most common element in the Earth's crust, dissolved iron concentrations in open ocean surface waters usually aren't higher than 0.2 nM, with higher concentrations at depth due to remineralization processes that take place throughout the water column [34]. The availability of electron donors, the reactivity of Fe(III) minerals, or a combination of both limit the amount of iron that can be reduced. The competition between iron reduction and sulfate reduction is primarily regulated by the accessibility of iron minerals because the sulfate reducers close to the sediment surface are only constrained by their electron donor and not by sulfate. Some sulfate-reducing bacteria, such as *Desulfotomaculum reducens*, can reduce Fe(III) in a catabolic metabolism that supports growth. Microbial iron reduction is well recognized from the genera *Geobacter* and *Shewanella*. In comparison to the more specialized metal reducers, the role of sulfate-reducing microbes in iron reduction is still little understood. Henceforth, the combination of C, N, S, and Fe biogeochemical cycles is influenced by the environmental stress and resilience of the natural and anthropogenic effects on the marine microbial communities. The metagenomic evolution of microbes and microbiomes significantly modulates global food web components and climate change parameters.

BIOACTIVE COMPOUNDS FROM MARINE ORGANISMS

A potential source of commercially significant bioactive substances such as exopolysaccharides, polyhydroxy-alkanoates, osmolytes, carotenoid pigments, bacterio-rhodopsin, *etc.*, is marine bacteria [35, 36]. Additionally, it has been thought that marine microorganisms that inhabit extremely salty conditions provide good study subjects for ecological and evolutionary processes. Remarkably, the capacity of extremophile bacteria from marine hot springs and hydrothermal vents to create enzymes, polymers, osmolytes, *etc.*, is drawing interest. Marine ecosystems are the most significant source of novel genes, enzymes, and natural products, which may enhance the efficiency and sustainability of industrial processes and result in more environmentally friendly and economically viable goods [37]. Identifying and characterizing marine microbial bioactive compounds would be easier if more automated and cheap methods were developed [38].

Different thermophilic bacteria were extracted from their marine environments and evaluated. Species belonging to *Thermotoga, Aquifex, Pyrococcus, Methanococcus, Thermococcus, Pyrodictium, Igneococcus,* and *Archeoaglobus* are all able to thrive in hydrothermal systems that are located in shallow water as well as in deep water. Coastal hydrothermal systems have been used to isolate members of the genus *Thermococcus*, specifically *T. celer* and *T. litoralis* [37]. *Fervidobacteriumpennivorans* V5 (recombinant) and *F. pennivorans* identified from the Hot springs, Azores islands capable of producing Amylase debranching and Serine peptidase respectively, assist in the poultry industry; detergent; fish firm [39]. *T. litoralis*, an organism that produced serine protease and was discovered in deep-sea hydrothermal springs and oil wells, is also used in the poultry industry and the production of detergent [40]. Another Metallo carboxy-peptidase-producing bacterium, *Caldithrixabyssi*, was isolated from the deep-sea hydrothermal chimneys on the Mid-Atlantic Ridge [41].

Recently, several remarkable instances of exopolysaccharides (EPSs) from extremophilic microbes that are both physiologically and biotechnologically beneficial have been described. These EPSs come from organisms that live in extremely cold environments. Two marine bacteria obtained from a deep-sea hydrothermal vent have had their EPSs studied for possible uses in regenerative medicine [42]. *Bacillus thermodenitrificans* strain B3-72 and *B. licheniformis* strain B3-15 are two EPS-producing strains isolated from the marine vents of Vulcanoislandin, Italy [43]. Both of these strains have interesting antiviral activities. An additional strain of *B. licheniformis* was also discovered on the Italian island of Panarea. This strain created a fructo-glucan EPS with anti-cytotoxic action [44]. *Alvinella pompejana*, a polychaete annelid, was the source of isolation of the marine bacteria *Alteromonas macleodii* subsp. *fijiensis*. EPS produced by these strains has been commercialized for use in cosmetics to protect delicate skin from chemical, mechanical, and UV-B aggressions [45].

Halophilic or halotolerant bacteria, as well as eukaryotes, accumulate neutral osmolytes. Due to the carboxylate or phosphate groups in their cell membranes, hyperthermophilic archaea, halophilic or halotolerant archaea, and thermophilic bacteria tend to accumulate negatively charged solutes [46]. For the salt stress response, hyperthermophilic microbes secrete mannosyl-glyceramide, di-glycero--phosphate, mannosyl glycerate, and α-glutamate. Contrarily, the levels of cyclic-2,3-bisphosphoglycerate and di-myo-inositol-1,1'-phosphate rise predominantly in response to high temperatures. A greater buildup of di-myo-inositol-1-1'-phosphate is seen in *Pyrococcus furiosus, Thermotoga maritima*, and, *Thermotoganeapolitana* at a temperature of 101°C; in particular, a 20-fold rise in DIP is seen in the species *Pyrococcus furiosus*. The majority of the osmolytes retrieved from the anaerobic thermohalophile (*T. Methanothermococcus*) were made up of the substance α-glutamate, also known as α-amino glutaric acid, which made up 37% of the total osmolyte [47].

In addition, the marine-derived fungus known as *Aspergillus fumigatus* isolated from Bohai sea sediments was reported to produce novel pseurotin analogs 4 and 5 [48]. Unfortunately, the spectroscopic data analyzed in such depth to identify these metabolites revealed that they did not have any antibacterial activity in the tests conducted for this study. In a related development, the well-known chemical helvolic acid was found to be particularly active against *Staphylococcus aureus* and methicillin-resistant *S. aureus*, with a minimum inhibitory concentration (MIC) of 0.78 g/mL. This compound was isolated from the same fungus. In the South China Sea, silt was collected at a depth of 2403 meters, and it was there that the fungus *Geosmithia pallida* FS140 was discovered. This fungus was used to isolate twelve different diketopiperazines, including three newly discovered thiodiketopiperazines designated geospallins A–C type [49].

USING MARINE MICROBES TO AMELIORATE ENVIRONMENTAL DETERIORATION

Marine microorganisms have a wide range of bioremediation potentials that are advantageous from an economic and environmental point of view. In order to degrade, modify, or accumulate hazardous materials, such as hydrocarbons, heterocyclic compounds, medicinal drugs, radionuclides, and toxic metals, marine microorganisms have been used in bioremediation and biotransformation techniques [50]. Marine bacteria and related species frequently use biofilm formation, enzyme production, biosurfactant generation, and heavy metal ions oxidation as bioremediation methods (Fig. **5**).

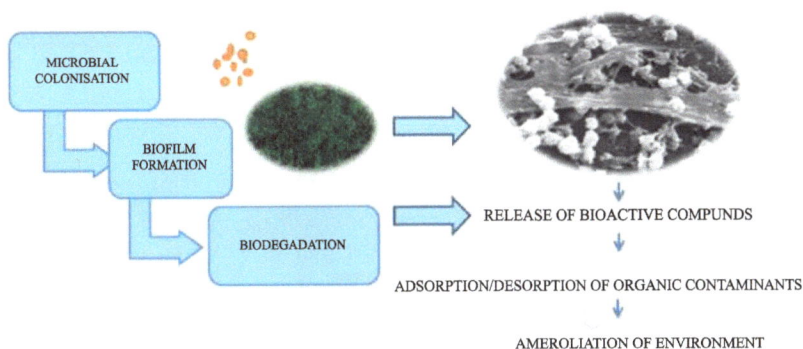

Fig. (5). Microbial interactions on environmental contaminants.

Polymers are one of the leading environmental pollutants because they are so resistant to biodegradation. The marine bacteria produce a variety of enzymes that target and break down plastics, including peroxidase, PHB depolymerase, polyurethanases, monooxygenase MHETase, PEG dehydrogenase, lipase, and PETase [51]. A PETase-like enzyme that could break down polycaprolactone was found in the actinomycete *Streptomyces* sp. associated with the marine sponge [52]. Several authors reported bacterial degradation of plastic in *in-vitro* conditions (30 days incubation): *Kocuriapalustris* (1%) and *Bacillus pumilus* (1.5%) [53]; *Pseudomonas* sp. (15%), and *Arthrobacter* sp. (12%) [54].

Marine bacteria use oxygen, nitrate, or sulfates as electron acceptors in aerobic or anaerobic processes to break down polyaromatic hydrocarbons (PAH) in marine environments. Moreover, synthesizing biosurfactants or the lipase enzyme by marine bacteria aids in breaking insoluble compounds [55]. *Alcanivorax, Exiguobacterium, Halomonas, Rhodococcus, Pseudomonas, Acinetobacter, Bacillus*, and *Streptomyces* were among the 55 marine bacterial communities that [56] reported producing biosurfactants and using n-hexadecane as their primary source of carbon. In a crude preparation, the lipase enzyme secreted by the *Alcanivoraxborkumensis* effectively degraded petroleum pollutants (6000 ppm) at a rate of 88.52%, indicating the possibility of bioremediation of hydrocarbon contamination [57]. Moreover, *Bacillus cereus* isolated from a marine sample contaminated with petroleum was discovered to synthesize an anionic biosurfactant that efficiently removed the toxicity of oil spills [58].

The rapid expansion of industrialization and mining in recent years has resulted in heavy metal contamination of aquatic water bodies. Heavy metal buildups can be crucial to the survival of microbes. Examples of such accumulations include calcium, cobalt, chromium, copper, iron, potassium, magnesium, manganese, sodium, nickel, and zinc. According to [59], the marine organism *Vibrio harveyi* can bioaccumulate cadmium up to 23.3 mg in its cells. According to a study [60], certain purple non-sulfur marine bacterial isolates, such as *Rhodobium marinum* and *Rhodobacter sphaeroides*, are able to bio-sorb or biotransform heavy metals like copper, zinc, cadmium, and lead from contaminated habitats.

FUTURE PERSPECTIVES AND LIMITATIONS

There are still many unanswered research concerns despite the past few years having seen remarkable technical advancements that have given us a wealth of information about the most common forms of marine microorganisms and their activity. Hence, learning about the ecology of marine microbial diversity is essential for comprehending marine ecosystems, improving and sustainably using ocean resources, and predicting changes in ecosystem functioning. Next generation sequencing now permits screening for biodiversity for any "marker gene" at unprecedented temporal and spatial resolutions. That could serve as the basis for figuring out the dynamics of microbial communities and identifying recurrent patterns of microbial connection and potential interactions. The age-old question of whether everything is truly everywhere may now have an answer.

After a baseline of ocean biodiversity has been established, anthropogenic factors that alter biodiversity can be separated from seasonal variations. That would also make it possible to track the growth of germs that could become harmful due to the warming of the ocean's surface.

More species must be researched to learn more about the changes brought on by ocean circulation, explore the connections between ecosystems, and construct models based on sampling and analytical techniques. Future research may be aided by taking into account information such as the fact that complex organisms and microorganisms in polar ecosystems are both susceptible to the effects of climate change, the crucial role played by bacteria in carbon flux and biogeochemical cycles, and a higher diversity of species is necessary for the ecosystems with changing surroundings to survive. Global warming may preferentially benefit species that have a range of genotypes among populations, quick generation rates, and wide thermal windows. Even a slight temperature increase has an impact on stenothermal marine creatures. In these circumstances, it is vital to comprehend how complicated responses to temperature fluctuations and their effects affect the entire ecosystem. Identifying species that exhibit phenotypic plasticity is essential so that organisms that are likely to suffer from the effects of global warming can be targeted. Biomarkers must be developed to gain access to information on stress experienced by microorganisms at the ecosystem level and to understand why some species can endure high temperatures better than others. Further field research and experimental studies are required to fully understand how different climatic changes affect the emergence of microbes, shaping ocean ecosystems.

CONCLUSION

Almost 90% of the biomass in the marine ecosystem is made up of microorganisms, representing a vast variety of living things. Most biogeochemical cycles in the ocean are controlled and dominated by a limited number of bacteria. While immeasurable microorganisms are uncommon, they represent a practically limitless genetic resource that marine bacteria draw upon to adapt to ecological changes and keep the ecosystem stable. Prokaryotes like bacteria and archaea and eukaryotes like microalgae, protista, zooplankton, and various other higher biota are abundant in seawater, which also contains a sizable number of viruses. Since they are the primary agents of genetic exchange between microbes, viruses control the biomass of microorganisms by "killing the winner" and also help to produce and maintain biodiversity. The stability of marine biogeochemical cycles and their constituents determines the dynamics of every ocean biome. With the industrial and lucrative application of microbial variety valued at millions of pounds, the organization and usage of marine microorganisms play a crucial role in sustainable growth. The biotechnology industry has a vast, largely untapped reserve for discovering new biomolecules and unique processes to deliver new medications, chemicals, and revolutionary technologies. Even though marine microbes play a vital role in sustainable development, little is known about the critical contribution that microbial diversity makes to the construction of prosperity, the national economy, and the primacy of human existence. The presentation that the sustainable use of microbial diversity has good financial value has a favorable impact on changing public and governmental perceptions of marine microorganisms.

REFERENCES

[1]   Nemergut DR, Schmidt SK, Fukami T, *et al.* Patterns and processes of microbial community assembly. Microbiol Mol Biol Rev 2013; 77(3): 342-56.
        [http://dx.doi.org/10.1128/MMBR.00051-12] [PMID: 24006468]

[2]   Doney SC, Ruckelshaus M, Emmett Duffy J, *et al.* Climate change impacts on marine ecosystems. Annu Rev Mar Sci 2012; 4(1): 11-37.
        [http://dx.doi.org/10.1146/annurev-marine-041911-111611] [PMID: 22457967]

[3]   Baltar F, Palovaara J, Vila-Costa M, *et al.* Response of rare, common and abundant bacterioplankton to anthropogenic perturbations in a Mediterranean coastal site. FEMS Microbiol Ecol 2015; 91(6): fiv058.
        [http://dx.doi.org/10.1093/femsec/fiv058] [PMID: 26032602]

[4]   Gupta VK, Paul S, Dutta C. Geography, ethnicity or subsistence-specific variations in human microbiome composition and diversity. Front Microbiol 2017; 8: 1162.
        [http://dx.doi.org/10.3389/fmicb.2017.01162] [PMID: 28690602]

[5]   Gifford SM, Sharma S, Moran MA. Linking activity and function to ecosystem dynamics in a coastal bacterioplankton community. Front Microbiol 2014; 5: 185.
        [http://dx.doi.org/10.3389/fmicb.2014.00185] [PMID: 24795712]

[6]   Bartkowski B, Lienhoop N, Hansjürgens B. Capturing the complexity of biodiversity: A critical review of economic valuation studies of biological diversity. Ecol Econ 2015; 113: 1-14.
        [http://dx.doi.org/10.1016/j.ecolecon.2015.02.023]

[7]   Salazar G, Sunagawa S. Marine microbial diversity. Curr Biol 2017; 27(11): R489-94.
        [http://dx.doi.org/10.1016/j.cub.2017.01.017] [PMID: 28586685]

[8]   Cognetti G, Maltagliati F. Strategies of genetic biodiversity conservation in the marine environment. Mar Pollut Bull 2004; 48(9-10): 811-2.
        [http://dx.doi.org/10.1016/j.marpolbul.2003.12.016] [PMID: 15111028]

[9]   Bindoff NL, Cheung WWL, Kairo JG, *et al.* Changing ocean, marine ecosystems, and dependent communities. In: Pörtner HO, Roberts DC, Masson-Delmotte V, PIPCC Special Report on the Ocean and Cryosphere in a Changing Climate. 2019.

[10]   Asch RG, Cheung WWL, Reygondeau G. Future marine ecosystem drivers, biodiversity, and fisheries maximum catch potential in Pacific Island countries and territories under climate change. Mar Policy 2018; 88: 285-94.
        [http://dx.doi.org/10.1016/j.marpol.2017.08.015]

[11]   Heiskanen AS, Berg T, Uusitalo L, *et al.* Biodiversity in marine ecosystems—European developments toward robust assessments. Front Mar Sci 2016; 3: 184.
        [http://dx.doi.org/10.3389/fmars.2016.00184]

[12]   Whitman WB, Coleman DC, Wiebe WJ. Prokaryotes: The unseen majority. Proc Natl Acad Sci USA

1998; 95(12): 6578-83.
[http://dx.doi.org/10.1073/pnas.95.12.6578] [PMID: 9618454]

[13] Teira E, Logares R, Gutiérrez-Barral A, *et al.* Impact of grazing, resource availability and light on prokaryotic growth and diversity in the oligotrophic surface global ocean. Environ Microbiol 2019; 21(4): 1482-96.
[http://dx.doi.org/10.1111/1462-2920.14581] [PMID: 30838751]

[14] Mayol E, Arrieta JM, Jiménez MA, *et al.* Long-range transport of airborne microbes over the global tropical and subtropical ocean. Nat Commun 2017; 8(1): 201.
[http://dx.doi.org/10.1038/s41467-017-00110-9] [PMID: 28779070]

[15] Kirchman DL, Gasol JM. Microbial ecology of the oceans. John Wiley & Sons 2018.

[16] Hagström Å, Pommier T, Rohwer F, *et al.* Use of 16S ribosomal DNA for delineation of marine bacterioplankton species. Appl Environ Microbiol 2002; 68(7): 3628-33.
[http://dx.doi.org/10.1128/AEM.68.7.3628-3633.2002] [PMID: 12089052]

[17] Curtis TP, Sloan WT, Scannell JW. Estimating prokaryotic diversity and its limits. Proc Natl Acad Sci USA 2002; 99(16): 10494-9.
[http://dx.doi.org/10.1073/pnas.142680199] [PMID: 12097644]

[18] Riley MA, Lizotte-Waniewski M. Population genomics and the bacterial species concept. Horiz Gene Transf Genomes Flux 2009; pp. 367-77.
[http://dx.doi.org/10.1007/978-1-60327-853-9_21]

[19] Zehr JP, Weitz JS, Joint I. How microbes survive in the open ocean. Science 2017; 357, 646-7.
[http://dx.doi.org/10.1126/science.aan5764]

[20] Zhu D, Sethupathy S, Gao L, *et al.* Microbial diversity and community structure in deep-sea sediments of South Indian Ocean. Environ Sci Pollut Res Int 2022; 29(30): 45793-807.
[http://dx.doi.org/10.1007/s11356-022-19157-3] [PMID: 35152353]

[21] Varliero G, Bienhold C, Schmid F, Boetius A, Molari M. Microbial diversity and connectivity in deep-sea sediments of the South Atlantic Polar Front. Front Microbiol 2019; 10: 665.
[http://dx.doi.org/10.3389/fmicb.2019.00665] [PMID: 31024475]

[22] Gao P, Qu L, Du G, Wei Q, Zhang X, Yang G. Bacterial and archaeal communities in deep sea waters near the Ninetyeast Ridge in Indian Ocean. J Oceanol Limnol 2021; 39(2): 582-97.
[http://dx.doi.org/10.1007/s00343-020-9343-y]

[23] Grossart HP, Massana R, McMahon KD, Walsh DA. Linking metagenomics to aquatic microbial ecology and biogeochemical cycles. Limnol Oceanogr 2020; 65(S1): S2-S20.
[http://dx.doi.org/10.1002/lno.11382]

[24] Fuhrman JA. Microbial community structure and its functional implications. Nature 2009; 459(7244): 193-9.
[http://dx.doi.org/10.1038/nature08058] [PMID: 19444205]

[25] Amado AM, Roland F. Microbial role in the carbon cycle in tropical inland aquatic ecosystems. Front Microbiol 2017; 8: 20.
[http://dx.doi.org/10.3389/fmicb.2017.00020] [PMID: 28154556]

[26] Peixoto RB, Marotta H, Enrich-Prast A. Experimental evidence of nitrogen control on pCO_2 in phosphorus-enriched humic and clear coastal lagoon waters. Front Microbiol 2013; 4: 11.
[http://dx.doi.org/10.3389/fmicb.2013.00011] [PMID: 23390422]

[27] Liengaard L, Nielsen LP, Revsbech NP, *et al.* Extreme emission of n(2)o from tropical wetland soil (pantanal, South america). Front Microbiol 2013; 3: 433.
[http://dx.doi.org/10.3389/fmicb.2012.00433] [PMID: 23293634]

[28] Martin-Pozas T, Cuezva S, Fernandez-Cortes A, *et al.* Role of subterranean microbiota in the carbon cycle and greenhouse gas dynamics. Sci Total Environ 2022; 831: 154921.

[http://dx.doi.org/10.1016/j.scitotenv.2022.154921] [PMID: 35364174]

[29]   Varadharajan C. Magnitude and spatio-temporal variability of methane emissions from a eutrophic freshwater lake. Ph.D. Thesis, 2009.

[30]   Deutzmann JS, Stief P, Brandes J, Schink B. Anaerobic methane oxidation coupled to denitrification is the dominant methane sink in a deep lake. Proc Natl Acad Sci USA 2014; 111(51): 18273-8.
[http://dx.doi.org/10.1073/pnas.1411617111] [PMID: 25472842]

[31]   Pajares S, Ramos R. Processes and microorganisms involved in the marine nitrogen cycle: knowledge and gaps. Front Mar Sci 2019; 6: 739.
[http://dx.doi.org/10.3389/fmars.2019.00739]

[32]   Jørgensen BB, Findlay AJ, Pellerin A. The biogeochemical sulfur cycle of marine sediments. Front Microbiol 2019; 10: 849.
[http://dx.doi.org/10.3389/fmicb.2019.00849] [PMID: 31105660]

[33]   Maia LB, Moura JJG. How biology handles nitrite. Chem Rev 2014; 114(10): 5273-357.
[http://dx.doi.org/10.1021/cr400518y] [PMID: 24694090]

[34]   Gledhill M, Buck KN. The organic complexation of iron in the marine environment: a review. Front Microbiol 2012; 3: 69.
[http://dx.doi.org/10.3389/fmicb.2012.00069] [PMID: 22403574]

[35]   Poli A, Anzelmo G, Nicolaus B. Bacterial exopolysaccharides from extreme marine habitats: production, characterization and biological activities. Mar Drugs 2010; 8(6): 1779-802.
[http://dx.doi.org/10.3390/md8061779] [PMID: 20631870]

[36]   Klippel B, Sahm K, Basner A, *et al.* Carbohydrate-active enzymes identified by metagenomic analysis of deep-sea sediment bacteria. Extremophiles 2014; 18(5): 853-63.
[http://dx.doi.org/10.1007/s00792-014-0676-3] [PMID: 25108363]

[37]   Elleuche S, Schäfers C, Blank S, Schröder C, Antranikian G. Exploration of extremophiles for high temperature biotechnological processes. Curr Opin Microbiol 2015; 25: 113-9.
[http://dx.doi.org/10.1016/j.mib.2015.05.011] [PMID: 26066287]

[38]   Hedlund BP, Dodsworth JA, Murugapiran SK, Rinke C, Woyke T. Impact of single-cell genomics and metagenomics on the emerging view of extremophile "microbial dark matter". Extremophiles 2014; 18(5): 865-75.
[http://dx.doi.org/10.1007/s00792-014-0664-7] [PMID: 25113821]

[39]   Friedrich AB, Antranikian G. Keratin degradation by Fervidobacterium pennavorans, a novel thermophilic anaerobic species of the order Thermotogales. Appl Environ Microbiol 1996; 62(8): 2875-82.
[http://dx.doi.org/10.1128/aem.62.8.2875-2882.1996] [PMID: 16535379]

[40]   Atomi H. Recent progress towards the application of hyperthermophiles and their enzymes. Curr Opin Chem Biol 2005; 9(2): 166-73.
[http://dx.doi.org/10.1016/j.cbpa.2005.02.013] [PMID: 15811801]

[41]   Lloyd KG, Schreiber L, Petersen DG, *et al.* Predominant archaea in marine sediments degrade detrital proteins. Nature 2013; 496(7444): 215-8.
[http://dx.doi.org/10.1038/nature12033] [PMID: 23535597]

[42]   Rougeaux H, Kervarec N, Pichon R, Guezennec J. Structure of the exopolysaccharide of Vibriodiabolicus isolated from a deep-sea hydrothermal vent. Carbohydr Res 1999; 322(1-2): 40-5.
[http://dx.doi.org/10.1016/S0008-6215(99)00214-1] [PMID: 10629947]

[43]   Arena A, Maugeri TL, Pavone B, Iannello D, Gugliandolo C, Bisignano G. Antiviral and immunoregulatory effect of a novel exopolysaccharide from a marine thermotolerant Bacillus licheniformis. Int Immunopharmacol 2006; 6(1): 8-13.
[http://dx.doi.org/10.1016/j.intimp.2005.07.004] [PMID: 16332508]

[44]   Gugliandolo C, Spanò A, Maugeri T, Poli A, Arena A, Nicolaus B. Role of bacterial exopolysaccharides as agents in counteracting immune disorders induced by herpes virus. Microorganisms 2015; 3(3): 464-83.
        [http://dx.doi.org/10.3390/microorganisms3030464] [PMID: 27682100]

[45]   Cambon-Bonavita MA, Raguénès G, Jean J, Vincent P, Guezennec J. A novel polymer produced by a bacterium isolated from a deep-sea hydrothermal vent polychaete annelid. J Appl Microbiol 2002; 93(2): 310-5.
        [http://dx.doi.org/10.1046/j.1365-2672.2002.01689.x] [PMID: 12147080]

[46]   Roberts MF. Organic compatible solutes of halotolerant and halophilic microorganisms. Saline Syst 2005; 1(1): 5.
        [http://dx.doi.org/10.1186/1746-1448-1-5] [PMID: 16176595]

[47]   Wilson ZE, Brimble MA. Molecules derived from the extremes of life. Nat Prod Rep 2009; 26(1): 44-71.
        [http://dx.doi.org/10.1039/B800164M] [PMID: 19374122]

[48]   Xu X, Han J, Wang Y, *et al.* Two new spiro-heterocyclic γ-lactams from a marine-derived Aspergillus fumigatus strain CUGBMF170049. Mar Drugs 2019; 17(5): 289.
        [http://dx.doi.org/10.3390/md17050289] [PMID: 31091673]

[49]   Sun ZH, Gu J, Ye W, *et al.* Geospallins A–C: new thiodiketopiperazines with inhibitory activity against angiotensin-converting enzyme from a deep-sea-derived fungus Geosmithia pallida FS140. Mar Drugs 2018; 16(12): 464.
        [http://dx.doi.org/10.3390/md16120464] [PMID: 30477129]

[50]   Karigar CS, Rao SS. Role of microbial enzymes in the bioremediation of pollutants: a review. Enzyme Res 2011; 2011: 1-11.
        [http://dx.doi.org/10.4061/2011/805187] [PMID: 21912739]

[51]   Ganesh Kumar A, Anjana K, Hinduja M, Sujitha K, Dharani G. Review on plastic wastes in marine environment – Biodegradation and biotechnological solutions. Mar Pollut Bull 2020; 150: 110733.
        [http://dx.doi.org/10.1016/j.marpolbul.2019.110733] [PMID: 31767203]

[52]   Almeida EL, Carrillo Rincón AF, Jackson SA, Dobson ADW. *In silico* screening and heterologous expression of a polyethylene terephthalate hydrolase (PETase)-like enzyme (SM14est) with polycaprolactone (PCL)-degrading activity, from the marine sponge-derived strain Streptomyces sp. SM14. Front Microbiol 2019; 10: 2187.
        [http://dx.doi.org/10.3389/fmicb.2019.02187] [PMID: 31632361]

[53]   Harshvardhan K, Jha B. Biodegradation of low-density polyethylene by marine bacteria from pelagic waters, Arabian Sea, India. Mar Pollut Bull 2013; 77(1-2): 100-6.
        [http://dx.doi.org/10.1016/j.marpolbul.2013.10.025] [PMID: 24210946]

[54]   Balasubramanian V, Natarajan K, Hemambika B, *et al.* High-density polyethylene (HDPE)-degrading potential bacteria from marine ecosystem of Gulf of Mannar, India. Lett Appl Microbiol 2010; 51(2): no.
        [http://dx.doi.org/10.1111/j.1472-765X.2010.02883.x] [PMID: 20586938]

[55]   Mohanrasu K, Rao RGR, Raja R, Arun A. Bioremediation process by marine microorganisms. Encycl Mar Biotechnol 2020; 4: 2211-28.
        [http://dx.doi.org/10.1002/9781119143802.ch100]

[56]   Cai Q, Zhang B, Chen B, Zhu Z, Lin W, Cao T. Screening of biosurfactant producers from petroleum hydrocarbon contaminated sources in cold marine environments. Mar Pollut Bull 2014; 86(1-2): 402-10.
        [http://dx.doi.org/10.1016/j.marpolbul.2014.06.039] [PMID: 25034191]

[57]   Kadri T, Magdouli S, Rouissi T, Brar SK. *Ex-situ* biodegradation of petroleum hydrocarbons using Alcanivorax borkumensis enzymes. Biochem Eng J 2018; 132: 279-87.

[http://dx.doi.org/10.1016/j.bej.2018.01.014]

[58]   Durval IJB, Mendonça AHR, Rocha IV, *et al.* Production, characterization, evaluation and toxicity assessment of a Bacillus cereus UCP 1615 biosurfactant for marine oil spills bioremediation. Mar Pollut Bull 2020; 157: 111357.
[http://dx.doi.org/10.1016/j.marpolbul.2020.111357] [PMID: 32658706]

[59]   Hanan AE, Gehan MAE, Nermeen AES. Cadmium resisting bacteria in Alexandria Eastern Harbor (Egypt) and optimization of cadmium bioaccumulation by Vibrio harveyi. Afr J Biotechnol 2011; 10(17): 3412-23.
[http://dx.doi.org/10.5897/AJB10.1933]

[60]   Panwichian S, Kantachote D, Wittayaweerasak B, Mallavarapu M. Removal of heavy metals by exopolymeric substances produced by resistant purple nonsulfur bacteria isolated from contaminated shrimp ponds. Electron J Biotechnol 2011; 14: 2.

CHAPTER 8

Microbiomes in Mangroves and Wetlands: Their Role and Importance in Ecosystem Sustainability

Zahra Haghani[1] and **Kamyar Amirhosseini**[1,*]

[1] *Department of Soil Science, Faculty of Agriculture, College of Agriculture and Natural Resources, University of Tehran, Tehran, Iran*

Abstract: Mangroves and wetlands are critical intermediary ecosystems between terrestrial and marine environments. These ecosystems offer a wide range of invaluable ecological and economic services. However, under the influence of natural and anthropogenic threats, mangroves and wetlands face rapid degradation. Microbes and microbiomes are integral components of a mangrove, playing key roles in the stability of the ecosystem. The present chapter compiles a comprehensive review of the classification and the role of microorganisms in the sustainability of mangrove and wetland ecosystems. The chapter discusses the most critical features of microbial groups, including archaea, bacteria, algae, and fungi in mangroves and wetlands. Bacterial groups under discussion consist of sulfur-related bacteria, nitrogen-related bacteria, phosphate-solubilizing bacteria, and photosynthetic bacteria. A separate section is dedicated to periphytic communities encompassing a microhabitat involving various prokaryotic and eukaryotic microorganisms. Moreover, biochemical transformations brought about by wetlands' microbial groups are explained. In addition, the following chapter emphasizes the degree of complexity in microbial interactions and draws attention to how alterations to these interactions ultimately impact ecosystems' health status. Furthermore, the role of wetland microorganisms in processes, such as detoxification, bioremediation, methanogenesis, carbon sequestration, nutrient cycling and transformations, and primary production is articulated.

Keywords: Biodiversity beneficial microorganisms, Ecosystem services, Mangrove restoration, Microbial processes, Wetland microbiology, Wetland ecology.

INTRODUCTION

Comprehensively, flooded and submerged habitats have tremendous ecologic and economic value, including but not limited to flood storage, drought prevention,

* **Corresponding author Kamyar Amirhosseini:** Department of Soil Science, Faculty of Agriculture, College of Agriculture and Natural Resources, University of Tehran, Tehran, Iran; E-mail: amirhosseini.k@ut.ac.ir

**Govindaraj Kamalam Dinesh, Shiv Prasad, Ramesh Poornima, Sangilidurai Karthika, Murugaiyan Sinduja &
Velusamy Sathya (Eds.)**

water purification, biodiversity preservation, climate change mitigation, nutrient accumulation, and fishery [1]. In particular, mangroves and wetlands are biologically-diverse and highly productive hotspots that offer a wide array of vital ecosystem services. Indeed, diverse biological communities that commonly thrive in estuaries play critical roles in realizing these services [2]. Despite their ecologic and economic importance, mangroves and wetlands face rapid degradation, and their overall decline approximates 50%. Natural environmental changes, anthropogenic activities, and the synergistic combination of the two threaten these ecosystems globally and are regarded as the primary agents driving ecological deterioration and wetland degradation [3]. Accordingly, studying the biological systems that are integral to such ecosystems proves helpful in implementing effective restoration programs that maintain or enhance mangroves' and wetlands' health and sustainability.

Characteristics of mangroves and wetlands have been adequately described in limnological sciences. Based on physiochemical limnology studies, such ecosystems are characterized by low oxygen levels with oxygen in-flow and out-flow controlled by molecular diffusion [4]. Soils of wetlands are saturated with water and possess low aeration porosity. The diffusion process is 10,000 times slower in an aqueous medium than in gas-filled pores [4]. Consequently, anaerobic condition prevails in the wetland ecosystem. This condition has essential consequences on oxidation-reduction reactions.

Furthermore, all forms of life comprising bacteria, fungi, cyanobacteria, microalgae, macroalgae, and fungus-like protists have been reported in these ecosystems [5]. In particular, microorganisms contribute to organic matter turnover by generating detritus by degrading mangrove organic residues of plant and animal origins [6]. A slow rate of organic matter oxidation commonly accompanies the anaerobic condition of the wetland environment. Therefore, such ecosystems are typically rich in organic matter [4]. Although some wetlands and mangroves may suffer nutrient deficiencies [7], nutrient regeneration from decomposing mangrove detritus and nutrient transformations by microbial activity establish an efficient nutrient recycling system in these habitats [8].

Bogland microbiomes and their processes are integral to ecosystem sustainability and restoration. As the basis of wetland ecosystems, the soil is a crucial indicator of the changes in wetland ecological characteristics. In wetlands research, especially mangrove restoration, soil physicochemical properties and soil enzymes are key soil attributes that are often regularly altered with the number of wetland restoration years [9]. In such ecosystems, free-living bacteria, fungi, and yeasts have been reported to play a significant role in the formation of detritus [10]. Bacteria perform various activities such as photosynthesis, nitrogen fixation, methanogenesis, carbon sequestration, and regulation of nutrient and energy flow [11]. Fungi species synthesize a wide array of enzymes catalyzing various biochemical processes [12]. Despite the wealth of systematic information, little knowledge of their role in bioremediation, restoration, and sustainability exists. Accordingly, ecological restoration of degraded wetlands is necessary for achieving ecological balance and food security. Despite numerous studies on biogeography, botany, zoology, ichthyology, environmental pollution, and the economic impact of mangroves, little is known about the diversity and activities of microbes in mangroves and wetlands [8].

The present chapter presents a comprehensive assessment of the microbial biodiversity of submerged habitats. Additionally, the following chapter explains the principal mechanisms involved in the nutrient-recycling system of microbiomes. It aims to highlight the critical roles of various microbes and microbiomes in ecosystem restoration and sustainability of mangroves and wetlands.

MANGROVE AND WETLAND MICROBIOMES

Mangroves and wetlands encompass unique physicochemical properties whose intertwining culminates in an ecology of considerable complexity. As remarked earlier, mangroves and wetlands are characterized by specific properties, including low oxygen levels, high organic matter content, and salinity, among which anoxic conditions and accumulation of carbon compounds appear to be the most influential characteristics impacting microbiomes (Table **1**). With undeniable importance as intermediary ecosystems between terrestrial and marine environments, mangroves and wetlands are hospitable hosts to many biological species ranging from unicellular microorganisms to macroscopic members of the Plantae kingdom. The scrupulous interplay among the living and non-living components of mangroves and wetlands is the key determinant of such ecosystems' health and sustainability. Table **1** outlines some of the most critical properties of wetland ecosystems and their impact on microbial communities.

Table 1. Critical properties of wetland ecosystems and their impact on the microbiomes.

Properties	Impact on Microbiome
pH	Alters microbial community composition and activity.
Temperature	Regulates bacterial community diversity and function.
Geographic gradients	Influences global distribution of microbes across habitats.
Soil Particle Size	Impacts soil bacterial community structure in wetlands.
Redox Potential	Regulates the abundance and species diversity of microbial communities along with their biochemical functions.

(Table 1) cont.....

Properties	Impact on Microbiome
Nutrients	Affects microbial growth and activity.
Salinity	Impacts ecosystem structure and function as well as abiotic and biotic processes.
Organic Matter	Influences bacterial community and function *via* nutrient cycles and stimulates extracellular enzyme activity.
Pollution Status	Microorganisms increase the remediation process efficiency. They are able to remove pollutants (*e.g.,* heavy metals) by biosorption, absorption, and/or biodegradation.

The water layer, sediments, submerged rocks, and plant parts provide a favorable medium with suitable substrates and colonizing surfaces for microbial groups to thrive. Moreover, mangrove sediments consist of a thin, overlying aerobic layer extending to layers beneath that are mainly anaerobic [13]. Bodelier and Dedysh (2013) pointed out that the coupled existence of oxic and anoxic conditions stimulate aerobic and anaerobic microbial group activity [14]. In addition, some studies speculate that the heterogeneity in microsite variations (*i.e.,* zonation) in a mangrove or wetland ecosystem further increases overall species richness. More specifically, it is stated that different compartments in a wetland ecosystem, having various physicochemical properties, enrich communities with distinct compositions. For example, in a study on the mangrove tree (*Rhizophora stylosa*) microbiome, different microbial groups colonized various soil and plant compartments: the rhizosphere, leaves endosphere, stem, and roots. In this study, bacteria were primarily detected in the rhizosphere, while fungal groups dominated soil and plant compartments and wildly proliferated in the leaf endosphere [15]. Similarly, Zhuang *et al.* (2020) observed niche differentiation across four root compartments—non-rhizosphere, rhizosphere, episphere, and endosphere—in natural mangroves of Guangdong, China [16]. The factors discussed work hand-in-hand to make mangrove and wetland ecosystems invaluable biodiversity hotspots.

Fundamental components of mangroves and wetlands microbiomes include archaea, bacteria, fungi, and algae. Archaea take part in essential aspects of the carbon cycle in wetlands. Archaea also contribute to the links between wetlands and global warming through methane emissions during methanogenesis. Bacteria and fungi comprise a large portion of the microbial biomass—up to 91% in tropical mangroves [17]. In this regard, the bacterial population in mangroves is many times greater than that of fungi [18]. Versatile oxygen requirements and the relative ability of bacteria to better withstand alkaline pH than fungi can explain the staggering bacterial population in alkaline wetlands of arid regions. Notably, fluctuations in properties, such as pH possibly alter the balance between microbial populations. For example, fungal species may become the dominant microbial group under acidic conditions.

While bacteria are involved in decomposition and other biogeochemical processes (*e.g.,* nutrient cycling), fungal activity primarily focuses on decomposition. Thatoi *et al.* (2013) introduced bacteria as primary decomposers of mangrove ecosystems that use up dissolved organic compounds and soluble inorganic substances [5]. Other studies suggested that fungi commonly colonize fallen debris before bacteria and are the first microbial group initiating detritus decomposition [19, 20]. In all cases, it is critical to note that while the diverse metabolism of bacteria allows them to harness energy from a wide range of substrates, the robust hydrolytic enzyme systems of fungal species entitle them to be the sole decomposers of hard-to-digest compounds such as lignin. Algae and photosynthetic bacteria require special attention as primary producers of mangrove ecosystems [5]. These primary producers power the entire mangrove and wetland food chain by harnessing light energy and converting it into usable chemical energy.

Apart from specialized processes specified to individual microorganisms, the interconnected nature of biological activities and the interactions among microbial groups ultimately decide ecosystem health. In other words, delicate interactions between microorganisms wove the fabric of a sustainable mangrove or wetland ecosystem. Such interactions are especially prominent in the mangrove rhizosphere microbiome. According to Purahong *et al.* (2018), plant-associated microorganisms are critical in plants' health and productivity [15]. For example, nutrient exchanges between plants' roots and growth-promoting rhizobacteria (PGPR) enhance plant nutrition and increase productivity—a concept commonly called phytostimulation.

Furthermore, PGPR can defend plants against pathogens and parasites (or biocontrol) and strengthen plants' resistance to abiotic stresses. From another perspective, biochemical transformations of elements like carbon (C), nitrogen (N), phosphorus (P), sulfur (S), iron (Fe), and manganese (Mn) in boglands are closely regulated by microbial activity. All of the elements mentioned are essential nutrients for plant growth, and their bioavailability is crucial for ecosystem sustainability and the successful establishment of vegetation in restoration programs. Studying mangrove and wetland microbiomes will shine a light on key biogeochemical processes that make mangrove ecosystems one of the most productive ecosystems on the planet. The necessity of mangrove and wetland conservation amid a changing climate further emphasizes the critical significance of wetland microbiological studies that hold promising potential for ecosystem restoration. As brilliantly put by the world-renewed chemist Louis Pasteur, "The role of the infinitely small in nature is infinitely large". The following sub-sections elucidate the most important microbial groups of mangroves and wetlands.

Archaea in Mangroves and Wetlands

Members of the archaea domain are commonly referred to as ancient forms of life that are especially well-suited to living under extreme conditions. Archaea represent an intermediate phylogeny between eubacteria and eukaryotes. In specific terms, archaea have some genetic similarities to eukaryotes while sharing properties with eubacteria. For example, like eukarya, archaea have histone protein frameworks in their genetic material. On the other hand, biochemical compounds, namely pseudo peptidoglycan resembling bacteria cell walls in structure and function, are also found in archaeal walls.

Archaea better withstand harsh environments than other microbial groups. Consequently, archaea have been successfully isolated from wetlands under various conditions. Cavicchioli (2006) reported the presence of archaea in permafrost and tundra wetlands [21]. Zhou *et al.* (2014) also isolated a methanogenic Archaean from the frozen soils of the Madoi wetland [22]. Another study points to reports documenting thermotolerant archaea living in a mangrove forest's sediments [5]. Conclusively, archaea are present in almost all waterlogged ecosystems regardless of the prevailing conditions. Accordingly, some pioneering studies stated that the archaeal population in mangroves ranges from 3.6×10^2 to 1.1×10^5 cfu.g^{-1} depending on various external factors [23].

Methanogens are the most abundant archaea dwelling in the saturated soils of mangroves and wetlands. These microorganisms are obligate anaerobes that reduce carbon dioxide (CO_2) to methane (CH_4) in a process known as methanogenesis (Eqn. 1). Under the anoxic condition of wetlands, compounds other than molecular oxygen (O_2) function as electron acceptors. Carbon dioxide plays this role once redox potentials fall to -240 mV and below. Hence, redox potential is an essential factor in controlling methanogens activity. The role of sulfate-reducing bacteria (SRB) in limiting the archaeal population in wetlands is a well-documented scientific fact [24]. This limitation is because SRB competes with hydrogenotrophic methanogens over H_2 to be utilized by the microorganisms as electron donors. However, SRB uses up available H_2 with higher efficiency than methanogens. Once H_2 levels are lowered below a threshold value, methanogen activity is inhibited [25]. Sulfate (SO_4^{2-}) is an electron acceptor under a more positive redox potential than CO_2. Therefore, under non-limiting sulfate concentration, H_2 is entirely used up by SRB [25]. Examples of archaeal genera in wetlands include *Methanobacterium*, *Methanosarcina*, *Methanosaeta* [26], *Methanopyrus*, *Methanothermococcus* [27], and *Methanococcoides* [28].

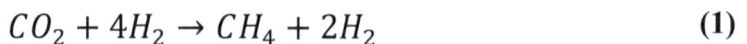

$$CO_2 + 4H_2 \rightarrow CH_4 + 2H_2 \qquad\qquad (1)$$

While methanogens appear to be the most important members of the archaea domain, other archaea detected in the waterlogged habitat also play critical roles in the overall ecosystem sustainability. For example, ammonia-oxidizing archaea (AOA) are integral to the N cycle in mangroves and wetlands. Recent studies emphasize the considerable capability of AOA to contribute to soil fertility [29]. These microorganisms are responsible for the initial stage of nitrification, converting ammonia (NH_4) into nitrite (NO_2^-). In a recent systematic review, Lai *et al.* (2022) highlighted the dominance of AOA under acidic conditions [30].

Bacteria in Mangroves and Wetlands

Bacteria comprise a significant portion of the microbial biomass in mangroves and wetlands. Researchers refer to bacteria as the second most important floral group in mangrove ecosystems, with the largest biomass and highest productivity next to mangrove trees [18]. A staggering population of varying bacterial species takes part in different processes in the wetland ecosystem. Here, we have categorized these integral components of mangroves and wetlands microbiomes according to their niche.

Sulfur-related Bacteria

Sulfur-related bacteria take part in highly complex reactions and interactions in submerged ecosystems. Considering that S-related bacteria influence—either directly or indirectly—cycles of elements such as S, N, P, C, and even Fe transformations, one may speculate that these microorganisms are the most critical microbiomes of mangroves and wetlands. Bacteria that regulate S transformations in wetland ecosystems possess varying assimilation routes and versatile O2 requirements. Sulfur-related bacteria encompass chemolithotrophs, chemoorganotrophs, aerobes, and anaerobic bacterial species. Photolithotrophs responsible for anaerobic S oxidation are discussed in a separate section.

Within the S-related bacteria population, the activity of one microbial group supplies the necessary substrate for the other group's activity. For instance, S-oxidizing bacteria (SOB) convert sulfides into sulfates that are consequently utilized by sulfate-reducing bacteria (SRB) as electron acceptors under anaerobic conditions (Eqn. 2). SRB, in return, reduce sulfates and recycle sulfides for use by SOB and anoxygenic photolithotrophs (Eqn. 3) [31]. SOB are generally chemolithotrophic microorganisms that mainly use reduced inorganic S compounds (hydrogen sulfide, thiosulfates, sulfites, elemental sulfur) and, in some cases, organic S compounds (*e.g.,* methanethiol, dimethyl sulfide, dimethyl disulfide) as electron donors [32]. By oxidizing S-containing compounds, SOB fixes CO_2 into their structural components, often using O_2 or, less commonly, oxidized forms of N (such as nitrate and nitrite) as electron acceptors [25].

Versatile metabolic behavior toward electron acceptors may result in niche differentiation among SOB, with those able to utilize O_2 occupying the aerobic surface layers and those using nitrate or nitrite being more active in the deeper anaerobic sediment layers. Observations of free-living and symbiotic SOB in mangrove and wetland ecosystems are reported in the literature [33]. Multiple S-oxidizing species belonging to Micrococcus, Bacillus, Pseudomonas, and Klebsiella were isolated from mangrove soil in Odisha, India [34].

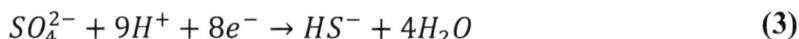

$$H_2S + 2O_2 \rightarrow SO_4^{2-} + 2H^+ \tag{2}$$

$$SO_4^{2-} + 9H^+ + 8e^- \rightarrow HS^- + 4H_2O \tag{3}$$

SRB is responsible for dissimilatory sulfate reduction under anaerobic conditions. Even though they are named according to their phenotypic feature of reducing SO_4^{2-}, SRB are also able to use a range of other inorganic S compounds (thiosulphate, sulfite, and sulfur) and organic compounds (*e.g.,* fumarate) as electron acceptors [35]. Although SRB is nutritionally versatile [8] and can oxidize a wide range of simple organic (chemoorganotrophy) and inorganic compounds (chemolithotrophy), they are unable to utilize complex polymeric organic compounds (*e.g.,* starch, cellulose, proteins, and nucleic acids). They depend on other microbial groups that break down these substrates [35]. Compounds typically used by SRB as substrates include, among others, simple carbohydrates, organic acids, alcohols, and aliphatic and aromatic hydrocarbons. In the chemolithotrophy nutrition mode, SRB uses molecular hydrogen and, thus, competes with hydrogenotrophic methanogens due to the overlapping of niches [36]. Notably, despite the popular representation of SRB as obligate anaerobes in the literature, specific isolates (*e.g., Desulfovibrio oxyclinae*) exhibit oxygen-dependent growth [37].

Organic matter degradation in anaerobic regions of mangrove or wetland ecosystems is primarily carried out *via* the sulfate-reducing action of SRB [13]. Accordingly, SRB's share of total organic matter decomposition in submerged ecosystem sediments ranges from 53% to 90% [8]. The rate of organic matter decomposition by SRB is affected by several factors, including, but not limited to, organic substrate availability, sediment wetting and mixing, sulfate availability, and redox potential [8]. SRB species isolated from Goa's mangrove were identified as *Desulfovibrio desulfuricans, Desulfovibrio desulfuricans aestuarii, Desulfovibrio salexigens, Desulfovibrio sapovorans, Desulfotomaculum orientis, Desulfotomaculum acetoxidans, Desulfosarcina variabilis, and Desulfococcus multivorans* [38].

Nitrogen-related Bacteria

The ability to fix N—converting the biologically unavailable N_2 into more valuable forms like ammonia and nitrate (Eqn. 4)—is exclusive to a particular group of prokaryotes called diazotrophs. Nitrogenase enzyme is a prerequisite for biological nitrogen fixation (BNF). This enzyme cleaves the strong triple bonds between N atoms in molecular nitrogen, exposing the compound to biochemical transformations. BNF is an energy-demanding process. In salt marshes and mangroves with sufficient energy sources (*i.e.,* environments with high organic matter accumulation), BNF rates reach up to 300 mg N. m^{-2} day^{-1} [8]. BNF is a significant source of bioavailable N in mangroves, contributing to 40%-60% of the ecosystem's total N requirement [39]. BNF is suggested to be the second most important microbial activity in mangroves following organic matter decomposition by SRB [8].

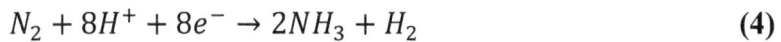

$$N_2 + 8H^+ + 8e^- \rightarrow 2NH_3 + H_2 \qquad (4)$$

Both free-living and symbiotic diazotrophs participate in BNF in mangroves and wetlands; however, their respective contributions to the total N fixed are not precisely known [8]. Difficulty in determining the exact shares of free-living and symbiotic diazotrophs in total N supply may lie in the fact that multiple and different micro- and macroorganisms benefit from BNF products. For instance, a study proposes that free-living N fixation is the primary source of BNF in systems [39] and provides an example in which N fixation by free-living diazotrophs in mangrove ecosystems is stated to rise up to 2-10 mg N. m^{-2} day^{-1} [40]. However, it should be noted that reporting overall estimates pertaining to 'entire' ecosystems might overshadow the importance of different BNF routes in more minor system components. For example, symbiotic diazotrophs fix more N per gram of cellular material than their free-living counterparts [41].

Moreover, one may infer that the survival of endophytic bacteria protected by a sustainable mutualistic symbiosis in the nodules faces less turbulence than the free-living diazotrophs dwelling in the dynamic soil environment. In addition, symbiotic diazotrophs are provided with a relatively constant supply of simple organic compounds from their hosts that may not always be readily available outside the nodules. BNF in mangrove sediments is susceptible to limitations due to insufficient energy sources [5]. Host plants also directly benefit from the bioavailable N in such symbioses [42]. Ultimately, with a fair degree of confidence, it may be speculated that symbiotic diazotrophs play more significant roles in providing mangrove trees with plant-available N forms than free-living N fixers. However, more research is required to conclusively delineate the contribution of each microbial group to N supply in mangrove and wetland ecosystems.

BNF is a process controlled by the genetic characteristics of the microbiome as well as external conditions. To name a few, the concentration of soluble N_2 in mangrove water, carbon sources availability, light intensity, temperature fluctuations, oxygen level, bacteria abundance, mangrove community composition, pH, rooting characteristics of host species, and microbe-microbe interactions are some of the factors affecting N fixation by free-living and symbiotic diazotrophs [5, 8, 39]. In general, diazotrophs of genera *Azospirillum*, *Azotobacter*, *Rhizobium*, *Clostridium*, and *Klebsiella* were observed in the sediments, rhizosphere, and rhizoplane of mangrove plants [5]. Some examples of N-fixing symbiotic bacteria genera isolated from mangrove tree roots include *Rhizobium*, *Sinorhizobium*, *Bradyrhizobium*, and *Frankia* [39]. Certain SRB, such as *Desulfovibrio* and *Desulfobulbus*, are regarded as the main free-living diazotrophs in the rhizosphere of mangrove trees [43].

Denitrifiers are another group of N-related bacteria that perform denitrification. In denitrification, nitrate is utilized as an electron acceptor under anoxic conditions and ultimately transformed to gaseous N forms (NO, N_2O, N_2) (Eqn. 5). Acidic wetlands are gaining attention as essential sources of N_2O [44] with estimations attributing 0.6% of global annual N_2O emissions to these ecosystems [45]. Key wetland denitrifiers include *Azospirillum*, *Bradyrhizobium*, *Mesorhizobium*, *Achromobacter*, *Ralstonia*, and *Pseudomonas* [46]. Ammonia-oxidizing bacteria (AOB) of genera *Vibrio* and *Methylophaga* are also essential as microbial components contributing to N transformations in mangroves [47].

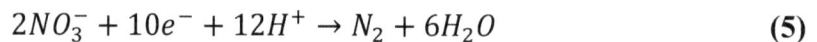

$$2NO_3^- + 10e^- + 12H^+ \rightarrow N_2 + 6H_2O \qquad (5)$$

Phosphate-solubilizing Bacteria

Phosphorus is an essential nutrient deemed vital for all forms of life. In higher plants, P is involved in the structural integrity of tissues and contributes to energy storage and energy transfer reactions [48]. Phosphorus is a critical component of essential biomolecules such as ATP, phospholipids, and nucleic acids. It is taken up as inorganic anions (HPO_3^{-2}, $H_2PO_3^-$) from the soil solution. A combination of the intrinsic properties of P and the inherent physicochemical properties of the ecosystem it is released into determines P availability for micro- and microorganisms.

The optimum pH for the highest bioavailability of P is between 6 and 7. In acidic and alkaline conditions, P forms insoluble complexes with cations such as Fe and Ca, respectively. Furthermore, a high degree of interaction between phosphate anions and clay particles enhances the adsorption and fixation of P by soil particles. For these reasons, muddy mangrove soils are known to have a high capacity for P adsorption [49]. Similarly, silty-clays soils of mangroves in China are shown to be poor in bioavailable P [50]. Processes controlling P transformations in wetlands are multifaceted and heavily regulated by various microbial interactions [51].

Although sulfide production by S-related bacteria promotes P dissolution under anoxic conditions of wetlands, phosphate-solubilizing bacteria (PSB) have become increasingly crucial in ecosystem regions where conditions are not always wholly anoxic [8]. PSB contribute to P bioavailability in several ways. One mechanism employed to improve the solubility of P-containing compounds involves the production of organic acids by PSB that acidifies the soil medium, releasing P from insoluble complexes. PSB can also secrete extracellular phosphate-solubilizing enzymes that dissociate phosphates from insoluble compounds, rendering them biologically available to other (micro)organisms. Enzymes secreted by PSB include alkaline phosphatases (EC 3.1.3.1) and acid phosphatases (EC 3.1.3.2) [51]. Phosphatases cleave lipid-phosphate bonds, accelerating hydrolytic reactions that increase P bioavailability [52].

It should be kept in mind that bacteria are not the only microbial group taking part in the P cycle in mangrove soils. For example, Euryarchaeota is a methanogen-related archaeon with some members having integral roles in the P cycle of mangrove sediments [53]. Fungal groups also contribute to P solubilization [51]. However, given that the bacterial population often outnumbers that of fungal groups, the role of PSB in increasing bioavailable P concentration has attracted more attention. Aerobic and anaerobic PSB are reported in the rhizosphere of wetlands [51]. Endophytic PSB has also been isolated from leaf samples of mangrove plants [54]. Common PSB sp. observed in wetland sediments, like *Achromobacter, Agrobacterium, Bacillus, Burkholderia, Micrococcus, Pseudomonas, and Rhizobium* [51]. The phosphate-solubilizing capability of bacteria was initially studied by halo formation around colonies growing in mediums supplemented with an insoluble inorganic P source. A study remarks that the amount of P solubilization by a PSB (*Bacillus amyloliquefaciens*) under controlled conditions was shown to theoretically fulfill the daily P requirement of a small terrestrial plant [8].

Photosynthetic Bacteria

Photosynthetic bacteria are a frequently detected ecological functional group in the mangrove microbiome [15]. The process of photosynthesis in wetlands—reducing CO_2 to high-energy compounds using light energy (Eqn. 6)—takes two routes. Namely, oxygenic photosynthesis and anoxygenic photosynthesis. As suggested by their names, the key difference between these processes is whether oxygen is produced as a product of photosynthesis. During oxygenic photosynthesis, water serves as the principal electron donor for the reduction of CO_2 and ultimately breaks down to liberate oxygen. In anoxygenic photosynthesis, reduced compounds other than water serve as electron donors. Thus, oxygen is not liberated during this type of photosynthesis. While oxygenic photosynthesis is a shared ability among prokaryotes, plants, and algae, anoxygenic photosynthesis is unique to certain bacteria. Cyanobacteria carry out oxygenic photosynthesis in mangroves. Bacterial groups able to perform anoxygenic photosynthesis in wetland ecosystems include green sulfur bacteria, purple sulfur bacteria, green non-sulfur bacteria, and purple non-sulfur bacteria. Anoxyphotobacteria are considered predominant photosynthetic organisms in anaerobic environments.

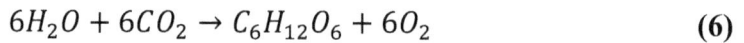

$$6H_2O + 6CO_2 \rightarrow C_6H_{12}O_6 + 6O_2 \tag{6}$$

Cyanobacteria are commonly found in the water column of wetlands [51]. *Lyngbya* and *Anacystis* are examples of cyanobacteria observed in mangroves [18]. These bacteria are aerobes; therefore, their activity is limited to the ecosystem's surface layers or oxic microsites. Cyanobacteria carry out oxygenic photosynthesis with various photosynthetic pigments, among which chlorophyll and phycobilins stand out. Phycobilins comprise phycocyanin and phycoerythrin that absorb light of different wavelengths. Phycocyanin is responsible for the blue-green coloration of cyanobacteria. Some cyanobacteria can fix N; thus, they can also be considered free-living diazotrophs [39].

In green and purple sulfur bacteria, sulfides (or other inorganic S compounds) are used as electron donors [8]. Consequently, sulfur bacteria produce S due to anoxygenic photosynthesis (Eqn. 7). Purple and green sulfur bacteria have bacteriochlorophyll a and b and bacteriochlorophyll c and d, respectively. Representatives of these bacteria in wetlands belong to the families Chromatiaceae (Purple S bacteria) and Chlorobiaceae (Green S bacteria). Anaerobic conditions in mangrove systems with high S content provide a suitable environment for these bacteria to thrive [8].

$$2H_2S + CO_2 + Light \rightarrow 2S^0 + CH_2O + H_2O \tag{7}$$

Green and purple non-sulfur bacteria do not rely on S to perform photosynthesis. Instead, they exhibit a remarkable ability to utilize a vast range of carbon compounds, such as malate, butyrate, and acetate, as electron donors [5]. Predominant non-sulfur bacteria in wetland ecosystems belong to *Chloroflexaceae* (green non-sulfur bacteria) and *Rhodospirillaceae* (purple non-sulfur bacteria) families. Certain anoxyphotobacteria are also diazotrophs [8]. For example, *Rhodospirillaceae* bacteria were responsible for a significant portion of N_2 fixation in a mangrove ecosystem [55]. Anoxyphotobacteria belonging to genera *Rhodopseudomonas* and *Rhodobacter* are commonly detected in the rhizosphere and root surface of mangrove trees [56].

Fungi in Mangroves and Wetlands

Fungi are vital components of the mangrove microbiome, maintaining essential functions such as decomposition and mineralization [30]. Fungal communities also benefit plants by aiding with nutrient acquisition [57]. A study provides evidence of fungal communities supporting plants under fluctuating soil moisture conditions [58]. Wetlands are hosts to diversified fungal communities [59], and wetland biogeography is considered an essential factor responsible for this diversity [51]. Accordingly, wetland biogeographic pattern strongly drives fungal community composition [58]. With 74000 species of fungi identified from various wetlands of the world [60], these microorganisms are found in all compartments of the mangrove or wetland ecosystem, including above- and below-ground parts, soil, rhizosphere, roots, phyllosphere, endosphere, and spermosphere.

Most fungal genera in mangroves are aerobes adapted to the prevailing moisture conditions [58]. However, the anoxic condition of mangrove and wetland ecosystems inevitably lowers fungal activity compared to terrestrial environments. Fungi have robust enzyme systems enabling them to decompose mangrove detritus effectively. For example, ligninolytic, cellulolytic, and amylolytic fungi are abundant in mangrove root environments [5]. Additionally, fungi isolated from dead leaves in a mangrove ecosystem exhibited pectinolytic and proteolytic activity [61].

It has been argued that external conditions, such as pH and C:N ratio, can alter fungal to bacterial dominance in mangroves [30]. In addition, fungi may be the predominant microbial group in specific ecosystem compartments. For example, while both bacteria and fungi were present in the rhizosphere of mangrove trees (*R. stylosa*), fungal classes belonging to Ascomycota (Dothideomycetes, Eurotiomycetes, and Sordariomycetes) dominated the endosphere microbiome [15]. Another study highlighted a moderate to high abundance of fungi in the root and rhizosphere of *Camassia* communities of the pacific northwest wetlands [58]. Moreover, this study observed Ascomycota, Basidiomycota, and Zygomycota associated with plant seeds in the wetland ecosystem [58].

Arbuscular mycorrhizal (AM) fungi forming symbiotic relationships with plant hosts are also widely reported in wetlands. Some common AM fungi identified in wetlands include *Acaulospora*, *Glomus*, and *Archaeospora* [62]. Despite saline and anaerobic conditions, AM fungi survive in wetland ecosystems. Two possible survival mechanisms have been proposed: a) indigenous AM fungi have developed adaptations to saline conditions and low oxygen levels, and b) aerating roots of mangrove plants supply the high O2 requirements of AM fungi, supporting their growth and activity [63]. Lichens—consisting of a photobiont (algae, sometimes accompanied by cyanobacteria) and a mycobiont—offer another example of symbioses in mangroves and wetlands. Lichens usually occupy the unsubmerged upper parts of mangrove trees [5]. Phytopathogenic fungal species, such as those belonging to the genera Fusarium and Leptosphaeria, have also been documented in wetlands [64].

Algae in Mangroves and Wetlands

Microalgae form another systemic group in wetlands that play vital roles in the ecosystem's physical, chemical, and biological processes [51]. They impact the survival of other organisms in the aquatic environment, regulate oxygen levels, and contribute to ecosystem maintenance [65]. Microalgal groups also reflect the average ecological condition of their habitat, and therefore, they may be relied upon as indicators of environmental quality [66]. Although the algal population in a tropical mangrove may only account for about 7% of the total microbial biomass [17], they serve as primary producers contributing to nutrient cycling and the wetland food web [51].

Algal species in a mangrove ecosystem may be differentiated as planktonic, benthic, and periphytic [5]. Favorable habitats in mangroves that promote the establishment and proliferation of algae include mangrove tree roots, rigid substrates, and soft mud [18]. Environmental factors such as biological and chemical oxygen demand, pH, temperature, electrical conductivity, and nutrient concentration influence algal and phytoplanktonic species composition and distribution [67]. Given the photosynthesizing capability of algae, other factors, including light intensity and shading, are also expected to impact their activity.

Analyses of 146 wetlands in Brazil identified 107 genera of algae belonging to Cyanophyta, Heterokontophyta, Dinophyta, Euglenophyta, and Chlorophyta [68]. Assessment of phytoplankton diversity in the Gandoman wetland in the west of Iran revealed 95 species that belonged to multiple divisions. This study identified phytoplankton taxa, including Bacillariophyta, Chlorophyta, Chrysophyta, and Dinophyta [67]. Another study confirmed the dominance of Bacillariophyta over other phytoplanktonic taxa in a wetland ecosystem [65]. Periphytic communities are primarily composed of algal species. However, they form a heterogenous microhabitat in which multiple life forms belonging to different domains may symbiotically exist. Hence, the periphyton's complex nature—as an independent yet integral microhabitat within the wetland ecosystem—is discussed in a separate section.

Periphyton in Mangroves and Wetlands

In submerged habitats, it is favorable for microbes to inhabit matrix-enclosed biofilms. However, coupling these biofilms' structure and dynamics to ecosystem function has yet to be addressed for the most part [69]. Periphyton biofilms are considered one of the most precious structural and functional indicators of ecosystem health status in waterlogged ecosystems [70, 71]. This attached microalgae-bacteria consortium consists of a complex mucopolysaccharide matrix, a variety of phototrophs, heterotrophs, and chemoautotrophs, as well as organic and inorganic substances. Periphytic microhabitats are distributed between sediments and the overlying water [72]. Periphytic biofilm is recognized as an integral hotspot for biogeochemical processes in the wetland ecosystem [69]. Critical variables that may affect the growth and distribution of the periphytic biofilm community include many biotic and abiotic factors. The alteration of these variables has been identified as one of the leading causes of changes in this microbiome's taxonomic composition and functioning.

Abiotic factors can be further categorized into hydrological (*e.g.,* the intensity and regime of water flow), physical (*e.g.,* light, temperature, substrate), and chemical (*e.g.,* the concentration of nutrients and pH) factors [73]. Living components of the periphyton are in a dynamic equilibrium with each other and the external environment. Microalgal biofilm species are attached to the substrate's surface by mucilage production and form aggregates [74]. Microalgae and bacteria species in biofilms secrete a mixture of polymers collectively called extracellular polymeric substances (EPS). Such polymers are woven to form an extensive net-like structure [75]. In an established biofilm, the consortium assimilates sufficient light and nutrient resources for growth during the growth stage. During the maturation stage, periphyton participates in various nutrient transformation processes such as ammonification, nitrification, and denitrification. Biomass disentanglement occurs when a maximum biomass-carrying capacity is reached in the stationary phase. Hence, deteriorating biofilm is replaced by a newly synthesized biomass at the fading stage [76].

Periphyton contains rich and diverse communities. Regarding taxonomic composition, algal groups include Diatomea, Chlorophyta, Charophyta, Chrysophyceae, Porphyndiophyceae, Eustigmatales, and the most abundant prokaryotic communities in periphytic biofilms consist of *Cyanobacteria, Bacteroidetes, Proteobacteria, SBR 1093, Actinobacteria, Planctomycetes,* and *Acidobacteria* [77]. Due to its dynamicity and high adaptability, multiple factors such as substrate type and nutrient status may alter periphytic community composition. It should be noted that one of the most common classifications of this microbiome has been rooted in substrate type. In this classification, periphytic biofilm is divided into two main groups: herpobenthos and haptobenthos. Herpobenthos includes metaphyton, epipelon, endopsammon, and endopelon. Haptobenthos classification is more extensive than herpobenthos and consists of epilithon, epixylon, epipsammon, epizoon, epiphyton, endolithon, endozoon, and endophyton.

Periphytic biofilms contribute to vital processes as primary producers and an energy source for higher trophic levels. In addition, this microbiome increases transient hydrodynamic storage, enhances the uptake of organic molecules with low bioavailability, and acts as an essential short-term sink for nutrients like nitrogen and phosphorus [77]. Nevertheless, most research on bogland environments focuses only on two phases, namely the water layer and soil. Therefore, the importance of microbial aggregates as the third phase between these layers has yet to be noticed [78].

Micropores on the surface of periphyton can act as adsorption sites that affect nutrient turnover and transformation. For instance, algal constituents of periphytic biofilms can temporarily remove P from the soluble phase, storing it in the insoluble complexes of Ca, Fe, and Al phosphates, affecting phosphorus cycling in the wetland ecosystem. Meanwhile, the disintegration of algal biomass during senescence releases this P back into the soil solution, rendering it bioavailable for plants and other organisms [79]. Besides, the organic matter released upon algae degradation can stimulate the growth and activity of heterotrophic bacteria [80]. This short-term storage mechanism can significantly reduce the harmful consequences of eutrophication or nutrient run-off events, preventing non-point source pollution. Aside from this vital mechanism and based on the few studies available, periphytic biofilms can fix CO_2 and increase the redox potential of their environment during the growth stage. Moreover, periphytic biofilm decay during the fading phase would diminish redox potential and increase soil organic carbon availability, accelerating methanogenic microbial community growth [81]. Furthermore, periphyton presents an environmentally friendly approach to the bioremediation of contaminated boglands.

Discussions on the principal role of periphytic communities and other microbiomes in wetland ecosystems, along with the contributions of other researchers, deepen the current understanding of the integral function of microorganisms—regarding such mechanisms as nutrient cycle regulation and ecosystem restoration *via* bioremediation capability—in achieving ecosystem sustainability.

Ecological Importance of Microbiomes in Mangrove and Wetland Sustainability

Previous sections characterized microbial niches—including microbial community structure and functional mechanisms—in a mangrove or wetland ecosystem. Under natural conditions, all components of mangrove and wetland microbiomes interact with one another to establish a sustainable climax community. Such a community comprises various biochemical processes closely regulated by microbial species that significantly influence the overall quality of the entire ecosystem. Fig. (**1**) displays a schematic representation of the major processes that occur in a wetland or mangrove ecosystem.

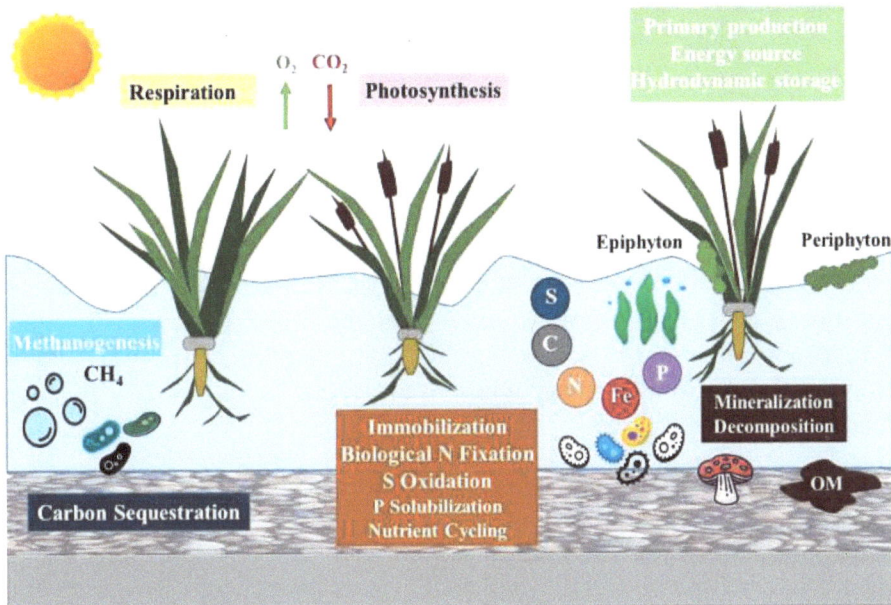

Fig. (1). Key processes occurring in a wetland or mangrove ecosystem.

The interplay among microorganisms in a submerged habitat involves multiple scenarios of commensalism and mutualistic symbioses. However, natural changes and human interference can cause drastic changes in a mangrove or wetland ecosystem. Restoration of a damaged ecosystem again relies on microbial interactions to regain ecosystem health. The following section further details microbiomes' ecological significance in the sustainability of mangroves and wetlands. The role of microbiomes in essential processes such as methanogenesis, carbon sequestration, nutrient transformations, primary production, global warming, and ecosystem restoration are discussed.

Circumventing the Threats to the Ecology of Mangroves and Wetlands

Assessing the negative impacts of natural and anthropogenic events on mangroves and wetlands is outside the scope of the present chapter. Nevertheless, a few examples of human alterations that influence wetlands' microbiological processes are outlined to emphasize the susceptibility and adaptability of such ecosystems. Submerged ecosystems encompass a rich diversity of habitats confronted with various threats [82]. Anthropogenic pressure alters a wetland ecosystem's critical biological, chemical, and physical properties. These alterations may impact microbial activity and limit their contribution to ecosystem processes. For example, a study showed that disturbed mangroves (those subjected to human interference, invasive species, and pollution) were relatively more acidic than well-persevered mangroves that were mainly alkaline. Moreover, more than 88 ppt salinity levels were reported in disturbed mangroves. The highest salinity level in well-preserved mangroves reached 64 ppt [30]. The study further elaborates that the AOB population—an integral component of the N cycle in mangrove ecosystems—decreases with increasing salinity and acidic pH [30]. Salinity and pH are essential drivers of microbial community composition [83, 84] influenced by human interference. It is noteworthy that salinity also negatively impacts mangrove vegetation through osmotic imbalances and specific ion toxicities. Undress such conditions, the role of PGPR in increasing mangrove plants' resistance to salinity stress is paramount. Additionally, pH fluctuations stagnate specific microbial activity while intensifying the rate of others, disrupting the dynamic balance of the ecosystem.

Polluted run-off may degrade wetlands' health, leading to gradual carbon depletion in wetland soils. Discharging agricultural run-offs containing fertilizers into transitional ecosystems such as mangroves promotes eutrophication [85]. Distortion of nutrient dynamics and increasing rate of algae blooms degenerates water quality. Moreover, agricultural fertilizers, especially P fertilizers, often contain considerable heavy metal impurities. Therefore, agricultural run-offs not only cause nutrient imbalances but also result in heavy metal pollution in mangroves and wetlands. Studies infer that the concentration and distribution of heavy metals affect the restoration time of boglands [9, 86]. Some review studies and field experiments indicate that algae, particularly periphyton, can act as bioremediation tools for submerged habitats [87 - 89]. Periphyton was shown to have the capability to use, convert, and modify a variety of other pollutants, such as heavy metals, microplastics, pharmaceutical waste, persistent organic pollutants, and personal care products [90, 91]. Moreover, there is evidence of the hydrocarbon degradation capability of S-related bacteria (SOB and SRB) in oil-contaminated mangroves [35, 92].

Human activities distorting the biogeochemical systems of mangroves and wetlands may exacerbate global warming by increasing wetlands' contribution to greenhouse gas (GHG) emissions. Wetlands contribute to about just under 40% of global CH4 emissions [93]. Human interference increases the CH4-emitting capacity of mangrove soils [94]. SRB are another microbial group playing roles in GHG emissions from mangrove and wetland ecosystems. The sulfate reduction mechanism carried out by these microorganisms accounts for almost the entire CO_2 emissions from mangrove sediments [95]. In-depth studies on microbial populations of wetlands are essential for understanding and monitoring natural processes that cause wetlands to contribute to global warming. Preserved mangroves and wetlands are potential carbon sources that could reverse global warming. This function of mangroves and wetlands is elaborated in the subsequent sub-section.

Carbon Sequestration

Wetlands and mangroves can effectively regulate the exchange of GHGs, such as carbon dioxide, methane, nitrous oxide, and sulfur dioxide, to and from the atmosphere. Submerged ecosystems act as sinks for carbon and nitrogen and are considered sources of methane and sulfur compounds. Alterations to wetland properties may change dynamic gas exchange equilibriums [96]. For example, the anaerobic conditions of submerged ecosystems favor methane and nitrous oxide production. When wetlands are flooded and anaerobic conditions exist, methane is produced.

On the contrary, wetlands may act as methane sinks when dried. Salinity also suppresses methane production in wetlands. In this context, coastal wetlands would have lower methane emission rates than freshwater wetlands. Disruptions in wetland processes also impact their capacity as GHG exchangers. Nitrous oxide production in undisturbed wetlands is lower than in terrestrial soil environments [97]. Portions of the methane generated in submerged ecosystems could be trapped at great depths. Firm and hard soils with hardpans have a reservoir of "entrapped methane". Under the pressure of the upper layers, methane liquifies, which may remain in the same state or be consumed by microbes. Trapped methane can be extracted in liquid form and utilized as fuel.

High capacity for sequestering and storing carbon, wetlands, and mangroves is critical to mitigating climate change. In particular, coastal wetlands are known as "blue carbon" ecosystems [98]. Mangroves contribute to carbon dioxide sequestration by burying biomass in sediments (long-term sink) and the net growth of forest biomass (shorter-term sink). In coastal and estuarine wetlands, salinity and anoxic conditions inhibit the disintegration of organic residue and greatly slow down decomposition processes. In addition, the tidal regime and sediment input from streams and rivers authorize the burial of organic matter. Collectively, these mechanisms result in the accumulation of carbon in wetland ecosystems. Therefore, even though methanogenesis is an integral process in wetlands, carbon storage in submerged ecosystems is proposed as the longest-term climate mitigation solution [98]. These processes (*i.e.,* carbon sequestration and methanogenesis) are in a dynamic equilibrium in undisturbed boglands. Microbiological communities such as bacteria, algae, and fungi sustain the ecosystem's function by balancing the mechanisms mentioned earlier. Similarly, the restoration of wetlands has led to enhanced sequestration or reduced gas emissions. On the contrary, human interference has negatively affected the balance of microbiome functions in a way that undermines the sequestration potential of wetlands [99].

Nutrient Transformations

Specific microbial composition and unique biogeochemical properties of wetlands make these ecosystems highly-active hotspots for nutrient cycling. Nutrient transformations and cycling in mangroves and wetlands give rise to numerous cases of commensalism and symbioses between microorganisms and microbiomes. The sustainability of mangrove ecosystems ultimately relies on such interactions. Microbial interactions in a wetland ecosystem mainly occur as direct exchanges of nutrients and growth factors between microorganisms or by improving one or more quality parameters of the environment by one microbial group for the benefit of the other. An example of the latter is sulfide detoxification by various S-related bacteria in the wetland ecosystem. Hydrogen sulfide produced by SRB is toxic to plants and some microbial groups. Sulfur-oxidizing bacteria are crucial in detoxifying sulfides in mangrove sediments [5], making the environment more hospitable for plants and susceptible microorganisms. A similar detoxification mechanism is observed in green and purple S bacteria activity that uses up H_2S during anoxygenic photosynthesis. Moreover, purple non-sulfur bacteria may alleviate organic acid toxicities in wetlands by consuming these compounds as growth substrates.

Furthermore, microorganisms with oxygenic photosynthesis increase the redox potential by producing O_2. An increase in redox eliminates the necessity for other inorganic electron acceptors and, as a result, prevents Fe and Mn toxicity in the wetland. Oxygen release and ventilation of the ecosystem are essential for plants' roots and other organisms that live in the ecosystem. Waterlogging limits oxygen diffusion into the sediment profile, creating anaerobic conditions and slowing decomposition rates. Subsequently, the pervasiveness of wetland anaerobic conditions significantly impacts their biogeochemistry, with important implications for C, N, P, Fe, Mn, and S transformations. Wetlands can function as sources, sinks, or transformers of these nutrients, depending on inflows, out-flows, and internal cycling rates. Particularly, N and P are the limiting nutrients in the submerged ecosystem [100]. These nutrients have an essential role in the food web's primary production and microbial pathways [101], the deficiency of which threatens mangroves' ecological functioning and increases the fragility of the entire ecosystem [39]. Biological nitrogen fixation by diazotrophs is considered a critical source of N in mangrove ecosystems, contributing to up to 60% of the total N requirement of mangroves [39]. Arbuscular mycorrhizal fungi are another ecologically important microbial group that maintains mangrove ecosystem stability by supplying bioavailable P to plants. These fungi solubilize inorganic P by secreting organic acids and phosphatases. The advantages of AM symbiosis are not limited to P nutrition. AM fungi also provide Zn, Cu, Mg, K, and N [52]. Plants that engage in AM symbiosis absorb more nutrients from the surrounding environment and possess a competitive advantage due to overall fitness. Consequently, AM symbiosis plays a significant role in establishing vegetation and reconstructing severely disturbed ecosystems [52].

Microbial processes rarely occur as stand-alone reactions without influencing or being influenced by other processes. In reality, the outcome of one microbial process may positively or negatively influence other processes or transformations. For example, SRB activity may impact Fe and P availability in mangrove sediments. Sulfides produced by SRB reduce Fe in Fe-P complexes, liberating soluble P [8]. It is noteworthy that high P concentrations in the wetland environment inhibit the formation of mycorrhizal symbiosis. One possible reason for such inhibition could be the reduction of AMF root colonization resulting from suppressing external hyphal growth and spread rates by increasing P concentration in the soil [52]. Recent studies highlighting the interplay between Fe availability and BNF [39] further demonstrate the complexity of microbial processes and interactions in mangrove and wetland ecosystems. Fe is an essential nutrient for nitrogenase biosynthesis by the diazotrophic community [39]. As previously stated, SRB impact Fe availability in the wetland environment. Thus, SRB activity indirectly affects BNF by regulating Fe availability in the wetland ecosystem.

Another vital role of wetland microbiomes in ecosystem sustainability is related to organic matter turnover in the soil. Bacteria and fungi contribute to the disintegration and recycling of organic compounds. Studies have demonstrated the production of several cellulolytic, proteolytic, amylolytic, lipolytic, chitinolytic, and phosphate-solubilizing enzymes that contribute to the biodegradation of organic matter and improving ecosystem productivity *via* nutrient cycling [102]. Additionally, studies have shown that microbial alteration of detritus enhances the nutritional value of organic compounds for subsequent utilization by other groups of micro and macroorganisms [8]. Besides, certain bacterial groups improve the growth of juvenile fish, shrimp, and prawns in the ecosystem by producing secondary metabolites [102].

Primary Production and the Food Chain

As a critical process, primary production displays an ecosystem's health and trophic status. Primary production directly responds to light and nutrient changes in the surrounding environment. Regardless, sustaining the wetland food web depends on natural levels of Primary production [103]. Aquatic plants, algae, and photosynthesizing bacteria convert atmospheric CO_2 to organic C in mangroves and wetlands through photosynthesis. The liberation of organic carbon is also stimulated by the aqueous medium and accumulation of leaves, sticks, barks, and grasses on submerged floors. Plants and animals subsequently absorb the carbon dissolved into the water.

Also, mold and bacteria get nourishment from dissolved carbon and other nutrients released from plant litter, forming biofilms on the decaying residue. These biofilms can act as a source of food for larger organisms. Accordingly, particulate organic compounds may be attached to soil particles, plants, algae, leaf pieces, or seeds. Small wetland animals would consume these particulates. Ultimately, larger animals like fish become a food source for birds, frogs, and other fishes in the ecosystem. In this context, the speed of biofilm formation—as a preliminary step in the transfer of nutrients along the wetland food chain—depends on the trophic status of the ecosystem (*i.e.,* primary production) [104].

Comprehensively, the trophic status of a freshwater or estuarine system could be described as oligotrophic, mesotrophic, and eutrophic, which are defined as low nutrients and productivity, moderate nutrients and productivity, and high nutrients and productivity situations, respectively [105]. Generally, the nutrient levels mentioned earlier are not considered stressors for aquatic communities. This fact illustrates the significant potential of submerged communities for adaptivity and sustainability of the ecosystem by possessing the ability to function equally at high and low nutrient levels.

Most commonly, algae and cyanobacteria are introduced as primary producers in the wetland food chain, harnessing solar energy and transforming inorganic substances into food for themselves and other organisms. Benthic and pelagic organisms utilize the food, and their rapid assimilation of carbon sources emphasizes the crucial role of algal species in the C cycle, food chain, and improving productivity in estuaries [5]. Organic matter impacts the mangrove ecosystem's biological, chemical, and physical properties. Therefore, photosynthesizing micro and macroorganisms that add organic matter is critical to ecosystem sustainability. Additionally, aquatic ecosystems' vegetation can alter soil's physicochemical properties by releasing secretions and O_2 from the rhizosphere and generating plant litter [106, 107].

CONCLUDING REMARKS

The underlining complex microbial interactions and processes in an ecosystem are critical determinants of the ecosystem's health and productivity. The microbial community governs fundamental biogeochemical processes such as primary production, nutrient cycling, nutrient transformations, carbon sequestration, methanogenesis, biological nitrogen fixation, sulfides detoxification, and organic matter turnover. Moreover, the entire food web of the wetland ecosystems depends on microbial processes. In addition, microorganisms actively contribute to plants' health and performance *via* phytostimulation and biocontrol. Therefore, any effort toward preserving or restoring disturbed mangroves and wetlands must address the viability of the ecosystems' microbiomes. Wetlands' microbiomes are dynamic and quickly respond to alterations brought about by natural events and anthropogenic activities. Studying the microbiology of wetlands and mangroves is crucial in understanding how external forces affect critical ecological processes and how these processes can be utilized to provide practical solutions to serious global issues such as climate change and environmental pollution.

REFERENCES

[1] Chen H, Su H, Guo P, *et al.* Effects of planting patterns on heavy metals (Cd, As) in soils following mangrove wetlands restoration. Int J Phytoremediation 2019; 21(8): 725-32.
[http://dx.doi.org/10.1080/15226514.2018.1556587] [PMID: 31037962]

[2] Zhou Y, Zhao B, Peng Y, Chen G. Influence of mangrove reforestation on heavy metal accumulation and speciation in intertidal sediments. Mar Pollut Bull 2010; 60(8): 1319-24.
[http://dx.doi.org/10.1016/j.marpolbul.2010.03.010] [PMID: 20378130]

[3] Nicholls RJ, Cazenave A. Sea-level rise and its impact on coastal zones. Science 2010; 328(5985): 1517-20.
[http://dx.doi.org/10.1126/science.1185782] [PMID: 20558707]

[4] Ponnamperuma FN. The chemistry of submerged soils. Adv Agron 1972; 24: 29-96.
[http://dx.doi.org/10.1016/S0065-2113(08)60633-1]

[5] Thatoi H, Behera BC, Mishra RR, Dutta SK. Biodiversity and biotechnological potential of microorganisms from mangrove ecosystems: a review. Ann Microbiol 2013; 63(1): 1-19.
[http://dx.doi.org/10.1007/s13213-012-0442-7]

[6] Bano N, Nisa MU, Khan N, *et al.* Significance of bacteria in the flux of organic matter in the tidal creeks of the mangrove ecosystem of the Indus River delta, Pakistan. Mar Ecol Prog Ser 1997; 157: 1-12.
[http://dx.doi.org/10.3354/meps157001]

[7] Sengupta A, Chaudhuri S. Ecology of heterotrophic dinitrogen fixation in the rhizosphere of mangrove plant community at the Ganges river estuary in India. Oecologia 1991; 87(4): 560-4.

[http://dx.doi.org/10.1007/BF00320420] [PMID: 28313699]

[8]   Holguin G, Vazquez P, Bashan Y. The role of sediment microorganisms in the productivity, conservation, and rehabilitation of mangrove ecosystems: an overview. Biol Fertil Soils 2001; 33(4): 265-78.
[http://dx.doi.org/10.1007/s003740000319]

[9]   Ye C, Butler OM, Chen C, Liu W, Du M, Zhang Q. Shifts in characteristics of the plant-soil system associated with flooding and revegetation in the riparian zone of Three Gorges Reservoir, China. Geoderma 2020; 361: 114015.
[http://dx.doi.org/10.1016/j.geoderma.2019.114015]

[10]   Maria G, Sridhar K. Richness and diversity of filamentous fungi on woody litter of mangroves along the west coast of India. Curr Sci 2002; •••: 1573-80.

[11]   Das S, Lyla P, Khan SA. Spatial variation of aerobic culturable heterotrophic bacterial population in sediments of the continental slope of western Bay of Bengal. 2007.

[12]   Bremer GB. Lower marine fungi (labyrinthulomycetes) and the decay of mangrove leaf litter. Hydrobiologia 1995; 295(1-3): 89-95.
[http://dx.doi.org/10.1007/BF00029115]

[13]   Sherman RE, Fahey TJ, Howarth RW. Soil-plant interactions in a neotropical mangrove forest: iron, phosphorus and sulfur dynamics. Oecologia 1998; 115(4): 553-63.
[http://dx.doi.org/10.1007/s004420050553] [PMID: 28308276]

[14]   Bodelier PL, Dedysh SN. Microbiology of wetlands. Frontiers Media, SA 2013; p. 79.
[http://dx.doi.org/10.3389/978-2-88919-144-4]

[15]   Purahong W, Sadubsarn D, Tanunchai B, *et al.* First insights into the microbiome of a mangrove tree reveal significant differences in taxonomic and functional composition among plant and soil compartments. Microorganisms 2019; 7(12): 585.
[http://dx.doi.org/10.3390/microorganisms7120585] [PMID: 31756976]

[16]   Zhuang W, Yu X, Hu R, Luo Z, Liu X, Zheng X, et al. Diversity, function and assembly of mangrove root-associated microbial communities at a continuous fine-scale. npj Biofilms and Microbiomes. 2020; 6(1): 52.

[17]   Alongi DM. Bacterial productivity and microbial biomass in tropical mangrove sediments. Microb Ecol 1988; 15(1): 59-79.
[http://dx.doi.org/10.1007/BF02012952] [PMID: 24202863]

[18]   Kathiresan K, Qasim SZ. Biodiversity of mangrove ecosystems. Hindustan Publishing, New Delhi(India) 251. 2005:2005.

[19]   Matondkar S, Mathani S, Mavinkurve S. Studies on Mangrove Swamps of Goa 1. Heterotrophic Bacterial Flora from Mangrove Swamps. Mahasagar. 1981.

[20]   Newell SY, Fell JW. Competition among mangrove oomycotes, and between oomycotes and other microbes. Aquat Microb Ecol 1997; 12(1): 21-8.
[http://dx.doi.org/10.3354/ame012021]

[21]   Cavicchioli R. Cold-adapted archaea. Nat Rev Microbiol 2006; 4(5): 331-43.
[http://dx.doi.org/10.1038/nrmicro1390] [PMID: 16715049]

[22]   Zhou L, Liu X, Dong X. Methanospirillum psychrodurum sp. nov., isolated from wetland soil. International journal of systematic and evolutionary microbiology. 2014; 64: 638-41.

[23]   Mohanraju R, Natarajan R, Eds. Methanogenic bacteria in mangrove sediments. Proceedings of the International Symposium held at Mombasa Kenya. Springer. 1992.24–30 September 1990; 1992.

[24]   Ramamurthy T, Raju RM, Natarajan R. Distribution And Ecology Of Methanogenic Bacteria. Mangrove Sediments Of Pitchavaram. East-Coast Of India 1990.

[25]   Pokorna D, Zabranska J. Sulfur-oxidizing bacteria in environmental technology. Biotechnol Adv 2015; 33(6): 1246-59.
[http://dx.doi.org/10.1016/j.biotechadv.2015.02.007] [PMID: 25701621]

[26]   Deng Y, Cui X, Hernández M, Dumont MG. Microbial diversity in hummock and hollow soils of three wetlands on the Qinghai-Tibetan Plateau revealed by 16S rRNA pyrosequencing. PLoS One 2014; 9(7): e103115.
[http://dx.doi.org/10.1371/journal.pone.0103115] [PMID: 25078273]

[27]   Taketani RG, Yoshiura CA, Dias ACF, Andreote FD, Tsai SM. Diversity and identification of methanogenic archaea and sulphate-reducing bacteria in sediments from a pristine tropical mangrove. Antonie van Leeuwenhoek 2010; 97(4): 401-11.
[http://dx.doi.org/10.1007/s10482-010-9422-8] [PMID: 20195901]

[28]   Lyimo TJ, Pol A, Jetten MSM, Op den Camp HJM. Diversity of methanogenic archaea in a mangrove sediment and isolation of a new *Methanococcoides* strain. FEMS Microbiol Lett 2009; 291(2): 247-53.
[http://dx.doi.org/10.1111/j.1574-6968.2008.01464.x] [PMID: 19146579]

[29]   Huang L, Chakrabarti S, Cooper J, *et al.* Ammonia-oxidizing archaea are integral to nitrogen cycling in a highly fertile agricultural soil. ISME Communications 2021; 1(1): 19.
[http://dx.doi.org/10.1038/s43705-021-00020-4] [PMID: 37938645]

[30]   Lai J, Cheah W, Palaniveloo K, Suwa R, Sharma S. A Systematic Review of the Physicochemical and Microbial Diversity of Well-Preserved, Restored, and Disturbed Mangrove Forests: What Is Known and What Is the Way Forward? Forests 2022; 13(12): 2160.
[http://dx.doi.org/10.3390/f13122160]

[31]   Holmer M, Storkholm P. Sulphate reduction and sulphur cycling in lake sediments: a review. Freshw Biol 2001; 46(4): 431-51.
[http://dx.doi.org/10.1046/j.1365-2427.2001.00687.x]

[32]   Cattaneo C, Nicolella C, Rovatti M. Denitrification Performance of *Pseudomonas denitrificans* in a Fluidized Bed Biofilm Reactor and in a Stirred Tank Reactor. Eng Life Sci 2003; 3(4): 187-92.
[http://dx.doi.org/10.1002/elsc.200390026]

[33]   Liang JB, Chen YQ, Lan CY, Tam NFY, Zan QJ, Huang LN. Recovery of novel bacterial diversity from mangrove sediment. Mar Biol 2007; 150(5): 739-47.
[http://dx.doi.org/10.1007/s00227-006-0377-2]

[34]   Behera B, Patra M, Dutta S, Thatoi H. Isolation and characterization of Sulphur oxidizing bacteria from mangrove soil of Mahanadi river delta and their Sulphur oxidizing ability. J Appl Environ Microbiol 2014; 2(1): 1-5.

[35]   Muyzer G, Stams AJM. The ecology and biotechnology of sulphate-reducing bacteria. Nat Rev Microbiol 2008; 6(6): 441-54.
[http://dx.doi.org/10.1038/nrmicro1892] [PMID: 18461075]

[36]   Liamleam W, Annachhatre AP. Electron donors for biological sulfate reduction. Biotechnol Adv 2007; 25(5): 452-63.
[http://dx.doi.org/10.1016/j.biotechadv.2007.05.002] [PMID: 17572039]

[37]   Sigalevich P, Meshorer E, Helman Y, Cohen Y. Transition from anaerobic to aerobic growth conditions for the sulfate-reducing bacterium Desulfovibrio oxyclinae results in flocculation. Appl Environ Microbiol 2000; 66(11): 5005-12.
[http://dx.doi.org/10.1128/AEM.66.11.5005-5012.2000] [PMID: 11055956]

[38]   Bharathi P, Oak S, Chandramohan D. Sulfate-reducing bacteria [rom mangrove swamps II: Their ecology and physiology. Oceanol Acta 1991; 14(2): 163-71.

[39]   Liu X, Yang C, Yu X, *et al.* Revealing structure and assembly for rhizophyte-endophyte diazotrophic community in mangrove ecosystem after introduced Sonneratia apetala and Laguncularia racemosa. Sci Total Environ 2020; 721: 137807.

[http://dx.doi.org/10.1016/j.scitotenv.2020.137807] [PMID: 32179356]

[40]   Reis CRG, Nardoto GB, Oliveira RS. Global overview on nitrogen dynamics in mangroves and consequences of increasing nitrogen availability for these systems. Plant Soil 2017; 410(1-2): 1-19. [http://dx.doi.org/10.1007/s11104-016-3123-7]

[41]   Mulder E. Physiology and ecology of free-living, nitrogen-fixing. Nitrogen fixation by free-living microorganisms. 1975; 6:3.

[42]   James EK. Nitrogen fixation in endophytic and associative symbiosis. Field Crops Res 2000; 65(2-3): 197-209. [http://dx.doi.org/10.1016/S0378-4290(99)00087-8]

[43]   Zhang Y, Yang Q, Ling J, *et al.* Diversity and structure of diazotrophic communities in mangrove rhizosphere, revealed by high-throughput sequencing. Front Microbiol 2017; 8: 2032. [http://dx.doi.org/10.3389/fmicb.2017.02032] [PMID: 29093705]

[44]   Marushchak ME, Pitkämäki A, Koponen H, Biasi C, Seppälä M, Martikainen PJ. Hot spots for nitrous oxide emissions found in different types of permafrost peatlands. Glob Change Biol 2011; 17(8): 2601-14. [http://dx.doi.org/10.1111/j.1365-2486.2011.02442.x]

[45]   Repo ME, Susiluoto S, Lind SE, *et al.* Large N2O emissions from cryoturbated peat soil in tundra. Nat Geosci 2009; 2(3): 189-92. [http://dx.doi.org/10.1038/ngeo434]

[46]   Kolb S, Horn MA. Microbial CH4 and N2O consumption in acidic wetlands. Front Microbiol 2012; 3: 78. [http://dx.doi.org/10.3389/fmicb.2012.00078] [PMID: 22403579]

[47]   Yin Y, Yan Z. Variations of soil bacterial diversity and metabolic function with tidal flat elevation gradient in an artificial mangrove wetland. Sci Total Environ 2020; 718: 137385. [http://dx.doi.org/10.1016/j.scitotenv.2020.137385] [PMID: 32092526]

[48]   Marschner H. Marschner's mineral nutrition of higher plants. Academic press 2011.

[49]   Hesse PR. Phosphorus fixation in mangrove swamp muds. Nature 1962; 193(4812): 295-6. [http://dx.doi.org/10.1038/193295b0]

[50]   Reef R, Feller IC, Lovelock CE. Nutrition of mangroves. Tree Physiol 2010; 30(9): 1148-60. [http://dx.doi.org/10.1093/treephys/tpq048] [PMID: 20566581]

[51]   De Mandal S, Laskar F, Panda AK, Mishra R. Microbial diversity and functional potential in wetland ecosystems Recent Advancements in Microbial Diversity. Elsevier 2020; pp. 289-314. [http://dx.doi.org/10.1016/B978-0-12-821265-3.00012-8]

[52]   Xie X, Weng B, Cai B, Dong Y, Yan C. Effects of arbuscular mycorrhizal inoculation and phosphorus supply on the growth and nutrient uptake of Kandelia obovata (Sheue, Liu & Yong) seedlings in autoclaved soil. Appl Soil Ecol 2014; 75: 162-71. [http://dx.doi.org/10.1016/j.apsoil.2013.11.009]

[53]   Yadav AN, Sharma D, Gulati S, *et al.* Haloarchaea endowed with phosphorus solubilization attribute implicated in phosphorus cycle. Sci Rep 2015; 5(1): 12293. [http://dx.doi.org/10.1038/srep12293] [PMID: 26216440]

[54]   Gayathri S, Saravanan D, Radhakrishnan M, Balagurunathan R, Kathiresan K. Bioprospecting potential of fast growing endophytic bacteria from leaves of mangrove and salt-marsh plant species 2010.

[55]   Gotto JW, Taylor BF. N2 fixation associated with decaying leaves of the red mangrove (Rhizophora mangle). Appl Environ Microbiol 1976; 31(5): 781-3. [http://dx.doi.org/10.1128/aem.31.5.781-783.1976] [PMID: 16345161]

[56]   Shoreit A, El-Kady I, Sayed W. Isolation and identification of purple non-sulfur bacteria of mangal

and non-mangal vegetation of Red Sea Coast, Egypt. Bulletin of the Faculty of Science, Assiut Univ (Egypt). 1992.

[57]   van der Valk A. The biology of freshwater wetlands. Oxford University Press 2012.
[http://dx.doi.org/10.1093/acprof:oso/9780199608942.001.0001]

[58]   Freed G, Schlatter D, Paulitz T, Dugan F. Mycological insights into wetland fungal communities: the mycobiome of Camassia in the Pacific Northwest. Phytobiomes J 2019; 3(4): 286-99.
[http://dx.doi.org/10.1094/PBIOMES-04-19-0022-R]

[59]   Wu B, Tian J, Bai C, Xiang M, Sun J, Liu X. The biogeography of fungal communities in wetland sediments along the Changjiang River and other sites in China. ISME J 2013; 7(7): 1299-309.
[http://dx.doi.org/10.1038/ismej.2013.29] [PMID: 23446835]

[60]   Kennedy N, Clipson N. Fingerprinting the fungal community. Mycologist 2003; 17(4): 158-64.
[http://dx.doi.org/10.1017/S0269915X04004057]

[61]   Raghukumar S, Sharma S, Raghukumar C, Sathe-Pathak V, Chandramohan D. Thraustochytrid and fungal component of marine detritus. IV. Laboratory studies on decomposition of leaves of the mangrove Rhizophora apiculata Blume. J Exp Mar Biol Ecol 1994; 183(1): 113-31.
[http://dx.doi.org/10.1016/0022-0981(94)90160-0]

[62]   Xu Z, Ban Y, Jiang Y, Zhang X, Liu X. Arbuscular mycorrhizal fungi in wetland habitats and their application in constructed wetland: a review. Pedosphere 2016; 26(5): 592-617.
[http://dx.doi.org/10.1016/S1002-0160(15)60067-4]

[63]   Allaway WG, Curran M, Hollington LM, Ricketts MC, Skelton NJ. Gas space and oxygen exchange in roots of Avicennia marina (Forssk.) Vierh. var. australasica (Walp.) Moldenke ex NC Duke, the grey mangrove. Wetlands Ecol Manage 2001; 9(3): 221-8.
[http://dx.doi.org/10.1023/A:1011160823998]

[64]   You YH, Lee MC, Kim JG. Diversity of Endophytic Fungi Isolated from Hydrophytes in Wetland of Nakdong River. Hanguk Kyun Hakoe Chi 2015; 43(1): 13-9.
[http://dx.doi.org/10.4489/KJM.2015.43.1.13]

[65]   Shams M, Karimian Shamsabadi . Identification of algae as pollution bioindicators in Shakh-Kenar, Gavkhouni Wetland, Isfahan. Journal of Phycological Research 2019; 3(2): 386-94.
[http://dx.doi.org/10.29252/JPR.3.2.386]

[66]   Saha SB, Bhattacharyya S, Choudhury A. Diversity of phytoplankton of a sewage polluted brackishwater tidal ecosystem. J Environ Biol 2000; 21(1): 9-14.

[67]   Cheraghpour J, Afsharzadeh S, Sharifi M, Ghadi RR, Masoudi M. Phytoplankton diversity assessment of Gandoman wetland, west of Iran. Iran J Bot 2013; 19(2): 153-6.

[68]   Matsubara CP, Maltchik L, Torgan LC. Diversity and distribution of algae in wetlands of the Rio Grande do Sul, Brazil. Neotropical: biology and conservation. 2008; 3(1): 21-7.

[69]   Battin TJ, Kaplan LA, Denis Newbold J, Hansen CME. Contributions of microbial biofilms to ecosystem processes in stream mesocosms. Nature 2003; 426(6965): 439-42.
[http://dx.doi.org/10.1038/nature02152] [PMID: 14647381]

[70]   McCormick PV, Stevenson RJ. Periphyton as a tool for ecological assessment and management in the Florida Everglades. J Phycol 1998; 34(5): 726-33.
[http://dx.doi.org/10.1046/j.1529-8817.1998.340726.x]

[71]   Feio MJ, Alves T, Boavida M, Medeiros A, Graça MAS. Functional indicators of stream health: a river basin approach. Freshw Biol 2010; 55(5): 1050-65.
[http://dx.doi.org/10.1111/j.1365-2427.2009.02332.x]

[72]   Bere T, Chia MA, Tundisi JG. Effects of Cr III and Pb on the bioaccumulation and toxicity of Cd in tropical periphyton communities: Implications of pulsed metal exposures. Environ Pollut 2012; 163: 184-91.

[http://dx.doi.org/10.1016/j.envpol.2011.12.028] [PMID: 22249022]

[73]   Azim ME, Verdegem MC, van Dam AA, Beveridge MC. Periphyton: ecology, exploitation and management: CABI; 2005.

[74]   Sekar R, Venugopalan V, Nandakumar K, Nair K, Rao V, Eds. Early stages of biofilm succession in a lentic freshwater environment. Proceeding of The Second Asian Pacific Phycological Forum, held in Hong Kong China. 2004.2004.
[http://dx.doi.org/10.1007/978-94-007-0944-7_13]

[75]   Nadell CD, Drescher K, Wingreen NS, Bassler BL. Extracellular matrix structure governs invasion resistance in bacterial biofilms. ISME J 2015; 9(8): 1700-9.
[http://dx.doi.org/10.1038/ismej.2014.246] [PMID: 25603396]

[76]   Liu J, Wu Y, Wu C, *et al.* Advanced nutrient removal from surface water by a consortium of attached microalgae and bacteria: A review. Bioresour Technol 2017; 241: 1127-37.
[http://dx.doi.org/10.1016/j.biortech.2017.06.054] [PMID: 28651870]

[77]   Su J, Kang D, Xiang W, Wu C. Periphyton biofilm development and its role in nutrient cycling in paddy microcosms. J Soils Sediments 2017; 17(3): 810-9.
[http://dx.doi.org/10.1007/s11368-016-1575-2]

[78]   Lu H, Liu J, Kerr PG, Shao H, Wu Y. The effect of periphyton on seed germination and seedling growth of rice (Oryza sativa) in paddy area. Sci Total Environ 2017; 578: 74-80.
[http://dx.doi.org/10.1016/j.scitotenv.2016.07.191] [PMID: 27503628]

[79]   Lu H, Yang L, Shabbir S, Wu Y. The adsorption process during inorganic phosphorus removal by cultured periphyton. Environ Sci Pollut Res Int 2014; 21(14): 8782-91.
[http://dx.doi.org/10.1007/s11356-014-2813-z] [PMID: 24728572]

[80]   Grossart HP, Czub G, Simon M. Algae–bacteria interactions and their effects on aggregation and organic matter flux in the sea. Environ Microbiol 2006; 8(6): 1074-84.
[http://dx.doi.org/10.1111/j.1462-2920.2006.00999.x] [PMID: 16689728]

[81]   Wang S, Sun P, Zhang G, *et al.* Contribution of periphytic biofilm of paddy soils to carbon dioxide fixation and methane emissions. Innovation 2022; 3(1): 100192.
[http://dx.doi.org/10.1016/j.xinn.2021.100192] [PMID: 34950915]

[82]   Geist J. Integrative freshwater ecology and biodiversity conservation. Ecol Indic 2011; 11(6): 1507-16.
[http://dx.doi.org/10.1016/j.ecolind.2011.04.002]

[83]   Jackson CR, Vallaire SC. Effects of salinity and nutrients on microbial assemblages in Louisiana wetland sediments. Wetlands 2009; 29(1): 277-87.
[http://dx.doi.org/10.1672/08-86.1]

[84]   Ceccon DM, Faoro H, Lana PC, Souza EM, Pedrosa FO. Metataxonomic and metagenomic analysis of mangrove microbiomes reveals community patterns driven by salinity and pH gradients in Paranaguá Bay, Brazil. Sci Total Environ 2019; 694: 133609.
[http://dx.doi.org/10.1016/j.scitotenv.2019.133609] [PMID: 31400683]

[85]   Barcellos D, Queiroz HM, Nóbrega GN, *et al.* Phosphorus enriched effluents increase eutrophication risks for mangrove systems in northeastern Brazil. Mar Pollut Bull 2019; 142: 58-63.
[http://dx.doi.org/10.1016/j.marpolbul.2019.03.031] [PMID: 31232342]

[86]   Bhattacharya BD, Nayak DC, Sarkar SK, Biswas SN, Rakshit D, Ahmed MK. Distribution of dissolved trace metals in coastal regions of Indian Sundarban mangrove wetland: a multivariate approach. J Clean Prod 2015; 96: 233-43.
[http://dx.doi.org/10.1016/j.jclepro.2014.04.030]

[87]   Jöbgen A, Palm A, Melkonian M. Phosphorus removal from eutrophic lakes using periphyton on submerged artificial substrata. Hydrobiologia 2004; 528: 123-42.
[http://dx.doi.org/10.1007/s10750-004-2337-5]

[88]   Wu Y, Zhang S, Zhao H, Yang L. Environmentally benign periphyton bioreactors for controlling cyanobacterial growth. Bioresour Technol 2010; 101(24): 9681-7.
[http://dx.doi.org/10.1016/j.biortech.2010.07.063] [PMID: 20702088]

[89]   Wu N, Dong X, Liu Y, Wang C, Baattrup-Pedersen A, Riis T. Using river microalgae as indicators for freshwater biomonitoring: Review of published research and future directions. Ecol Indic 2017; 81: 124-31.
[http://dx.doi.org/10.1016/j.ecolind.2017.05.066]

[90]   Wu Y, Xia L, Yu Z, Shabbir S, Kerr PG. *In situ* bioremediation of surface waters by periphytons. Bioresour Technol 2014; 151: 367-72.
[http://dx.doi.org/10.1016/j.biortech.2013.10.088] [PMID: 24268508]

[91]   Faheem M, Shabbir S, Zhao J, *et al.* Multifunctional periphytic biofilms: Polyethylene degradation and Cd2+ and Pb2+ bioremediation under high methane scenario. Int J Mol Sci 2020; 21(15): 5331.
[http://dx.doi.org/10.3390/ijms21155331] [PMID: 32727088]

[92]   dos Santos HF, Cury JC, do Carmo FL, *et al.* Mangrove bacterial diversity and the impact of oil contamination revealed by pyrosequencing: bacterial proxies for oil pollution. PLoS One 2011; 6(3): e16943.
[http://dx.doi.org/10.1371/journal.pone.0016943] [PMID: 21399677]

[93]   Conrad R, Babbel M. Effect of dilution on methanogenesis, hydrogen turnover and interspecies hydrogen transfer in anoxic paddy soil. FEMS Microbiol Lett 1989; 62(1): 21-7.
[http://dx.doi.org/10.1111/j.1574-6968.1989.tb03654.x]

[94]   Giani L, Bashan Y, Holguin G, Strangmann A. Characteristics and methanogenesis of the Balandra lagoon mangrove soils, Baja California Sur, Mexico. Geoderma 1996; 72(1-2): 149-60.
[http://dx.doi.org/10.1016/0016-7061(96)00023-7]

[95]   Kristensen E, Holmer M, Bussarawit N. Benthic metabolism and sulfate reduction in a southeast Asian mangrove swamp. Mar Ecol Prog Ser 1991; 73: 93-103.
[http://dx.doi.org/10.3354/meps073093]

[96]   Pritchard D. Reducing emissions from deforestation and forest degradation in developing countries (REDD): the link with wetlands. Reducing emissions from deforestation and forest degradation in developing countries (REDD): the link with wetlands. 2009.

[97]   Page KL, Dalal RC. Contribution of natural and drained wetland systems to carbon stocks, CO2, N2O, and CH4 fluxes: an Australian perspective. Soil Res 2011; 49(5): 377-88.
[http://dx.doi.org/10.1071/SR11024]

[98]   Aliance LT. Wetlands United States of America https://climatechange.lta.org/wetlands/

[99]   Solutions N. Protected areas helping people cope with climate change. Gland, Switzerland 2010.

[100]   Robson BJ, Bukaveckas PA, Hamilton DP. Modelling and mass balance assessments of nutrient retention in a seasonally-flowing estuary (Swan River Estuary, Western Australia). Estuar Coast Shelf Sci 2008; 76(2): 282-92.
[http://dx.doi.org/10.1016/j.ecss.2007.07.009]

[101]   Human LRD, Snow GC, Adams JB, Bate GC, Yang SC. The role of submerged macrophytes and macroalgae in nutrient cycling: A budget approach. Estuar Coast Shelf Sci 2015; 154: 169-78.
[http://dx.doi.org/10.1016/j.ecss.2015.01.001]

[102]   Sivakumar K, Sahu MK, Thangaradjou T, Kannan L. Research on marine actinobacteria in India. Indian J Microbiol 2007; 47(3): 186-96.
[http://dx.doi.org/10.1007/s12088-007-0039-1] [PMID: 23100666]

[103]   Munn MD. Understanding the influence of nutrients on stream ecosystems in agricultural landscapes. United States Department of the Interior 2018.
[http://dx.doi.org/10.3133/cir1437]

[104] Sinduja, M., Avudainayagam, S., Davamani, V., & Suganthi, R. (2018). Uptake of mercury by marigold and amaranthus on spiked soil.

[105] Dodds WK. Trophic state, eutrophication and nutrient criteria in streams. Trends Ecol Evol 2007; 22(12): 669-76.
[http://dx.doi.org/10.1016/j.tree.2007.07.010] [PMID: 17981358]

[106] Zhang H, Cui B, Zhang K. Surficial and vertical distribution of heavy metals in different estuary wetlands in the Pearl River, South China. Clean (Weinh) 2012; 40(10): 1174-84.
[http://dx.doi.org/10.1002/clen.201100730]

[107] Karppinen EM, Payment J, Chatterton S, *et al.* Distribution and abundance of Aphanomyces euteiches in agricultural soils: effect of land use type, soil properties, and crop management practices. Appl Soil Ecol 2020; 150: 103470.
[http://dx.doi.org/10.1016/j.apsoil.2019.103470]

CHAPTER 9

Forest Microbiomes: Their Role and Importance in Ecosystem Sustainability and Restoration

Jerome O. Ihuma[1,*], Malgwi T. Doris[2], Tayo I. Famojuro[3], R. Raveena[4] and Govindaraj Kamalam Dinesh[5,6,7]

[1] *Department of Biological Science, Faculty of Science and Technology, Bingham University, Karu Nasarawa State, Nigeria*

[2] *Department of Community Medicine, Nnamdi Azikiwe University, Nnewi, Anambra State, Nigeria*

[3] *Department of Pharmacognosy, Faculty of Pharmaceutical Sciences, Bingham University, Karu Nasarawa State, Nigeria*

[4] *Department of Environmental Sciences, Tamil Nadu Agricultural University, Coimbatore, Tamil Nadu, India*

[5] *Division of Environment Science, ICAR-Indian Agricultural Research Institute, New Delhi-110012, India*

[6] *Division of Environmental Sciences, Department of Soil Science and Agricultural Chemistry, SRM College of Agricultural Sciences, SRM Institute of Science and Technology, Baburayanpettai-603201, Chengalpattu, Tamil Nadu, India*

[7] *INTI International University, Persiaran Perdana BBN, Putra Nilai, 71800 Negeri Sembilan, Malaysia*

Abstract: A forest is a large area of land covered with big trees of different species, approximately covering one-third of the Earth's surface. Forest ecosystems are more than what can be seen physically (aboveground); below the ground level, they are extraordinarily diverse and have unique communities of microbiomes with a large population of bacteria and fungi species. These microorganisms are essential to how plants interact with the soil environment and are necessary to access critically limiting soil resources. This book chapter focuses on the ecosystems below and above ground level of a forest microbiome, including the soil microorganisms, their importance, and the diverse interrelationships among soil microorganisms (parasitism, mutualism, commensalism). The aboveground part of a plant is known as the phyllosphere, harboring diverse microorganisms, such as viruses, bacteria, filamentous fungi, yeast, algae, and rarely protozoa and nematodes with a role in disease resistance that is critical to plant health and development. The rhizosphere is the soil region immediately adjacent to and affected by plant roots where plants, soil, microorganisms, nutrients,

---

* **Corresponding author Jerome O. Ihuma:** Department of Biological Science, Faculty of Science and Technology, Bingham University, Karu Nasarawa State, Nigeria;
E-mail: jeromeihuma@binghamuni.edu.ng

Govindaraj Kamalam Dinesh, Shiv Prasad, Ramesh Poornima, Sangilidurai Karthika, Murugaiyan Sinduja & Velusamy Sathya (Eds.)

and water meet and interact. In this region, plants and microbes coordinate and show a symbiotic relationship by fulfilling each other's nutrient requirements, roles, and functions. The endosphere is the plant interior and is colonized by endophytes, and their functions range from mutualism to pathogenicity. Archaebacteria, anaerobic bacteria, aerobic prokaryotes, fungi, and viruses exist as forest biomes. Examples of fungi include *Trichoderma harzianum and* obligate parasites *Puccinia striiformis*and*Gremmeniella abietina.* Plants, fungal endophytes, mycoviruses, and the environment all participate in a four-way interactive system.

Keywords: Abiotic, Archaea, Association, Biotic, Bacteria, Endosphere, Ecosystem, Forest, Forest microbiomes, Fungi, Hubs, Microorganisms, Microbial communities, Rhizosphere, Trees, Rree pest, Plants, Phyllosphere, Soil, Virus.

INTRODUCTION

Forest, in its simplest definition, is a large area of land dominated by trees (mostly comprised of different species) and associated fauna that inhabits a specific land area. According to expert estimations, approximately thirty percent of the surface of the planet is covered with forests. According to Pan and co-workers [1], forests are the predominant terrestrial ecosystem and are evenly distributed on the surface of the Earth. Forests cover a total land area of approximately 4 billion hectares, which is estimated to be 31% of the world's total area [2].

Nearly 80% of the Earth's plant biomass comprises forests, with a 75% gross primary production. The forest biomes are composed of tropical forests, temperate forests, and boreal forests, with a net direct annual output estimated at 21.9 gigatonnes, 8.1 gigatonnes, and 2.6 gigatonnes, respectively [1], and they have significant economic as well as ecological value. Forest ecosystems are more than what can be seen physically – they do not only comprise the plants and animals aboveground. Diverse and complex fungi and bacteria communities inhabit the Earth's surface underground. To get access to these scarce soil nutrients, plants rely on fungal and bacterial interactions with their soil environment.

A forest ecosystem is, therefore, a place that provides natural habitat to millions of microorganisms. It can be classified into three common types: tropical, boreal, and temperate. Forest ecosystems play essential roles in the environment by balancing the climate of the planet Earth, providing oxygen, maintaining the balance of carbon dioxide in the atmosphere and biogeochemical cycles, and preventing soil erosion. This is the general description of a forest aboveground; however, diverse fungi and bacteria communities (microbiomes) exist and cohabit below the ground level. These soil microorganisms are equally necessary for interdependence cycles.

Forest microbiomes are microbial organisms closely associated with tree species in a forest ecological community. Microbiota and microbiome are used interchangeably, describing the assembly of tree-associated microorganisms and the relationships between groups. The existence of these organisms is essential to the interaction between soil and plant environment and is required to access soil resources. The subterranean stratum of the organism establishes interconnected systems amongst trees, facilitating the exchange of nutrients and providing mutual support in the face of adverse conditions [3]. In addition, the existence of microorganisms in the soil plays a vital function in aiding the conversion of nutrients within forest ecosystems, therefore contributing to the overall stability and long-term viability of these ecosystems. These microbes enhance energy transformation processes and facilitate interactions between the many components of the ecosystem, both above and below the surface of the ground.

Soil microorganisms are essential to forest ecosystems as they are vital in nutrient transformations. Forest ecosystems depend on these nutrient transformations and interactions between above and underground components for stability and sustainable development that drive forest ecosystem processes [4].

IMPORTANCE OF SOIL ORGANISMS

Soil organisms are of importance in many ways and are outlined below:

- Accountable for cycling of K, P, C, N, and other nutrients in the soil.
- Increase and strengthen soil structure
- Replace and decompose organic materials
- Support soil health and quality
- Enhance soil penetrability and soil aeration

FOREST ECOSYSTEM

As described earlier, a forest ecosystem provides a natural habitat to millions of plants, animals, fungi (Unicellular and multi-cellular), and microbial species. It can be classified into three general types: temperate, tropical, and boreal [1]. This division is a result of the climatic condition of a particular locality. The three types of forest ecosystems are:

- Temperate Forest
- Tropical Forest
- Boreal Forest

Temperate Forest

Temperate forests constitute Eurasia and eastern North America. The recorded temperatures ranged from -30 degrees Celsius (-22°F) to 30 degrees Celsius (86°F), with an average mean of 10 degrees Celsius (50 degrees Fahrenheit) on an average day. The temperature in these forests varies throughout the year [1, 2]. Nonetheless, abundant precipitation contributes to the fertility of the soil, which supports a variety of plant life, such as birch, oak, and maple trees.

Tropical Forest

Tropical forests are mostly located in close proximity to the equator, including regions such as sub-Saharan Africa and Central America. The ambient temperature inside tropical forests exhibits a range of around 20 to 31°C (68 to 88°F); as a result, tropical rainforests have an abundant representation of biodiversity [1]. Tropical forests are characterized by the proliferation of arboreal and shrubby vegetation in saline or brackish aquatic environments, mostly occurring in regions within the subtropical and tropical latitudes.

Boreal Forest

Boreal forests, also referred to as taiga, are a substantial terrestrial biome spanning vast regions in Siberia, Scandinavia, and North America, including Alaska and Canada. The region has harsh winters and moderately warm summers, characterized by a temperature range of 21 °C during the summer months and -54 °C during the winter season. Boreal forests are substantially responsible for the sequestration of CO_2 from the Earth's atmosphere [1, 2]. It is the world's biggest terrestrial biome and is distinguished by thick, cold-tolerant, coniferous forests that are dominated by evergreen trees such as spruces, firs, and pines. These forests constitute the world's largest terrestrial ecosystem. The boreal forest, which comprises roughly 29% of the world's total forest cover, is an essential component in maintaining the planet's unique ecological and climatic balance. On average, temperatures inside boreal forests remain below the freezing point. The majority of avian species indigenous to the taiga biome rely on seasonal migration as a means to seek more favorable environmental circumstances within the challenging winter period inside the forest ecosystem.

FOREST MICROBIOMES

Forest microbiomes are microbial organisms closely associated with tree species in a forest ecological community. Microbiota and microbiome are used interchangeably to explain the environmental community of microorganisms.

Pathogenic Microbiomes

It is a parasitic relationship when one organism benefits from the interaction at the cost of another, which is subjected to harm. It is a symbiotic relationship wherein one organism thrives at the expense of the other. The organism that derives an advantage is often referred to as the "parasite", while the organism that experiences advantages is known as the "host" while also incurring disadvantages. The creature that derives advantages is referred to as the "parasite". An example of this relation is the *Armillaria*, a parasitic fungal species that benefits free-living organisms. The fungi can survive and stay in the woods without the need for any parasitic activity. These organisms are essential agents in recycling nutrients through microbial decomposition in the ecosystem [5].

Mutualistic Forest Microbiomes

In a mutualistic connection, members of both species gain from one another's presence. The potential advantages derived from the connection include pollination, defense, dispersion, and nutrition provisioning. Mutualism is a phenomenon seen in some animals whereby their longevity is contingent upon the existence of a mutually beneficial connection. In other instances, mutualism confers advantages while the survival of the species remains unaffected by the connection. The concept of facultative mutualism refers to a kind of symbiotic relationship between two organisms in which both parties may get benefits [6, 7].

Lichens exemplify the concept of obligatory mutualism when a symbiotic association is formed between a fungus and algae. The mutual relationship facilitates the survival of both species. The fungus uses its filamentous structures to provide protection and acquire nourishment and moisture from the surrounding environment. The algae reciprocally supply nourishment to the fungus *via* the process of photosynthesis, whereby organic matter is produced [8]. Mycorrhiza establishes symbiotic associations between fungus and plant roots, whereby the leaves provide carbon resources to the fungi, while the fungi reciprocate by supplying essential nutrients. Mycorrhizas further enhance water absorption and protection against pathogenic infections [7 - 9].

Commensalistic Forest Microbiomes

Commensalism refers to a kind of contact between many species in which one species derives advantages while the other species neither incurs damage nor gains any advantage from the association. Typically, it is a long-term association in which species interact with one another during their whole lives [8, 9].

Commensalism mainly occurs when a larger organism, referred to as the host, coexists with a smaller organism, referred to as the commensal. The species that provide benefits without experiencing any adverse effects are referred to as the host organisms. *Caenorhabditis remanei*, a nematode species, has restricted dispersion capabilities and relies on slugs, snails, and isopods for extended distances. The interactions serve only for the purpose of dissemination and are confined to the body of the slugs [9 - 11].

PHYLLOSPHERE, RHIZOSPHERE, AND ENDOSPHERE MICROBIOMES

Phyllosphere Microbiome

The phyllosphere encompasses the aerial portion of a plant and functions as a home for a diverse range of microorganisms residing in both epiphytic and endophytic ecological niches [12, 13]. Yeast, viruses, bacteria, filamentous fungi, algae, and even protozoa and nematodes may all be found living in the phyllosphere. Bacteria are the most common kind of microbe, whereas fungi and archaea are far less common [12, 13]. The phyllosphere is a dynamic ecological niche consisting of diverse microbiotas that play a crucial role in regulating the physiological interactions among plants, microbes, and the environment.

Phyllosphere Diversity and Function

The skin microbiome is widely recognized as the first line of defense against the infiltration of diseases. Similarly, the phyllosphere serves as a protective barrier for plant hosts, shielding them from the invasion of viruses [12]. The association between phyllosphere microbiomes and illness has been reported across a range of plant species, spanning from seedlings of short-lived plants to mature, long-lived trees. The phenomenon arises as a consequence of inherent microbial competition and the indirect manipulation of plant defense mechanisms [12]. The significance of the phyllosphere microbiota in relation to disease resistance is of utmost importance for the overall well-being of plants, their growth and maturation, as well as the functioning of ecosystems [12]. During a workshop facilitated by Isabelle Laforest-Lapointe, an examination was conducted on the phyllosphere microbiomes inside several experimental forest plots. This investigation revealed a correlation between the richness of bacterial leaf species and the overall productivity of the ecosystem. Additionally, it was discovered that the identification of tree species served as a predictor for the diversity of phyllosphere bacteria, and this association was proven statistically significant in relation to the productivity of the plant community [12].

Rhizosphere Microbiome

The term "rhizosphere" is used to describe the localized region of soil where plant roots have an immediate impact. Dynamic interactions between plants, microbes, soil, water, and nutrients characterize the ecosystem. The rhizosphere is distinct from the surrounding soil due to the interactions and impacts of soil organisms and plant roots [14]. The term "rhizosphere" refers to the region around the root system of actively developing plants, characterized by a high level of microbial activity, which extends up to a few millimeters. Microorganisms that thrive in the presence of roots exhibit notable qualitative and quantitative distinctions compared to those residing in the soil environment away from such impact [14].

The term "rhizosphere" refers to the small portion of soil that is positioned only a few millimeters away from the root system of a plant. Symbiotic relationships between bacteria and plants make nutrient intake easier for both in this region. There is a wide range of variations in the size of the rhizosphere zone based on the plant type and the microbiota (Fig. **1**). There are qualitative and quantitative variations in microorganisms across different plant species. Hence, the rhizosphere area may be classified into two distinct zones:

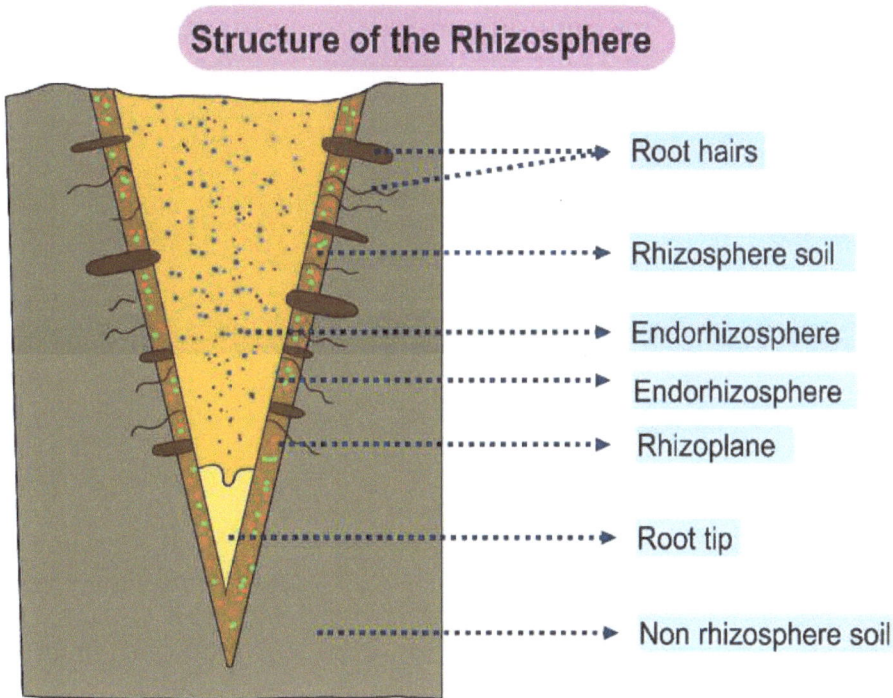

Fig. (1). Structure of the Rhizosphere (modified from [14]).

- The region in close proximity to the root surface, known as the inner rhizosphere, is referred to as the endo-rhizosphere. The concept is very dependent on the underlying foundation. The apoplastic space, which encompasses the cortical and epidermal regions, is an area inside the plant structure where microorganisms reside, occupying the intercellular gaps.
- The outer rhizosphere encompasses the soil directly next to it, which is often referred to as the exo-rhizosphere. The location of this layer is immediately next to the epidermis.

Microbial Activity in Rhizosphere Zone

In the interstitial space between the soil and the rhizosphere, a region of transition exists whereby the impact of roots gradually lessens as the distance increases. (Fig. **2**). The consensus in the scientific community is that rhizosphere soil refers to the thin layer that remains attached to a root subsequent to the removal of loose soil particles and aggregates by mechanical agitation [14]. This phenomenon undeniably impacts the 'rhizosphere effect'. The rhizosphere impact is a fundamental concept used to quantify the microbial population inside the rhizosphere zone. The text discusses the comparison between microorganisms present in rhizospheric soil and those found in edaphosphere or non-rhizospheric soil.

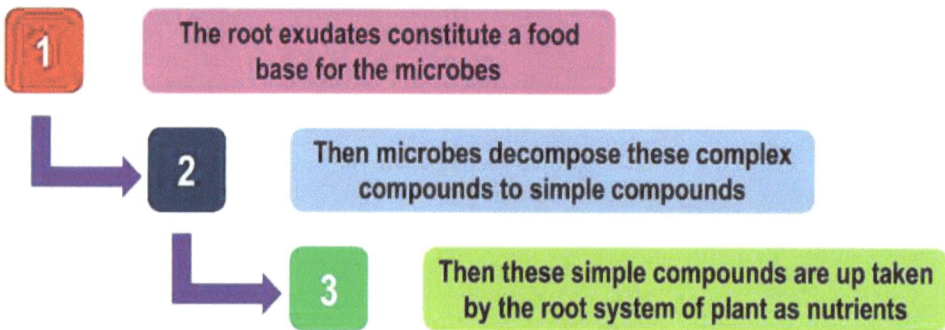

1 The root exudates constitute a food base for the microbes

2 Then microbes decompose these complex compounds to simple compounds

3 Then these simple compounds are up taken by the root system of plant as nutrients

Fig. (2). Microbial Activity in Rhizosphere Zone (modified from [14]).

Root Exudation

The process known as root exudation is responsible for the occurrence of extensive microbial activity. Root exudates refer to the diffusion of inorganic and organic substances from the root under adverse circumstances. Therefore, roots communicate with bacteria through exudates. Exudates from roots include both root secretions and diffusates [15].

Classification of Root Exudates

Root exudates may be classified into three categories based on their chemical composition:

- **Organic compounds**: These are chemical compounds composed chiefly of carbon atoms, which are covalently bonded to other atoms, such as hydrogen, nitrogen, and other elements. Some examples of essential macromolecules found in living organisms are proteins, vitamins, amino acids, carbohydrates, and enzymes.
- **Inorganic compounds**: These are chemical substances composed of elements other than carbon and hydrogen. Examples of such compounds include water, as well as other gases such as carbon dioxide and nitrogen gas.
- **Miscellaneous compounds**: The plant also emits several molecules known as miscellaneous substances, which have the potential to have detrimental consequences. Some examples of compounds that may be found in plants are glycosides, saponins, and hydrocyanic acids [15].

Role of Root Exudates

Root exudates have many important functions:

- The provision of a nutrition source facilitates the sustenance of microbial organisms.
- Root exudates serve as a protective mechanism for plants against desiccation.
- The moisture provided by the root exudates is ideal for the proliferation of bacteria.
- The release of defense proteins and antibacterial compounds serves to protect the root against pathogenic microorganisms.
- It elicits systemic tolerance to abiotic stressors like temperature, concentrations of salts, and pH levels.
- Mucilage secretion has a role in promoting soil aggregation.
- Osmotic pressure, ionic balance, and redox potential are only a few of the factors that may affect the soil's chemical and physical properties.

Root Exudation and Its Influencing Factors

- The existence of light.
- High temperature.
- The process of plant wilting results in the substantial release of amino acids.
- Certain bacteria can release secondary metabolites. The microorganisms present in the rhizosphere have the potential to influence both root permeability and metabolism [15].

ENDOSPHERE MICROBIOME

The plant endosphere is inhabited by diverse microbial communities and microorganisms, collectively known as endophytes, that reside inside the internal tissues of the plant for a certain period of time. Their roles range from pathogenicity to mutualism. Endophytes generally colonize plants' organs and tissues, and their diversity depends on the plant, its physiological conditions, the plant growth stage, and its environment [15, 16]. Endophytic microorganisms inhabit both the aerial and subterranean components of plants. Nevertheless, the aerial parts of a plant, including the leaves, stems, flowers, and mature seeds, inherently possess a lesser abundance of endophytes compared to the underground root area [17].

Microbial endophytes that inhabit the internal tissues of plant roots are commonly considered to be a subsection of rhizosphere microbes. This is because of the high quantities of root exudates, which function as chemo-attractants and provide an ideal habitat for a huge population of microorganisms in the rhizosphere zone. Consequently, this facilitates the unrestricted movement and infiltration of microbes from the rhizoplane into the roots of plants [18]. The rhizodeposits and root exudation contribute to an elevated microbial population inside the root zones relative to other regions of the plant [19]. The microorganisms have been identified and classified in relation to several economically significant plant species, including sunflower, maize, sugarcane, tomato, chickpeas, wheat, rice, corn cowpea, pearl millet, soybean, citrus, and others. These microorganisms have shown the capacity to improve crop productivity [20 - 22].

Endophytic microorganisms inhabit specific ecological niches inside plants, hence facilitating plant development and bolstering plant defense mechanisms [23]. In recent times, a wide range of endophytic organisms have been discovered; however, there are still plenty more that have not been successfully cultivated [24 - 26]. The process of extracting endophytic fungus from different plant components for bioprospecting in agricultural biotechnology has been documented [27, 28]. Endophytes establish a symbiotic relationship with plants by colonizing their interior tissues, demonstrating asymptomatic characteristics that contribute to maintaining plant development and overall health. There is a current understanding of about one million distinct endophytic fungus species that exhibit selectivity toward plants.

Endophytes may be able to produce bioactive compounds in proportion to the degree to which they have coevolved with their host plants by adopting the host's genetic information. This may enable endophytes to exercise adaptive mechanisms within the plant and to protect plants from pathogens and insect infections. Despite their biological origins, endophytes have the capacity to release chemicals inside the plants; hence, plants mediate endophyte selection for the production of nontoxic bioactive metabolites to higher species [29].

Endophytic fungi are classified into two major groups: ascomycetes and fungi imperfecti. There is widespread agreement that the mutually beneficial connection between endophytes and plant hosts is key in generating specialized metabolites with important implications for agricultural biotechnology [30]. Therefore, the use of endophytes in biotechnology may enhance the potential for investigating the varied range of endophytic resources found in distinct plant species. The interior environment of the plant is often described as sterile; however, it contains imperceptible microbial colonies with diverse capabilities to enhance production. Numerous studies have shown that almost all plants with structurally developed supportive tissues host endophytic microorganisms [31]. Despite the prevailing estimate of the existence of over three million plant species globally, it is important to note that each plant has diverse endophytic communities [32]. The isolation of various endophyte species from host plants, as seen *via* colonization patterns, clearly indicates the potential to extract novel and specific natural chemicals from microbial endophytes residing in diverse plant species within ecological niches [33].

The capacity of endophytes to synthesize bioactive compounds may be influenced by their coevolutionary relationship with host plants, wherein they acquire plant genetic information. This acquisition potentially enables endophytes to employ adaptive mechanisms within the plant, providing functional benefits such as protection against pathogen attacks and insect infestation [34]. Despite the inherent biological characteristics of endophytes, it is noteworthy that they can release chemical substances inside the host plants. This phenomenon highlights the crucial involvement of plants in selecting endophytes capable of producing bioactive metabolites that are non-toxic to higher species.

As a result of the direct host-linkage established by endophytic fungi, a condition of cell biomass equilibrium develops between the plant and the associated microorganisms [35]. It has been proven that these bacteria have a valuable consequence on plant growth by increasing agricultural output and shielding plants against pathogens [36]. Fungal endophytes have the capacity to inhabit plant niches without inducing any pathological manifestations, owing to their capability to produce distinct bioactive metabolites. These chemicals are potential antimicrobial agents that might be used to control harmful bacteria. It is possible for the symbiotic connections that exist among flora and endophytic pathogens to continue even in the absence of apparent symptoms of disease and by avoiding the activation of the host's defensive mechanisms [37]. In adverse circumstances, advantageous endophytes have the potential to exhibit pathogenic behavior as a result of genetic changes [22]. The interactions between plants and endophytes may include more than only maintaining an antagonistic balance. The release of biologically active chemicals reinforces the links between fungal endophytes and the vegetation that serves as their hosts [32]. Endophytes and vegetation engage in intricate contact with one another across several species, which results in the creation of secondary metabolites, which may be partly or wholly synthesized. The synthesis of these metabolites varies across different endophytes and plant species [38].

MICROORGANISMS IN THE RHIZOSPHERE

Numerous microorganisms proliferate and carry out their metabolic processes in the rhizosphere. Consequently, the region has a vast microbial diversity of Bacteria, protozoa, fungi, algae, and actinomycetes, while viruses, archaea, and arthropods are extremely uncommon. Based on their effect, microorganisms can be divided into three distinct categories:

Remunerative Microorganisms

Certain groups or species of microorganisms, such as mycorrhizal fungi, protozoa, and nitrogen-fixing bacteria, do not have any detrimental effects on the root system.

The Significance of Remunerative Microorganisms

- The process of nitrogen fixation facilitates the solubilization of inorganic phosphate, iron, and other essential elements.
- Microorganisms with remunerative properties play a significant role in the process of root nodulation.
- The root system is inhabited by microbes.
- Microbes of commercial value promote plant growth.

Pathogenic Microorganisms

Pathogenic fungi, bacteria, nematodes, and other types of microbes, among others, are known to cause damage to the root system of soil. The significance of harmful microorganisms in many contexts:

- Plant diseases are caused by several factors.
- Acquire essential nutrients

Neutral Microorganisms

These groups of microorganisms have zero effect; examples are actinomycetes, algae, *etc*.

The function of neutral microorganisms

- It contributes to the aggregation of the soil.
- It has hostile characteristics against other species.

ARCHAEA

The Archaebacteria kingdom comprises a collection of bacteria that exhibit both anaerobic and aerobic characteristics while also being classified as prokaryotes. These microbes are able to survive under extreme conditions, such as those found near volcanoes or at the depths of the ocean. These bacteria have undergone evolutionary adaptations that enable their survival in the absence of light or oxygen. Living creatures are classified under the five-kingdom system, which includes Plantae, Animalia, fungus, Protoctista, and Monera. In the not-too-distant past, before 1977, archaea were regarded as a collective of microorganisms belonging to the domain Bacteria.

Consequently, they were classified into the taxonomic category of Kingdom Plantae. Subsequently, they were assigned to the recently established kingdom Monera, named after the microorganisms known as bacteria. In 1977, scientists Carl Woese and George Fox proposed that archaebacteria be classified as a distinct kingdom [39].

During the 1990s, researchers made a significant finding on the dissimilarity of 16S rRNA and 18S rRNA sequences in archaea. The differentiation between bacteria and archaea was elucidated by the analysis of their genomes in the year 2003. The proportion of Archaea within forest ecosystems represents a minor fraction of prokaryotes' overall abundance [39].

Characteristics of Archaebacteria

The Greek word "achaio" means "ancient". The name appropriately describes archaebacteria, which are thought to share a common ancestor with bacteria and eukaryotes. Archaebacteria have a greater degree of structural similarity to eukaryotes than bacteria. Various traits of the archaebacteria kingdom aid in differentiating them from eubacteria. The following are the features of archaebacteria:

- Archaebacteria lack peptidoglycan inside their cellular walls.
- The cellular wall is composed of glycoproteins and polysaccharides.
- The cell wall envelopes exhibit a notable degree of resistance to antibiotics and lytic agents due to variations in their composition.
- Cell membranes are composed of a lipid bilayer that exhibits significant differences.
- The RNA polymerase found in archaea is highly similar to the RNA polymerase present in eukaryotes.
- The ribosomal proteins found in eukaryotes and archaea exhibit notable similarities.

Archaebacteria exhibit autotrophic characteristics by using atmospheric CO_2 as a carbon source for carbon fixation. Archaebacteria, known as extremophiles in scientific terminology, have a remarkable ability to flourish in environments that are characterized by adverse conditions. These organisms have the ability to thrive in aquatic environments characterized by high levels of acidity, alkalinity, and salinity. Certain organisms have the ability to endure temperatures above 100°C or 212°F. Nevertheless, only a limited number of organisms have the ability to endure extreme pressures exceeding 200 atmospheres and survive at the depths of the Earth. These organisms use many chemical processes to adapt and thrive in extreme environmental circumstances. Consequently, these microorganisms may be classified into three distinct subcategories: methanogens, extreme halophiles, and thermoacidophiles [39, 40].

Methanogens

Methanogens can catalyze the reduction of carbon dioxide (CO_2) into methane (CH_4). These organisms are classified as obligate anaerobes and are susceptible to mortality upon exposure to oxygen. The organisms generate gaseous substances that are visually detectable in still water as bubbles. Methanogens use CO_2 as an electron acceptor in the process of hydrogen oxidation, facilitated by coenzymes such as coenzyme M and methanofuran. The presence of these coenzymes is exclusive to archaebacteria.

Halophiles

Halophiles are a kind of bacteria that can thrive in conditions with salt concentrations that are ten times higher than that found in the sea. Halobacterium utilizes photophosphorylation as a metabolic process. Light-activated ion pumps such as bacteriorhodopsin and halorhodopsin are used to establish ion gradients that facilitate ion transport across the plasma membrane. ATP synthesis is facilitated by the conversion of energy contained within electrochemical gradients through the enzyme ATP synthase. Bacteriorhodopsin, a pigmented compound that exhibits red or orange hues, is present inside them.

Thermoacidophiles

Thermoacidophiles refer to a class of organisms that possess the ability to thrive in environments characterized by very high temperatures and low pH levels. These organisms have the ability to endure temperatures as high as 100 degrees Celsius and thrive in an environment with a pH level of 2. The organisms in question exhibit anaerobic characteristics.

Reproduction in Archaea

Archaea engage in asexual reproduction *via* many mechanisms, including binary or multiple fission, fragmentation, and budding.

The procedure of isolating cultures of Archaea from soil that is free of contamination is an intricate one, and all efforts made up to this point have been futile in terms of obtaining pure isolates that are acceptable for research into the essential functions that Archaea play in the ecosystems of forest soil [41]. In 2011, it was shown that nitrifying acid soil might effectively support the establishment of an obligatory acidophilic ammonia oxidizer [42]. Molecular ecology technologies have emerged as the most effective and rational approach for studying Archaeal diversity and function. The investigation of methane production in the soil inside boreal peatlands has led to the identification of a correlation between the presence of methane-producing Archaea sequences and the capacity for methane generation. This connection establishes a link between the variety of archaeal species and their functional role in methane production [43 - 45].

In 2002, researchers demonstrated the correlation between soil depth and methanogen diversity in Finland's boreal peatlands. Methanogens necessary for methane production were also identified for the first time in these peatlands using the genetic marker gene mcrA (methyl coenzyme M reductase) [46]. It was discovered through studying the link between methanogen diversity and boreal fen microsites that different fen features, such as hummocks, slopes, and lawns, support distinct populations of methane-producing archaea [47]. The 16S rRNA gene, which acts as a phylogenetic marker, and the mcrA gene, which acts as a functional marker, were the two molecular markers that were used in order to define the methanogenic bacteria that were found in a drained bog that was situated in Northern Finland. These genes were located in the same drained bog [45]. Ash fertilization has been used as a strategy to increase tree growth, impacting the architectures of methanogen communities. This has been accomplished *via* the employment of a variety of different techniques [48, 49].

Genome Sequences of Archaea

The analysis of extended genomic segments from uncultivated I.1b Crenarchaeota and the genome sequences of I.1a Crenarchaeote. The discovery of *Cenarchaeum symbiosum* has prompted the suggestion that Group I Crenarchaeota may represent a new taxonomic group. The taxonomic classification of the archaeal domain encompasses the phylum level [53]. Thaumarchaeota is the name that has been given to this newly discovered phylum [54]. Thaumarchaeota, sometimes referred to as collection I Crenarchaeota, is a diverse collection of organisms that were once classified under that name. There are aerobic ammonia-oxidizing marine Group I.1a Archaea, such as Cenarchaeum, Nitrosoarchaeum, Nitrosopumilales, and Nitrosopumilus. They are also included in this group. These organisms are known for their ecological versatility and may be found in many soil habitats. The first group, Group I.1b Archaea, consists of the taxa Nitrososphaerales and Nitrososphaera [55].

Methanobacterium   thermoautotrophicum   (Methanobacterium thermoautotrophicum str. Marburg) is classified as a hydrogenotrophic methanogen since it has the ability to generate methane by the reduction of carbon dioxide using hydrogen as a substrate. It is often seen in anaerobic habitats, such as geothermal regions and the gastrointestinal systems of animals. *Halobacterium salinarum*, also known as Halobacterium NRC-1, is a species of halophilic archaea. The above-mentioned species is classified as a halophile, indicating its ability to flourish in settings characterized by high salinity levels, such as salt flats and pans. This creature has a remarkable ability to withstand salt concentrations beyond those tolerated by most other species. *Pyrococcus furiosus*, an extremophilic archaeon, exhibits robust growth in settings characterized by elevated temperatures, such as deep-sea hydrothermal vents. The organism has the ability to thrive in environments with temperatures reaching up to 100°C (212°F), making it a subject of interest in the field of extremophile research.

The discovery of additional genetic material from newly found archaeal lineages has offered essential insights into most of Thaumarchaeota's untapped metabolic capacity. Thaumarchaeota has been seen in poplar plantations located in the area of China's northern region, which is characterized by a moderate climate. The amount of Archaea present in the different soil strata observed inside poplar plantings was found to be relatively low. However, it was observed that Thaumarchaeota exhibited a reduced relative abundance with increasing soil depth [56]. The Soil Crenarchaeotic Group (SCG) has been identified as the predominant class of Archaea sequences that have been uncovered. Genetic research findings have established a substantial association between Thaumarchaeota and the biological processes of ammonia oxidation and carbon sequestration.

Furthermore, previous research has shown that Thaumarchaeota has a substantial capability for denitrification in hypoxic conditions [57, 58]. With increasing depth in the first four layers of poplar plants, the relative abundance of Euryarchaeota was shown to grow significantly. The observed pattern has been ascribed to their preference for anaerobic environments. The Thaumarchaeota ammonia oxidizers, namely those classified under Groups 1.1a and 1.1b, mostly exhibit an aerobic metabolic preference. Nevertheless, it is important to acknowledge that Thaumarchaeota that are firmly buried may not consistently display aerobic traits, and there is still uncertainty around their metabolic capacity. The existence of both aerobic and anaerobic metabolic processes has been supported by the findings of a significant body of research that has offered empirical evidence in favor of this proposition. The phenomena being examined were further examined inside controlled mesocosms, where Scottish pine wood soil was separated into oxic and anoxic conditions [59].

ARCHAEA IN TROPICAL FOREST

Archaea perform crucial roles in biogeochemical cycling and nutrient recycling in tropical forests. Participate in processes including nitrogen fixation, nitrification, and methanogenesis. These processes are crucial for maintaining the ecosystem's balance of nutrients and gases. In the context of research into basic organisms found in forest soils, the archaeal domain is frequently juxtaposed with the bacterial domain. Methanogenic archaea are one of the most well-known archaea found in tropical forest environments. They flourish in anaerobic environments, such as flooded soils, marshes, and wetlands. As a consequence of their metabolic processes, methanogenic archaea generate methane gas. In tropical forests, these microorganisms contribute to the ecosystem's total methane emissions. When compared to the rhizosphere, it was discovered that the phylum and family levels had a stronger influence on the relative abundance of the bulk soil in the Amazonian rainforest.

The archaeal taxa that were found to be abundant in this study were Nitrosopumilus, which belongs to the Thaumarchaeota phylum. These bacteria, which can oxidize ammonia and contain nitrite reductase activity, are sometimes referred to as ammonia-oxidizing archaea. The use of aerobic oxidation of ammonia to nitrite for chemolithoautotrophic development, the mesophilic crenarchaeon Nitrosopumilus maritimus, was discovered. This discovery marked the first detection of nitrification inside the Archaea domain [60]. The involvement of Thaumarchaeota in geochemical cycles has been shown [54], and these microorganisms are recognized for possessing copper-dependent nitrite reductases (NirK). Furthermore, the presence of Desulphurococcaceae, a group of microorganisms capable of metabolizing sulfur, was observed in the Amazonian jungle. Several energy metabolism mechanisms are thought to rely heavily on these microbes. The phenomenon of the observed transition in the prevalence of archaea from the rhizosphere to the bulk soil may be elucidated by considering the variable of soil depth.

The relatively low prevalence seen in comparison to other taxonomic groups of fungus and bacteria may be attributed to a putative compartmentalized function within the microbiome. The presence of Archaea in bulk soil that exhibited a notable increase in abundance was associated with acidic soil conditions. This correlation was established due to the identification of acidophilic Archaea belonging to the Ferroplasmaceae and Picrophilaceae families. The conversion of the Amazonian rainforest into agricultural land was the primary focus of the study of microbial community structures. A comparison was conducted between forest and grassland locations in terms of their methane turnover. The forest location was found to have a negative methane flow due to methane consumption, even in the rainy season [61]. The study documented a positive methane flow in a cow pasture, indicating the occurrence of methane emissions specifically during the dry season [35].

According to the study, methane-cycling microorganisms in the Amazon showed a clear reaction to land-use changes. Specifically, the dominant reaction was an increased population of methane-consuming microorganisms. A notable discovery from the research indicated a statistically significant rise in the relative abundance of methanotrophs compared to methanogens inside the forest ecosystem. This showed a significant decrease in the number of genes encoding a particulate methane monooxygenase and a decrease in the relative abundance of methanotrophs. The significant differential in the number of methanotrophs compared to methanogens within the forest ecosystem was a discovery that stood out as particularly interesting [62]. Changes in the methanotroph community pointed to forest conversion for grazing as the cause of those changes. The alteration in land use was hypothesized to affect methane flow in the Amazon because of the highly sensitive nature of methane-cycling microorganisms. Methanogenic Archaea composition changed significantly between forest and grassland at the operational taxonomic unit level.

The ammonia-oxidizing archaea are a second type of archaea typically found in tropical forest soils. They play an essential role in the nitrogen cycle by converting ammonia into nitrite or nitrate, which is then converted to nitrogen dioxide by other microorganisms during the denitrification process. Archaea contribute considerably to the ecological processes that define the unique ecosystems of tropical forests. Their actions impact nutrient availability, greenhouse gas emissions, and the health of the ecosystem as a whole. Understanding and managing tropical forests, which are highly complex ecosystems, and their reactions to environmental changes like climate change may be aided by research on the archaea that live in these woods.

Archaea as a domain of life remains an enigma due to the paucity of actual isolates to examine. In this study, the relationship between Archaea and forests has been elucidated in order to clarify their functions and duties. Thermophiles, Halophiles, metabolic nitrifiers, acidophiles, and methanogens are the five major physiological categories of cultivated archaea. Bacterial occurrence in environmental gradients and alterations is frequently compared to Archaeal occurrence. It has been postulated that archaea and bacteria have different functional responses to prolonged circumstances of energy stress that include both the maintenance energy and the biological energy quantum [63].

The term "maintenance energy" refers to the minimum energy flow that is essential to maintain cellular function and is the product of catabolism. This energy flow is distinct from the energy that is required for growth or survival [64]. The term "biological energy quantum" refers to the minimum quantity of catabolic energy required for the continued existence of an organism. This maintenance process requires the utilization of a chemiosmotic potential, which is important for anaerobic organisms. Starvation occurs when an organism is unable to get the necessary maintenance energy, whereas the inability to acquire the biological energy quantum results in the decoupling of energy conservation from catabolism [61]. This summarizes the current knowledge emergence of life on Earth, emphasizing the theory that bacteria and archaea came first, with eukaryotes deriving from archaea. According to a study [65], energy stress served as a selection factor that drove the evolutionary split between archaea and bacteria.

In the context of tropical forests, archaea play a substantial role in influencing the ecological processes that are crucial in shaping the distinct characteristics of these ecosystems. The actions undertaken by individuals significantly impact the availability of nutrients, emissions of greenhouse gases, and the general health of ecosystems. Insights into the complex dynamics of tropical forest ecosystems and their responses to environmental changes, such as climate change, may be gained *via* the study of archaea within these ecosystems.

VIRUSES IN FOREST ECOSYSTEM

Viruses are integral components of forest ecosystems, similar to their significance in other ecological systems. Viruses, despite their diminutive size and relative simplicity compared to other biological entities, can substantially influence the well-being and ecological dynamics of forest ecosystems. Viruses have been identified in several fungal organisms, including endophytes such as *Trichoderma harzianum* and obligate parasites like *Puccinia striiformis* [66, 67]. Bao and Roossinck presented a comprehensive survey indicating that mycoviruses inside endophytes are widely postulated [67, 68].

Forests provide a considerable array of viral varieties. Viruses are widely distributed and have the ability to infect a diverse range of creatures, including plants, animals, fungi, and bacteria. Plant viruses have significant importance within the realm of forest ecosystems due to their potential to impact the overall well-being and viability of trees and other flora. A considerable proportion of the presently documented mycoviruses are found to infect grass endophytes, as shown by studies indicating that mycoviruses have been identified in 53 distinct grass endophytes. This represents a considerable fraction of the mycoviruses that are presently known to infect grass endophytes [68, 69]. The identification of *Epichloe festucae* virus 1, a double-stranded RNA virus, was explicitly associated with the grass endophyte *Epichloe festucae* [70].

Tree endophytes have the potential to function as hosts for mycoviruses as well. In this instance, the presence of a mycovirus was observed in *Colletotrichum gloeosporioides*, an endophytic organism known for its detrimental impact on cashew trees (*Anacardium occidentale* L.) by causing anthracnose infection. Specifically, viruses that are a part of the family Endornaviridae were found inside the root mycorrhizal Ceratobasidium fungus. These viruses have the capacity to infect plants, fungi, and oomycetes, and they were discovered within the fungi [71, 72]. In the study conducted on *Pterostylis sanguinea*, a native orchid species in the western part of Australia, the researchers discovered a total of 22 previously unidentified mycoviruses inside the fungal symbionts that were examined [73]. In addition, it should be noted that mycoviruses exhibit a broader range of host specificity beyond filamentous, phytopathogenic fungi since they have been shown to infect edible mushrooms as well. This observation underscores the economic importance of these viruses [74, 75]. Furthermore, it has been shown that mycoviruses are present inside the parasitic fungus associated with mushrooms [76 - 78].

Certain viruses are known to be plant pathogens, leading to the development of disease in many species of trees and plants. These diseases have the potential to result in diminished growth, decreased yields, and, in some cases, even the death of the affected plants. Typical symptoms are leaf browning, stunting, necrosis, and deformation. Plants, mycoviruses, the environment, and fungal endophytes together engage in a mutually interacting system. In natural ecosystems, the dynamics of interactions between different components are subject to continuous fluctuations. For instance, abiotic stressors, such as dryness, have the potential to selectively benefit some components, such as pathogens, hence promoting the infection of host plants [79 - 81].

Speculation has arisen over the capacity of a symbiotic virus to alter the pathogenicity of the host endophyte by eliciting modifications in the phenotypic or gene expression profile of the infected endophyte [82]. Prior studies have shown that these modifications may be heritable since mycoviruses have demonstrated the ability to inhibit the host's RNA-silencing defense systems, perhaps through epigenetic modifications [83, 84]. There has been speculation on the potential of a symbiotic virus to induce pathogenicity in the host endophyte by modifying the infected endophyte's phenotypic or gene expression profile. Previous research has shown that these modifications may be heritable since mycoviruses have demonstrated the ability to inhibit the host's RNA-silencing defense systems, perhaps through epigenetic modifications [83]. The effect of mycoviruses on the plant-endophyte connection may manifest in several ways, such as the augmentation or diminishment of host virulence (hyper- and hypovirulence, respectively), as well as alterations in host fitness in response to changes in environmental circumstances [67].

However, it should be mentioned that most endophytes do not cause plants to show any apparent symptoms. So, the presence of a mycovirus in a certain endophyte may have little effect on both the endophyte host and the plant with which it is linked. The degree to which this impact is exerted is directly proportional to the particular action method. Nevertheless, it is worth considering that mycoviruses could be utilized as viable bioagents for managing fungal diseases. This assertion is supported by existing evidence since it has been shown that mycoviral infections in several fungal species may induce a state of hypovirulence.

The mycovirus known as *Cryphonectria hypovirus* 1 (CHV1) is a particularly noteworthy example of this phenomenon since it reduces the virulence of the pathogenic fungus Cryphonectria parasitica in order to prevent the incidence of chestnut blight in the United States [83, 83]. Several other instances of mycoviruses that induce hypovirulence in the host fungus include *Cryphonectria hypovirus* 2 (CHV2) in *Cryphonectria parasitica* [85], *Sclerotinia sclerotiorum* hypovirus 1 in *Sclerotinia sclerotiorum* [86], and W370 double-stranded RNA in

Rosellinia necatrix [87]. Furthermore, substantial advancements have been achieved in identifying and characterizing mycoviruses and their associated domains responsible for inducing hypovirulence in the host organism [88].

The impact of climate change on viral dynamics is that the alterations in temperature patterns can potentially influence the spatial distribution and frequency of viral infections within forest ecosystems. The effects of increased temperatures and changes in precipitation on the population and behavior of viral vectors might have significant implications for the dynamics of diseases within forest ecosystems.

MICROBIOTA OF FOREST NURSERIES

The term "microbiota of forest nurseries" encompasses the vast array of microbes, such as bacteria, fungi, archaea, and other microorganisms, that reside in the soil, plant roots, and surrounding ecosystem within forest nursery environments. Microorganisms are crucial to the preservation of forest ecosystems and the prosperous development of tree seedlings in nurseries. Several microorganisms are often seen in the roots of seedlings grown on growth substrates.

The impact of these factors on reciprocal development and achievement might vary from no discernible correlation to a state of synergy or antagonism. The significance of these interactions is growing due to the prospective use of microorganisms as commercially viable biological soil additives [89]. The advantageous applications of these microorganisms include enhancing plant productivity and mitigating the impact of microbial diseases *via* implementing a biocontrol strategy (Fig. **3**).

TREE PESTS

Tree pests are living creatures that have the potential to inflict harm on trees, resulting in a range of detrimental consequences, including the impairment of leaves, branches, trunks, and roots. Insects, mites, fungi, bacteria, viruses, nematodes, and even rats are some species that might be considered pests. Fungal infections have the potential to diminish the structural integrity of trees, impede their development, and, in some cases, result in tree mortality if not adequately addressed. It is anticipated that both natural and agroecosystems within the African continent would exhibit susceptibility to the impacts of climate change [91]. The potential consequences of this phenomenon are expected to significantly influence nations located in the Sub-Saharan region, notably concerning ecological well-being, food accessibility, land administration, and the sustenance of rural communities [91-93].

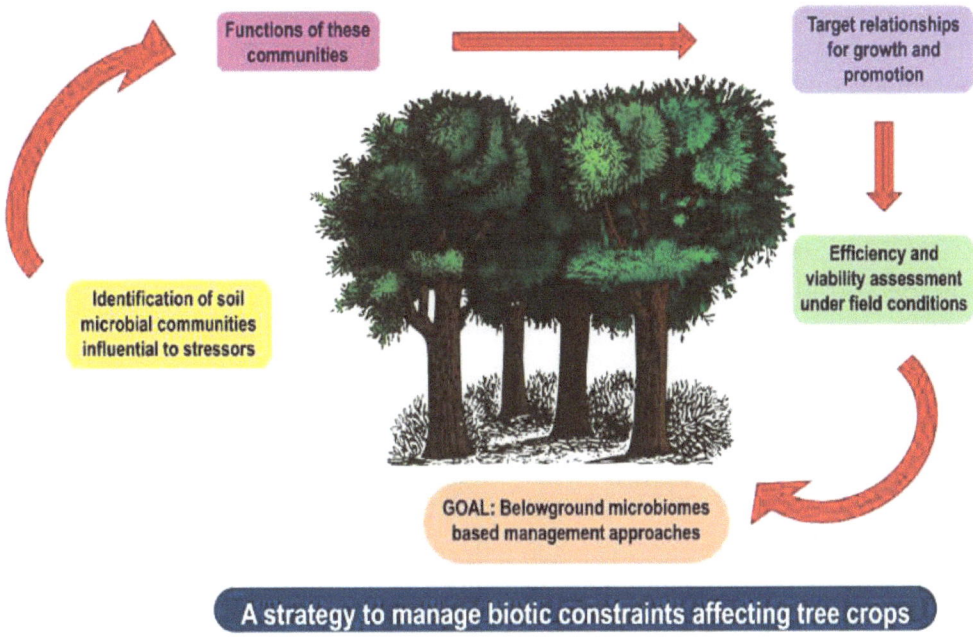

Fig. (3). Management of tree crop infections, pests, and invasive species *via* the study, classification, and use of soil and root microbiota [based on a conceptual framework by [90].

Trees and forest cover on farmland have a significant role in slowing global warming and storing carbon dioxide [94 - 96]. In plantations and agroforestry systems throughout Africa, evidence indicates that herbivorous pathogenic microorganisms, both local and invasive, are negatively impacting native tree species.

Herbivorous pathogenic microbes of indigenous and native species are damaging African plantations and agroforestry systems, and it has been observed. According to the genome sequences report by the FAO of the United Nations [1, 2], about a hundred different types of insects and diseases have been documented damaging trees in both artificial and natural forests throughout Africa. These regions include northern, western, eastern, central, and southern African countries such as Malawi, South Africa, Mauritius, Ghana, Morocco, Sudan, and Kenya.

Approximately 50% of the species under consideration have been recognized as native, whilst more than 33% may be classified as non-native, sometimes referred to as invaders. The remaining 15% have an undetermined origin. Invasive species account for more than 33% of insects that cause harm to trees, whilst over 66% of diseases are either non-native or their origin is unclear. Broadleaf trees are more susceptible to insect and disease species than conifers, where just one-third of such species are shown to cause damage. The observed pattern indicates the botanical makeup of the indigenous plant species, which primarily comprises a significant proportion of broad-leaved plants with a substantial presence of conifers that have been introduced *via* tropical planting efforts [98].

A total of 47 indigenous and 19 non-indigenous organisms that defoliate feed on sap, burrow into wood, and attack shoots are causing injury. Examples include plant species *Acacia mearnsii* from the Fabaceae family, several species of Eucalyptus from the Myrtaceae family, and different species of Pinus from the Pinaceae family. Furthermore, it has been shown that the teak tree, scientifically known as *Tectona grandis* (Lamiaceae), along with plantations [99], is susceptible to damage caused by 15 different kinds of diseases, which may significantly harm eucalyptus [100].

CHALLENGES AND POTENTIALS

The forest microbiome, which is made up of bacteria, fungi, archaea, and viruses, plays a crucial role in determining the health and functional dynamics of forest ecosystems. Soil represents a diverse ecosystem on our planet and is characterized by various biotic and abiotic constituents. Within this environment, many fungi, bacteria in the billions, and bigger creatures such as worms, ants, and moles find their dwelling [101]. Plants rely on soil bacteria to protect against biotic and abiotic stresses and recycle important nutrients like nitrogen and phosphorus [102, 103]. The implementation of intensive agricultural practices has resulted in notable enhancements in crop productivity. However, it is important to acknowledge that these practices have also adversely impacted soils' physical and biological characteristics [105, 106].

Using fertilizers in intensively managed agricultural systems may help slow the reduction in soil fertility. Conversely, the practice of tillage has the potential to disturb microbial populations [107]. This issue has significant relevance in the context of contemporary crop production methods, given the alarming deterioration seen in more than 50% of the world's agricultural land. Simultaneously, we have significant issues linked to the perturbation of nitrogen and phosphorus cycles. This condition is anticipated to exacerbate in the context of climate change [108, 109]. In order to lessen the damage caused by land degradation, the United Nations has advocated for the reinstatement of sustainable land management practices [110].

The phenomenon of nutrient cycling within a forest ecosystem pertains to the ongoing and iterative process through which vital nutrients are exchanged and reused among species and their surrounding environment. The process mentioned above is vital to the continued health and productivity of the forest ecosystem and plays a significant role.

Crop diversification, the use of locally adapted species or intercropping strategies to maintain the fertility of the soil, capturing carbon, cycling of nutrients, and controlling soil erosion are some of the practices that have been described above [109]. These measures have significantly improved the overall reduction of soil-borne diseases [110, 111]. Furthermore, the maintenance of microbial community variety, structure, and composition may contribute to preserving ecosystem services, such as regulating nutrient cycles. Overall, nutrient cycling is a multifaceted and interrelated phenomenon that maintains the sustainable operation of forest ecosystems *via* the preservation of a harmonious distribution of vital nutrients. The overall sustenance of the ecosystem needs to provide support for the growth and development of plants, animals, and bacteria in order to ensure the ecosystem's continued existence (Fig. **4**).

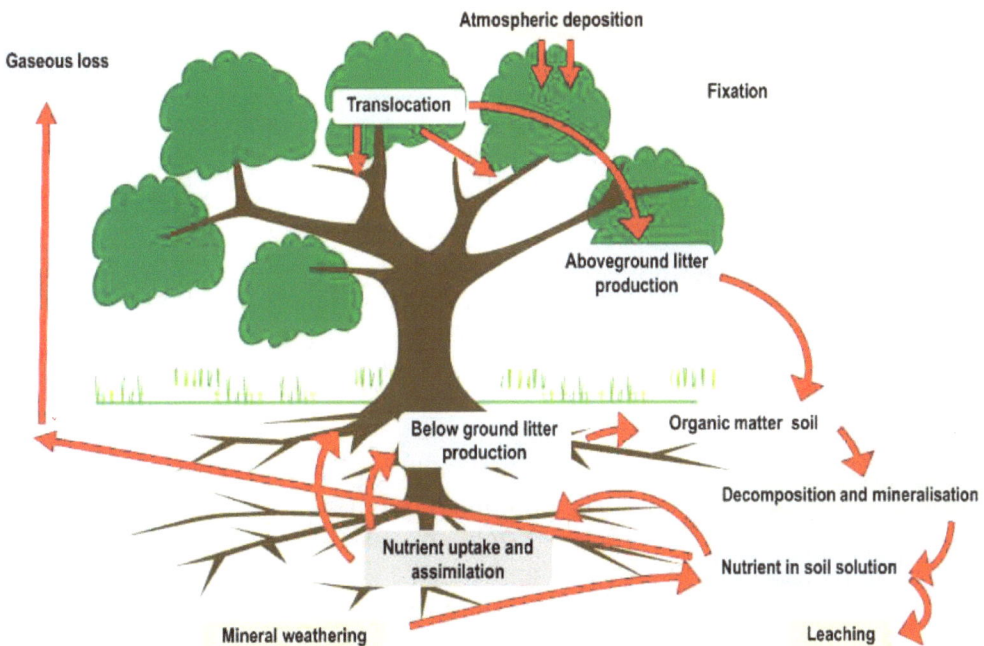

Fig. (4). A conceptual diagram of nutrient cycling of forest ecosystem [112].

TREES AND THE MICROBIAL COMMUNITIES

The interdependence between trees and microbial populations is a subject of great interest due to its complicated nature and significant impact on the functioning of ecosystems. This interaction may be comprehended from two distinct perspectives: firstly, the effect exerted by trees on microbial populations, and secondly, the impact of microbial communities on trees. As significant living organisms, trees have a significant impact on the microbial communities that are present in their immediate surroundings. This includes both the soil and the rhizosphere, which is the term given to the region of soil that is located around the tree's roots. Multiple lines of evidence suggest that alterations in the plant community have the potential to instigate significant modifications in the carbon and nitrogen cycle within ecosystems [113, 114]. Soil microbes help in soil nutrient cycle processes, serving as the primary drivers. They serve as the essential link between changes in the composition of the dominant vegetation and fundamental shifts in the way the ecosystem functions, which they mediate. The extent to which plant community features affect soil microbial activity and composition is a fascinating discovery. Changes in the plant community owing to disturbances like climate change and selective breeding of species for forestry objectives are only two examples of how the interaction between two components of an ecosystem may lead to unexpected variations in the functioning of the ecosystem.

Sugars, amino acids, and organic acids are only some of the compounds that trees secrete into the soil *via* their root exudates. Many microbial communities have formed around the area around the roots because the exudates provide food for many different kinds of bacteria. Various tree species have distinct root systems, hence generating distinct microhabitats inside the soil. For example, some trees establish mycorrhizal associations, a mutually beneficial partnership with certain fungi, increasing their nutrient absorption capabilities. The composition and diversity of soil microbes may be further altered as a result of these interactions [114].

The features of the plants that are currently growing in a location have the potential to impact the microbial community's makeup as well as its ability to perform its functions. That impact establishes itself in various ways, such as changes in the local climate brought on by shade and frost protection, the passage of time throughout autumn, and the absorption and loss of soil water *via* transpiration. In addition, the formation of litter by aboveground systems, as well as root systems, interactions with herbivores in both aboveground and belowground habitats, and the discharge of root exudates all play a role. Notwithstanding the intricate network of interconnections, the degree of robustness and immediacy in the connections between plant and microbial populations remains uncertain. The challenges associated with characterizing microbial communities have been improved *via* the use of molecular methods [115].

Research carried out on natural stands shows the influence of different types of trees on the microbial communities found in mineral soils. This research, for instance, analyzed polylactic acid (PLA) profiles in mineral soils from several plots in a mixed deciduous forest in Germany. These plots were characterized by dominant species compositions, including Fagus, Tilia, and Fraxinus, as well as plots with the additional presence of Carpinus and Acer [116]. Samples taken from depths ranging from 0 to 5 cm and 10 to 20 cm showed that the microbial communities differed depending on which of the three unique plot types they were taken from. The differentiation seen in this study was shown to be associated with the presence of Tilia and Acer compared to Fagus and with variations in pH levels. This suggests that the presence of species that generate litter with a high base content has the most significant impact on soil microbial communities, which aligns with the cedar effect previously discussed [117]. In the state of Colorado, this research looks at the different fungal and bacterial communities that may be found within the first 10 cm of soil in neighboring stands that are distinguished by the preponderance of pine, spruce, and aspen trees. Subsequent investigations have shown variations in soil microbial assemblages throughout distinct arboreal ecosystems, including disparate tree species, which were geographically disjunct and not situated inside shared botanical enclosures [118 - 120].

The process of leaf shedding in trees contributes to the provision of organic matter for soil bacteria *via* the decomposition of leaf litter. Because of the substantial role that this process plays in the cycling of nutrients and carbon sequestration, it affects the dynamics of the microbial population. Soil temperature, moisture, and nutrient availability are just a few of the internal soil environmental factors that trees may affect. In turn, these changes affect the composition and performance of microbial communities. In the conducted investigations, isolating the exclusive impact of tree species is challenging due to the presence of confounding variables such as soil composition and topographical variations. Therefore, the observed discrepancies may be ascribed not only to the intrinsic properties of the tree species but also to the influence of the site-species combination.

GENERALISTS AND SPECIALISTS MICROORGANISMS AMONG TREES

Based on their interactions with trees and the ecosystem, "generalists" and "specialists" refer to distinct categories of microorganisms among trees. Generalist microorganisms are those that can interact with multiple varieties of trees and have a broad spectrum of host species. These microorganisms are adaptable and able to flourish in a variety of ecological niches. Some fungi and microbes, for instance, can establish mutualistic relationships with various tree species, supplying them with essential nutrients in exchange for access to the tree's resources. Generalists may also include decomposers that can decompose organic matter from various tree species, playing a crucial role in the ecosystem's nutrient cycling. Specialist microorganisms, on the other hand, have a specific and frequently restricted host range. These microorganisms are highly adapted to particular tree species and may have evolved specialized interaction mechanisms with them.

Some fungi, for instance, create mycorrhizal associations that are highly specialized for specific tree species, meaning that they can only form mutually beneficial relationships with those trees. Likewise, some pathogens and diseases may only affect certain tree species or closely related species, rendering them specialists in their interactions. Microorganisms are ubiquitous and perform essential functions in numerous ecosystems, such as the biogeochemical cycles of the earth [121], human metabolism [122], and biotechnological processes [123]. Microbial generalists and specialists have distinct effects on microbial community dynamics [123, 124].

As a consequence of continual movement and competition with invasive species, microbes improve their capacity to survive by becoming either generalists (i.e., adapt to environments) or specialists (i.e., those that adapted to particular habitats). Specialists have a limited spectrum of habitats and specific environmental fitness compared to generalists [125]. The role of specialists and generalists in species interactions may be comparable to that of ubiquitous and rare microbes, respectively. This can be indicated by a higher degree of connections (that is, more links with other species in the network) and a bigger number of keystone species than generalists in the topologies of co-occurrence networks.

This is based on the hypothesis that experts are more reliant on the interactions between species, such as auxotrophy [126]. On top of that, it is predicted that specialists would have more strict development circumstances, which may include particular metabolic needs, while generalists are less influenced by or better buffered against the effects of environmental filtration [127]. The variety of microorganisms that live among the trees and their interactions with one another play a vital part in the health and operation of forest ecosystems. On the other hand, specialists have the potential to have a more focused and, in some cases, considerable influence on specific tree populations. Generalists are beneficial to the ecosystem as a whole because they help maintain its resilience and stability. Understanding these relationships is crucial for sustainable forest management and the conservation of biodiversity.

INFLUENCED OF FOREST COMPOSITION ON THE STRUCTURE OF MICROBIAL COMMUNITIES

Forest tree microbial communities are influenced by a wide variety of biotic variables, both in terms of their composition and their architecture. These are the organisms that interact directly or indirectly with the microbial populations associated with forest trees. Some of the most important biotic factors are tree species, tree age and health, root exudates, leaf detritus decomposition, plant pathogens, forest succession, human activities, competition and predation, biological interactions, and mycorrhizal associations. It is of the highest significance to examine the effect of abiotic factors on the make-up and organization of the microbial communities linked with forest trees. These characteristics are related to abiotic aspects of the environment that have an effect on the number, distribution, and interaction of microorganisms found in the soil and on the surfaces of the roots of trees in the forest. The impacts of various abiotic conditions on microbial communities in forest trees include many critical elements, including soil pH, texture, moisture content, temperature, soil organic matter, nutrient availability, oxygen and light availability, and soil compaction. In general, the interaction of these non-living components generates distinct microenvironments within the soil of the forest and on the surfaces of trees, resulting in varied and dynamic communities of microorganisms that have crucial functions in the processes of nutrient cycling, soil quality, and tree well-being.

The physiological control significantly influences plants' interaction with the inorganic parts of their environment that microorganisms exert over plant processes. In order to ensure their survival, trees need a sufficient supply of various resources, including water, light, carbon dioxide, oxygen, heat, nutrients, and other abiotic factors. These resources are essential for sustaining trees' life, growth, and reproductive processes. When these variables are present in insufficient or excessive quantities, they give rise to mortality. A certain quantity of trees must undergo mortality as the forest progresses in order to sustain a state of optimal health. Various abiotic conditions may lead to the demise of trees *via* distinct mechanisms, such as desiccation, uprooting, or stem fracture. The assessment of sustainability and productivity in relation to changing stand structures is contingent upon the analysis of mortality patterns and the subsequent response of the forest ecosystem.

THE CONCEPT OF "MICROBIAL "HUBS."

Microbial hubs, sometimes referred to as microbial communities or microbiomes, denote intricate ecosystems of microorganisms that inhabit many settings, including soil, water, flora, fauna, and even the human anatomy. These ecological communities comprise many microorganisms, including bacteria, archaea, fungi, viruses, and other microbes, which interact with one another and their surrounding environment. In recent years, there has been a noticeable increase in the attention given to the study of microbial hubs. The development of DNA sequencing technology has allowed scientists to explore and understand the diversity and functions of numerous microorganisms throughout many ecosystems, which may account for the present surge in interest. Through the examination of the genetic material of these microorganisms, researchers are able to discern the species that are present, elucidate their relationships, and ascertain their possible ecological functions [128].

The scientific world has come to acknowledge the idea that hosts and the microbial populations that coexist with them should be regarded as a single, indivisible entity. This idea is referred to as the meta-organism hypothesis. It is now widely acknowledged that the ecology and evolution of hosts and their associated microbial communities are intricately interconnected [129, 130]. Both the phyllosphere, which refers to the portions of plants that are aboveground, and the rhizosphere, which refers to the parts of plants that are belowground, act as homes for a wide variety of different microorganisms. Microorganisms like this significantly influence the health and development of plants, which means they play an essential part in determining the fate of plants [131].

In addition to this, they serve as a barrier against herbivores and contribute to the development of resistance against a variety of illnesses, which in turn shape the evolutionary path that plants take [132]. The comprehension of the plant holobiont, including both the plant itself and the creatures inhabiting it, has significant implications for crucial aspects such as human food security, biodiversity preservation, and the operation of ecosystems [133, 134]. Three significant processes influence microbial community architectures. Random colonization occurrences, species sorting based on local criteria like nutrition availability, host availability, and microbial interactions, and isolating factors like dispersion and distance are all factors that affect colonization patterns [135, 136].

The distribution and composition of bacteria and fungal communities in plants are influenced by a wide range of neutral, abiotic, and host factors [137 - 139]. These factors influence the presence of certain microorganisms that facilitate their colonization in particular plant host conditions [140, 141]. Nevertheless, while these adaptations have the ability to establish a connection between abiotic and biotic host characteristics and the efficacy of colonization, it is crucial to recognize that they cannot be comprehended in isolation. This occurs when the host, acting as a holobiont, is simultaneously colonized by a variety of prokaryotes and eukaryotes [142].

ECOSYSTEM SUSTAINABILITY AND RESTORATION

The notions of ecosystem sustainability and restoration have significant importance in the realm of environmental conservation and management. The primary objective of their efforts is the preservation of the vitality and operational integrity of natural ecosystems, as well as the restoration of ecosystems that have suffered deterioration or impairment due to human activities or natural calamities. The concept of ecosystem sustainability pertains to the capacity of an ecosystem to endure and uphold its fundamental processes, variety of species, and ability to recover from disruptions. Sustainable ecosystems have the capacity to sustainably maintain and cater to the needs of both present and future generations while concurrently avoiding the depletion of resources and the infliction of irreversible damage to the environment. The task entails preserving equilibrium between human activities and the ecosystem's ability to replenish and recuperate from disruptions.

Ecosystem sustainability and restoration include efforts to restore and rehabilitate systems that have experienced significant destruction or degradation while also emphasizing the conservation of existing systems. Ecosystems that exhibit enhanced health and possess more biodiversity are associated with a range of advantageous outcomes, including increased soil fertility, improved wood and fish yields, and bigger reservoirs of greenhouse gases. Ecological restoration, as defined by the Ecological Restoration Society, encompasses implementing several methods to facilitate the recuperation of ecosystems that have undergone degradation, damage, or complete destruction. The practical approach encompasses a diverse range of strategies and procedures to enhance these ecosystems' functionality and quality. In this way, we can affirm that within its primary goals, Ecological restoration serves to:

1. Ensuring the optimal operation of ecosystems, preservation of biodiversity, and implementation of sustainable management practices.

2. Enhance the physical, socio-economic, and cultural dimensions pertaining to the preservation of ecosystems.

3. Facilitate the development of constructive and progressive connections in individuals and the environments they occupy.

4. The objective is to mitigate the impacts of unfavorable climatic conditions, such as erosion and floods, to ensure the stability of soils as substrates and preserve hydrological systems.

The process of restoration involves the deliberate implementation of measures to either introduce or eliminate external factors that exert pressure on the natural environment, hence facilitating the autonomous recovery of ecological systems. Nevertheless, restoring an ecosystem to its former condition is not always feasible. For example, it is essential to acknowledge the ongoing need for arable land and infrastructure development in areas formerly covered by forests. Additionally, both ecosystems and civilizations are compelled to undertake adaptive measures in response to the dynamic shifts in climatic conditions [143].

CONCLUSION

All types of ecosystems, such as woods, farmlands, towns, wetlands, and seas, have the potential to undergo both destruction and restoration processes. Furthermore, the implementation of restoration programs may be undertaken by various actors, including governmental bodies, development organizations, corporate entities, local communities, and private stakeholders. The multitude and diversity of reasons contribute to the degradation phenomenon, which may manifest at various scales. Because of their vigor and diversity, the ecosystems of the Earth, which include forests, farmlands, lakes, oceans, and coasts, play an essential part in ensuring the prosperity and well-being of humankind. However, the erosion of these invaluable resources is occurring in concerning ways. The United Nations Decade on Ecosystem Restoration is a valuable opportunity to effectively address the current ecosystem challenges and foster a sustainable future for both humanity and the world.

REFERENCES

[1]   Pan YA, Freundlich T, Weissman TA, *et al.* Zebrabow: multispectral cell labeling for cell tracing and lineage analysis in zebrafish. Development 2013; 140(13): 2835-46.
[http://dx.doi.org/10.1242/dev.094631] [PMID: 23757414]

[2]   Food and Agricultural Organization. Forest area - Our World in DataState of the World's Forests 2020 (fao.org).

[3]   Baldrian P. Forest microbiome: diversity, complexity and dynamics. FEMS Microbiol Rev 2017; 41(2): 109-30.
[PMID: 27856492]

[4]   Balasubramanian A. Soil Microorganisms, Report number: 1: University of Mysore, 2017.
[http://dx.doi.org/10.13140/RG.2.2.27925.12008]

[5]   Young PA. Facultative Parasitism and Host Ranges of Fungi. Am J Bot 1926; 13(8): 502-20. [JSTOR].
[http://dx.doi.org/10.1002/j.1537-2197.1926.tb05899.x]

[6]   Leigh EG Jr. The evolution of mutualism. J Evol Biol 2010; 23(12): 2507-28.
[http://dx.doi.org/10.1111/j.1420-9101.2010.02114.x] [PMID: 20942825]

[7]   Bascompte J. Mutualism and biodiversity. Curr Biol 2019; 29(11): R467-70.
[http://dx.doi.org/10.1016/j.cub.2019.03.062] [PMID: 31163160]

[8] Loeschcke V, Christiansen FB. Evolution and Mutualism. In: Wöhrmann K, Jain SK, Eds. Population Biology. Berlin, Heidelberg: Springer 1990.
[http://dx.doi.org/10.1007/978-3-642-74474-7_14]

[9] Sourakov A. Inquilines and Cleptoparasites. In: Capinera JL, Ed. Encyclopedia of Entomology. Dordrecht: Springer 2008.
[http://dx.doi.org/10.1007/978-1-4020-6359-6_1536]

[10] Petersen C, Hermann RJ, Barg MC, *et al.* Travelling at a slug's pace: possible invertebrate vectors of *Caenorhabditis* nematodes. BMC Ecol 2015; 15(1): 19.
[http://dx.doi.org/10.1186/s12898-015-0050-z] [PMID: 26170141]

[11] White PS, Morran L, de Roode J. Phoresy. Curr Biol 2017; 27(12): R578-80.
[http://dx.doi.org/10.1016/j.cub.2017.03.073] [PMID: 28633022]

[12] Iqra B, Aadil FW, Iflah R, Zafar AR, *et al.* Shouche, Phyllosphere microbiome: Diversity and functions, Microbiological Research 2022; p 254 126888, ISSN 0944-5013.

[13] Koskella B. The phyllosphere. Curr Biol 2020; 30(19): R1143-6.
[http://dx.doi.org/10.1016/j.cub.2020.07.037] [PMID: 33022257]

[14] Sun X, Song B, Xu R, *et al.* Root-associated (rhizosphere and endosphere) microbiomes of the Miscanthus sinensis and their response to the heavy metal contamination. J Environ Sci (China) 2021; 104: 387-98.
[http://dx.doi.org/10.1016/j.jes.2020.12.019] [PMID: 33985741]

[15] Babalola OO, Adeleke BS, Ayangbenro AS. Whole genome sequencing of sunflower root-associated Bacillus cereus. EvolBioinform2021b; 17: 16.

[16] Wolfe ER, Ballhorn DJ. Do foliar endophytes matter in litter decomposition? Microorganisms 2020; 8(3): 446.
[http://dx.doi.org/10.3390/microorganisms8030446] [PMID: 32245270]

[17] Adeleke BS, Babalola OO. The endosphere microbial communities, a great promise in agriculture. Int Microbiol 2021; 24(1): 1-17. a
[http://dx.doi.org/10.1007/s10123-020-00140-2] [PMID: 32737846]

[18] Santoyo G, Moreno-Hagelsieb G, del Carmen Orozco-Mosqueda M, Glick BR. Plant growth-promoting bacterial endophytes. Microbiol Res 2016; 183: 92-9.
[http://dx.doi.org/10.1016/j.micres.2015.11.008] [PMID: 26805622]

[19] Yuan ZS, Liu F, Liu ZY, Huang QL, Zhang GF, Pan H. Structural variability and differentiation of niches in the rhizosphere and endosphere bacterial microbiome of moso bamboo (Phyllostachys edulis). Sci Rep 2021; 11(1): 1574.
[http://dx.doi.org/10.1038/s41598-021-80971-9] [PMID: 33452327]

[20] Adeleke BS, Babalola OO. Biotechnological overview of agriculturally important endophytic fungi. Hortic Environ Biotechnol 2021; 62(4): 507-20. a
[http://dx.doi.org/10.1007/s13580-021-00334-1]

[21] Chowdhary K, Kaushik N, Ganapathi TR. Fungal endophyte diversity and bioactivity in the Indian medicinal plant *Ocimum sanctum* Linn. PLoS One 2015; 10(11): e0141444.
[http://dx.doi.org/10.1371/journal.pone.0141444] [PMID: 26529087]

[22] Hiruma K, Kobae Y, Toju H. Beneficial associations between Brassicaceae plants and fungal endophytes under nutrient-limiting conditions: evolutionary origins and host–symbiont molecular mechanisms. Curr Opin Plant Biol 2018; 44: 145-54.
[http://dx.doi.org/10.1016/j.pbi.2018.04.009] [PMID: 29738938]

[23] dos Santos Souza CR, de Oliveira Barbosa AC, Fortes Ferreira C, *et al.* Diversity of microorganisms associated to *Ananas* spp. from natural environment, cultivated and *ex situ* conservation areas. Sci Hortic (Amsterdam) 2019; 243: 544-51.

[http://dx.doi.org/10.1016/j.scienta.2018.09.015]

[24]   Dastogeer K, Oshita Y, Yasuda M, *et al.* Host specificity of endophytic fungi from stem tissue of nature farming tomato (*Solanum lycopersicum* Mill.) in Japan. Agron 2020; 10: 10-9.

[25]   Dubey A, Saiyam D, Kumar A, Hashem A, Abd Allah EF, Khan ML. Bacterial root endophytes: Characterization of their competence and plant growth promotion in soybean (*Glycine max* (L.) Merr.) under drought stress. Int J Environ Res Public Health 2021; 18(3): 931.
[http://dx.doi.org/10.3390/ijerph18030931] [PMID: 33494513]

[26]   Fouda A, Eid AM, Elsaied A, *et al.* Plant growth-promoting endophytic bacterial community inhabiting the leaves of *Pulicaria incisa* (Lam.) DC Inherent to arid regions. Plants 2021; 10(1): 76.
[http://dx.doi.org/10.3390/plants10010076] [PMID: 33401438]

[27]   Shah S, Shrestha R, Maharjan S, Selosse MA, Pant B. Isolation and characterization of plant growth-promoting endophytic fungi from the roots of Dendrobium moniliforme. Plants 2018; 8(1): 5.
[http://dx.doi.org/10.3390/plants8010005] [PMID: 30597827]

[28]   Yang B, Huang J, Zhou X, *et al.* The fungal metabolites with potential antiplasmodial activity. Curr Med Chem 2018; 25(31): 3796-825.
[http://dx.doi.org/10.2174/0929867325666180313105406] [PMID: 29532754]

[29]   Adeleke BS, Babalola OO. The endosphere microbial communities, a great promise in agriculture. Int Microbiol 2021; 24(1): 1-17. a
[http://dx.doi.org/10.1007/s10123-020-00140-2] [PMID: 32737846]

[30]   Adeleke BS, Babalola OO. Biotechnological overview of agriculturally important endophytic fungi. Hortic Environ Biotechnol 2021; 62(4): 507-20. a
[http://dx.doi.org/10.1007/s13580-021-00334-1]

[31]   Cope-Selby N, Cookson A, Squance M, Donnison I, Flavell R, Farrar K. Endophytic bacteria in Miscanthus seed: implications for germination, vertical inheritance of endophytes, plant evolution and breeding. Glob Change Biol Bioenergy 2017; 9(1): 57-77.
[http://dx.doi.org/10.1111/gcbb.12364]

[32]   Baltrus DA. Adaptation, specialization, and coevolution within phytobiomes. Curr Opin Plant Biol 2017; 38: 109-16.
[http://dx.doi.org/10.1016/j.pbi.2017.04.023] [PMID: 28545003]

[33]   Foito A, Stewart D. Metabolomics: A high-throughput screen for biochemical and bioactivity diversity in plants and crops. Curr Pharm Des 2018; 24(19): 2043-54.
[http://dx.doi.org/10.2174/1381612824666180515125926] [PMID: 29766789]

[34]   Hartmann A, Fischer D, Kinzel L, *et al.* Assessment of the structural and functional diversities of plant microbiota: Achievements and challenges – A review. J Adv Res 2019; 19: 3-13.
[http://dx.doi.org/10.1016/j.jare.2019.04.007] [PMID: 31341665]

[35]   Fernandez JC, Burch-Smith TM. Chloroplasts as mediators of plant biotic interactions over short and long distances. Curr Opin Plant Biol 2019; 50: 148-55.
[http://dx.doi.org/10.1016/j.pbi.2019.06.002] [PMID: 31284090]

[36]   Fadiji AE, Babalola OO. Elucidating mechanisms of endophytes used in plant protection and other bioactivities with multifunctional prospects. Front Bioeng Biotechnol 2020; 8: 467.
[http://dx.doi.org/10.3389/fbioe.2020.00467] [PMID: 32500068]

[37]   Xu T, Cao L, Zeng J, *et al.* The antifungal action mode of the rice endophyte *Streptomyces hygroscopicus* OsiSh-2 as a potential biocontrol agent against the rice blast pathogen. Pestic Biochem Physiol 2019; 160: 58-69.
[http://dx.doi.org/10.1016/j.pestbp.2019.06.015] [PMID: 31519258]

[38]   Sahu PK, Singh S, Gupta A, Singh UB, Brahmaprakash GP, Saxena AK. Antagonistic potential of bacterial endophytes and induction of systemic resistance against collar rot pathogen *Sclerotium rolfsii* in tomato. Biol Control 2019; 137: 104014.

[http://dx.doi.org/10.1016/j.biocontrol.2019.104014]

[39]   Woese CR, Fox GE. Phylogenetic structure of the prokaryotic domain: The primary kingdoms. Proc Natl Acad Sci USA 1977; 74(11): 5088-90.
         [http://dx.doi.org/10.1073/pnas.74.11.5088] [PMID: 270744]

[40]   Woese CR, Kandler O, Wheelis ML. Towards a natural system of organisms: proposal for the domains Archaea, Bacteria, and Eucarya. Proc Natl Acad Sci USA 1990; 87(12): 4576-9.
         [http://dx.doi.org/10.1073/pnas.87.12.4576] [PMID: 2112744]

[41]   Lehtovirta LE, Prosser JI, Nicol GW. Soil pH regulates the abundance and diversity of Group 1.1c Crenarchaeota. FEMS Microbiol Ecol 2009; 70(3): 367-76.
         [http://dx.doi.org/10.1111/j.1574-6941.2009.00748.x] [PMID: 19732147]

[42]   Lehtovirta-Morley LE, Stoecker K, Vilcinskas A, Prosser JI, Nicol GW. Cultivation of an obligate acidophilic ammonia oxidizer from a nitrifying acid soil. Proc Natl Acad Sci USA 2011; 108(38): 15892-7.
         [http://dx.doi.org/10.1073/pnas.1107196108] [PMID: 21896746]

[43]   Merilä P, Galand P, Fritze H, Tuittila E, *et al.* development of methanogen communities during a primary succession of mire ecosystems. In: Program and Abstracts of the Joint International Symposia for Subsurface Microbiology (ISSM 2005) and Environmental Biogeochemistry (ISEB XVII). August 14–19, 2005, Jackson Hole, Wyoming, USA.

[44]   Juottonen H, Galand PE, Tuittila ES, Laine J, Fritze H, Yrjälä K. Methanogen communities and *Bacteria* along an ecohydrological gradient in a northern raised bog complex. Environ Microbiol 2005; 7(10): 1547-57.
         [http://dx.doi.org/10.1111/j.1462-2920.2005.00838.x] [PMID: 16156728]

[45]   Juottonen H, Galand PE, Yrjälä K. Detection of methanogenic Archaea in peat: comparison of PCR primers targeting the mcrA gene. Res Microbiol 2006; 157(10): 914-21.
         [http://dx.doi.org/10.1016/j.resmic.2006.08.006] [PMID: 17070673]

[46]   Galand PE, Saarnio S, Fritze H, Yrjälä K. Depth related diversity of methanogen Archaea in Finnish oligotrophic fen. FEMS Microbiol Ecol 2002; 42(3): 441-9.
         [http://dx.doi.org/10.1111/j.1574-6941.2002.tb01033.x] [PMID: 19709303]

[47]   Galand PE, Fritze H, Yrjälä K. Microsite□dependent changes in methanogenic populations in a boreal oligotrophic fen. Environ Microbiol 2003; 5(11): 1133-43.
         [http://dx.doi.org/10.1046/j.1462-2920.2003.00520.x] [PMID: 14641593]

[48]   Galand PE, Juottonen H, Fritze H, Yrjälä K. Methanogen communities in a drained bog: effect of ash fertilization. Microb Ecol 2005; 49(2): 209-17. b
         [http://dx.doi.org/10.1007/s00248-003-0229-2] [PMID: 15965727]

[49]   Bräuer SL, Cadillo-Quiroz H, Ward RJ, Yavitt JB, Zinder SH. Methanoregula boonei gen. nov., sp. nov., an acidiphilic methanogen isolated from an acidic peat bog. Int J Syst Evol Microbiol 2011; 61(1): 45-52.
         [http://dx.doi.org/10.1099/ijs.0.021782-0] [PMID: 20154331]

[50]   Bomberg M, Münster U, Pumpanen J, Ilvesniemi H, Heinonsalo J. Archaeal communities in boreal forest tree rhizospheres respond to changing soil temperatures. Microb Ecol 2011; 62(1): 205-17.
         [http://dx.doi.org/10.1007/s00248-011-9837-4] [PMID: 21394607]

[51]   Teh YA, Silver WL, Conrad ME. Oxygen effects on methane production and oxidation in humid tropical forest soils. Glob Change Biol 2005; 11(8): 1283-97.
         [http://dx.doi.org/10.1111/j.1365-2486.2005.00983.x]

[52]   Yrjälä K, Tuomivirta T, Juottonen H, *et al.* CH4 production and oxidation processes in a boreal fen ecosystem after long-term water table drawdown. Glob Change Biol 2011; 17(3): 1311-20.
         [http://dx.doi.org/10.1111/j.1365-2486.2010.02290.x]

[53]   Hallam SJ, Konstantinidis KT, Putnam N, *et al.* Genomic analysis of the uncultivated marine

crenarchaeote *Cenarchaeum symbiosum.* Proc Natl Acad Sci USA 2006; 103(48): 18296-301. [http://dx.doi.org/10.1073/pnas.0608549103] [PMID: 17114289]

[54]   Brochier-Armanet C, Boussau B, Gribaldo S, Forterre P. Mesophilic crenarchaeota: proposal for a third archaeal phylum, the Thaumarchaeota. Nat Rev Microbiol 2008; 6(3): 245-52. [http://dx.doi.org/10.1038/nrmicro1852] [PMID: 18274537]

[55]   Pester M, Schleper C, Wagner M. The Thaumarchaeota: an emerging view of their phylogeny and ecophysiology. Curr Opin Microbiol 2011; 14(3): 300-6. [http://dx.doi.org/10.1016/j.mib.2011.04.007] [PMID: 21546306]

[56]   Feng H, Guo J, Wang W, Song X, Yu S. Feng h, Guo J, Wang W, Song X, Yu S. Soil depth determines the composition and diversity of bacterial and archaeal communities in a poplar plantation. Forests 2019; 10(7): 550. [http://dx.doi.org/10.3390/f10070550]

[57]   Könneke M, Schubert DM, Brown PC, *et al.* Ammonia-oxidizing archaea use the most energy-efficient aerobic pathway for CO_2 fixation. Proc Natl Acad Sci USA 2014; 111(22): 8239-44. [http://dx.doi.org/10.1073/pnas.1402028111] [PMID: 24843170]

[58]   Walker CB, de la Torre JR, Klotz MG, *et al. Nitrosopumilus maritimus* genome reveals unique mechanisms for nitrification and autotrophy in globally distributed marine crenarchaea. Proc Natl Acad Sci USA 2010; 107(19): 8818-23. [http://dx.doi.org/10.1073/pnas.0913533107] [PMID: 20421470]

[59]   Biggs-Weber E, Aigle A, Prosser JI, Gubry-Rangin C. Oxygen preference of deeply-rooted mesophilic thaumarchaeota in forest soil. Soil Biol Biochem 2020; 148: 107848. [http://dx.doi.org/10.1016/j.soilbio.2020.107848]

[60]   Könneke M, Bernhard AE, de la Torre JR, Walker CB, Waterbury JB, Stahl DA. Isolation of an autotrophic ammonia-oxidizing marine archaeon. Nature 2005; 437(7058): 543-6. [http://dx.doi.org/10.1038/nature03911] [PMID: 16177789]

[66]   Liu C, Li M, Redda ET *et al.* Complete nucleotide sequence of a novel mycovirus from Trichoderma harzianum in China Arch Virol 2019; 164:1213–1216.

[67]   Zheng L, Zhao J, Liang X, Zhuang H, Qi T, Kang Z. Complete genome sequence of a novel mitovirus from the wheat stripe rust fungus *Puccinia striiformis.* Arch Virol 2019; 164(3): 897-901. [http://dx.doi.org/10.1007/s00705-018-04134-4] [PMID: 30600350]

[68]   Bao X, Roossinck MJ. Chapter two—multiplexed interactions: viruses of endophytic fungi. In: Ghabrial SA, Ed. Advances in virus research. Boston: Academic Press 2013; pp. 37-58.

[69]   Herrero N, Sánchez Márquez S, Zabalgogeazcoa I. Mycoviruses are common among different species of endophytic fungi of grasses. Arch Virol 2009; 154(2): 327-30. [http://dx.doi.org/10.1007/s00705-008-0293-5] [PMID: 19125219]

[70]   Herrero N, Zabalgogeazcoa I. Mycoviruses infecting the endophytic and entomopathogenic fungus *Tolypocladium cylindrosporum.* Virus Res 2011; 160(1-2): 409-13. [http://dx.doi.org/10.1016/j.virusres.2011.06.015] [PMID: 21736906]

[71]   Figueirêdo LC, Figueirêdo GS, Giancoli ACH, *et al.* Detection of isometric, dsRNA-containing viral particles in *Colletotrichum gloeosporioides* isolated from cashew tree. Trop Plant Pathol 2012; 37: 142-5. [http://dx.doi.org/10.1590/S1982-56762012000200007]

[72]   Ong JWL, Li H, Sivasithamparam K, Dixon KW, Jones MGK, Wylie SJ. Novel Endorna-like viruses, including three with two open reading frames, challenge the membership criteria and taxonomy of the Endornaviridae. Virology 2016; 499: 203-11. [http://dx.doi.org/10.1016/j.virol.2016.08.019] [PMID: 27677157]

[73]   Ong JWL, Li H, Sivasithamparam K, Dixon KW, Jones MGK, Wylie SJ. Novel and divergent viruses associated with Australian orchid-fungus symbioses. Virus Res 2018; 244: 276-83.

[http://dx.doi.org/10.1016/j.virusres.2017.11.026] [PMID: 29180114]

[74]   Song HY, Choi HJ, Jeong H, Choi D, Kim DH, Kim JM. Viral effects of a dsRNA mycovirus (PoV-ASI2792) on the vegetative growth of the edible mushroom *Pleurotusostreatus*. Mycobiology 2016; 44(4): 283-90.
[http://dx.doi.org/10.5941/MYCO.2016.44.4.283] [PMID: 28154486]

[75]   Lin YH, Fujita M, Chiba S, *et al.* Two novel fungal negative-strand RNA viruses related to mymonaviruses and phenuiviruses in the shiitake mushroom (Lentinula edodes). Virology 2019; 533: 125-36.
[http://dx.doi.org/10.1016/j.virol.2019.05.008] [PMID: 31153047]

[76]   Petrzik K, Siddique AB. A mycoparasitic and opportunistic fungus is inhabited by a mycovirus. Arch Virol 2019; 164(10): 2545-9.
[http://dx.doi.org/10.1007/s00705-019-04359-x] [PMID: 31317260]

[77]   Siddique AB, Khokon AM, Unterseher M. What do we learn from cultures in the omics age? High-throughput sequencing and cultivation of leaf-inhabiting endophytes from beech (*Fagus sylvatica* L.) revealed complementary community composition but similar correlations with local habitat conditions. MycoKeys 2017; 20: 1-16.
[http://dx.doi.org/10.3897/mycokeys.20.11265]

[78]   Rosseto P, Costa AT, Polonio JC, *et al.* Investigation of mycoviruses in endophytic and phytopathogenic strains of colletotrichum from different hosts. Genet Mol Res.2016;15: gmr.15017651
[http://dx.doi.org/10.4238/gmr.15017651]

[79]   Petrzik K, Siddique AB. A mycoparasitic and opportunistic fungus is inhabited by a mycovirus. Arch Virol 2019; 164(10): 2545-9.
[http://dx.doi.org/10.1007/s00705-019-04359-x] [PMID: 31317260]

[80]   Siddique AB, Khokon AM, Unterseher M. What do we learn from cultures in the omics age? High-throughput sequencing and cultivation of leaf-inhabiting endophytes from beech (*Fagus sylvatica* L.) revealed complementary community composition but similar correlations with local habitat conditions. MycoKeys 2017; 20: 1-16.
[http://dx.doi.org/10.3897/mycokeys.20.11265]

[81]   Rosseto P, Costa AT, Polonio JC *et al.* Investigation of mycoviruses in endophytic and phytopathogenic strains of colletotrichum from different hosts. Genet Mol Res. 2016; 15: gmr.15017651
[http://dx.doi.org/10.4238/gmr.15017651]

[82]   Nerva L, Turina M, Zanzotto A, *et al.* Isolation, molecular characterization and virome analysis of culturable wood fungal endophytes in esca symptomatic and asymptomatic grapevine plants. Environ Microbiol 2019; 21(8): 2886-904.
[http://dx.doi.org/10.1111/1462-2920.14651] [PMID: 31081982]

[83]   Muvea AM, Subramanian S, Maniania NK, Poehling HM, Ekesi S, Meyhöfer R. Endophytic colonization of onions induces resistance against *Viruliferous thrips* and virus replication. Front Plant Sci 2018; 9: 1785.
[http://dx.doi.org/10.3389/fpls.2018.01785] [PMID: 30574155]

[84]   Morsy MR, Oswald J, He J, Tang Y, Roossinck MJ. Teasing apart a three-way symbiosis: Transcriptome analyses of *Curvularia protuberata* in response to viral infection and heat stress. Biochem Biophys Res Commun 2010; 401(2): 225-30.
[http://dx.doi.org/10.1016/j.bbrc.2010.09.034] [PMID: 20849822]

[85]   Segers GC, Zhang X, Deng F, Sun Q, Nuss DL. Evidence that RNA silencing functions as an antiviral defense mechanism in fungi. Proc Natl Acad Sci USA 2007; 104(31): 12902-6.
[http://dx.doi.org/10.1073/pnas.0702500104] [PMID: 17646660]

[86]   Hammond TM, Andrewski MD, Roossinck MJ, Keller NP. *Aspergillus* mycoviruses are targets and

suppressors of RNA silencing. Eukaryot Cell 2008; 7(2): 350-7.
[http://dx.doi.org/10.1128/EC.00356-07] [PMID: 18065651]

[87]   Saikkonen K, Faeth SH, Helander M, Sullivan TJ. Fungal endophytes: a continuum of interactions with host plants. Annu Rev Ecol Syst 1998; 29(1): 319-43.
[http://dx.doi.org/10.1146/annurev.ecolsys.29.1.319]

[88]   Zabalgogeazcoa I, Benito EP, Ciudad AG, Criado BG, Eslava AP. Double-stranded RNA and virus-like particles in the grass endophyte *Epichloë festucae*. Mycol Res 1998; 102(8): 914-8.
[http://dx.doi.org/10.1017/S0953756297005819]

[89]   Caron M. Potential use of mycorrhizae in control of soil-borne diseases. Can J Plant Pathol 1989; 11(2): 177-9.
[http://dx.doi.org/10.1080/07060668909501135]

[90]   Kowalski KP, Bacon C, Bickford W, *et al.* Advancing the science of microbial symbiosis to support invasive species management: a case study on Phragmites in the Great Lakes. Front Microbiol 2015; 6: 95.
[http://dx.doi.org/10.3389/fmicb.2015.00095] [PMID: 25745417]

[91]   Wheeler T, von Braun J. Climate change impacts on global food security. Science 2013; 341(6145): 508-13.
[http://dx.doi.org/10.1126/science.1239402] [PMID: 23908229]

[92]   Müller C, Waha K, Bondeau A, Heinke J. Hotspots of climate change impacts in sub☐Saharan Africa and implications for adaptation and development. Glob Change Biol 2014; 20(8): 2505-17.
[http://dx.doi.org/10.1111/gcb.12586] [PMID: 24796720]

[93]   Schlenker W, Lobell D.B. Robust negative impacts of climate change on African agriculture. Environ Res.Lett 2010; 5: 014-010.
[http://dx.doi.org/10.1088/1748-9326/5/1/014010]

[94]   Albrecht A, Kandji ST. Carbon sequestration in tropical agroforestry systems. Agric Ecosyst Environ 2003; 99(1-3): 15-27.
[http://dx.doi.org/10.1016/S0167-8809(03)00138-5]

[95]   Bonan GB. Forests and climate change: forcings, feedbacks, and the climate benefits of forests. Science 2008; 320(5882): 1444-9.
[http://dx.doi.org/10.1126/science.1155121] [PMID: 18556546]

[96]   Zomer RJ, Neufeldt H, Xu J, *et al.* Global Tree Cover and Biomass Carbon on Agricultural Land: The contribution of agroforestry to global and national carbon budgets. Sci Rep 2016; 6(1): 29987.
[http://dx.doi.org/10.1038/srep29987] [PMID: 27435095]

[97]   Craib IJ. The silviculture of exotic conifers in South Africa. Journal of the South African Forestry Association 1947; 15(1): 11-45.
[http://dx.doi.org/10.1080/03759873.1947.9630547]

[98]   Hurley BP, Garnas J, Wingfield MJ, Branco M, Richardson DM, Slippers B. Increasing numbers and intercontinental spread of invasive insects on eucalypts. Biol Invasions 2016; 18(4): 921-33.
[http://dx.doi.org/10.1007/s10530-016-1081-x]

[99]   Wingfield MJ, Slippers B, Roux J, Wingfield BD. The worldwide movement of exotic forest fungi, especially in the tropics and the southern hemisphere. AIBS Bull 2001; 51(2): 134-40.

[100] Bardgett RD, van der Putten WH. Belowground biodiversity and ecosystem functioning. Nature 2014; 515(7528): 505-11.
[http://dx.doi.org/10.1038/nature13855] [PMID: 25428498]

[101] Bender SF, Wagg C, van der Heijden MGA. An underground revolution: biodiversity and soil ecological engineering for agricultural sustainability. Trends Ecol Evol 2016; 31(6): 440-52.
[http://dx.doi.org/10.1016/j.tree.2016.02.016] [PMID: 26993667]

[102] Lladó S, López-Mondéjar R, Baldrian P. Forest soil bacteria: diversity, involvement in ecosystem processes, and response to global change. Microbiol Mol Biol Rev 2017; 81(2): e00063-16.
[http://dx.doi.org/10.1128/MMBR.00063-16] [PMID: 28404790]

[103] Pimentel D, Harvey C, Resosudarmo P, et al. Environmental and economic costs of soil erosion and conservation benefits. Science 1995; 267(5201): 1117-23.
[http://dx.doi.org/10.1126/science.267.5201.1117] [PMID: 17789193]

[104] Bouwman AF, Beusen AHW, Billen G. Human alteration of the global nitrogen and phosphorus soil balances for the period 1970–2050. Global Biogeochem Cycles 2009; 23(4): 2009GB003576.
[http://dx.doi.org/10.1029/2009GB003576]

[105] Johnson NC, Graham J-H, Smith FA. Functioning of mycorrhizal associations along the mutualism–parasitism continuum. New Phytol 1997; 135(4): 575-85.
[http://dx.doi.org/10.1046/j.1469-8137.1997.00729.x]

[106] Rockström J, Steffen W, Noone K, et al. Planetary boundaries: exploring the safe operating space for humanity. Ecol Soc 2009; 14(2): art32.
[http://dx.doi.org/10.5751/ES-03180-140232]

[107] Yuan Z, Jiang S, Sheng H, et al. Human perturbation of the global phosphorus cycle: changes and consequences. Environ Sci Technol 2018; 52(5): 2438-50.
[http://dx.doi.org/10.1021/acs.est.7b03910] [PMID: 29402084]

[108] Sanz MJ, de Vente J, Chotte JL, Bernoux M, Kust G, Ruiz I, Eds. Sustainable land management contribution to successful land-based climate change adaptation and mitigation. 2017.

[109] Weller DM, Raaijmakers JM, Gardener BBM, Thomashow LS. Microbial populations responsible for specific soil suppressiveness to plant pathogens. Annu Rev Phytopathol 2002; 40(1): 309-48.
[http://dx.doi.org/10.1146/annurev.phyto.40.030402.110010] [PMID: 12147763]

[110] Bonilla N, Vida C, Martínez-Alonso M, et al. Organic amendments to avocado crops induce suppressiveness and influence the composition and activity of soil microbial communities. Appl Environ Microbiol 2015; 81(10): 3405-18.
[http://dx.doi.org/10.1128/AEM.03787-14] [PMID: 25769825]

[111] Barnes BV, Zak DR, Denton SR, Spurr SH. Forest Ecology. 4th ed., New York: John Wiley and Sons 1998.

[112] Hobbie SE. Temperature and plant species control over litter decomposition in the Alaskan tundra. Ecol Monogr 1996; 66(4): 503-22.
[http://dx.doi.org/10.2307/2963492]

[113] Mitchell RJ, Hester AJ, Campbell CD, Chapman SJ, et al. Is vegetation composition or soil chemistry the best predictor of the soil microbial community? J Ecol 2010; 98: 50-61.
[http://dx.doi.org/10.1111/j.1365-2745.2009.01601.x]

[114] Zak DR, Blackwood CB, Waldrop MP. A molecular dawn for biogeochemistry. Trends Ecol Evol 2006; 21(6): 288-95.
[http://dx.doi.org/10.1016/j.tree.2006.04.003] [PMID: 16769427]

[115] Thoms C, Gattinger A, Jacob M, Thomas FM. Gleixner. Direct and indirect effects of tree species diversity drive soil microbial diversity intemperate deciduous forests. Soil Biol Biochem 2010; 42: 1558-65.
[http://dx.doi.org/10.1016/j.soilbio.2010.05.030]

[116] Ayres E, Steltzer H, Berg S, Wallenstein MD, et al. In high-elevation forests, tree species traits influence soil physical, chemical, and biological properties. PLoS One 2009; 4: e5964. a
[http://dx.doi.org/10.1371/journal.pone.0005964] [PMID: 19536334]

[117] Myers RT, Zak DR, White DC, Peacock A. Landscape-level patterns of microbial community composition and substrate use in upland forest ecosystems. Soil Sci Soc Am J 2001; 65(2): 359-67.

[http://dx.doi.org/10.2136/sssaj2001.652359x]

[118] Hackl E, Pfeffer M, Donat C, Bachmann G, Zechmeisterboltenstern S. Composition of the microbial communities in the mineral soil under different types of natural forest. Soil Biol Biochem 2005; 37(4): 661-71.
[http://dx.doi.org/10.1016/j.soilbio.2004.08.023]

[119] Selvam A, Tsai SH, Liu CP, Chen IC, Chang CH, Yang SS. Microbial communities and bacterial diversity of spruce, hemlock and grassland soils of Tatachia Forest, Taiwan. J Environ Sci Health B 2010; 45(5): 386-98.
[http://dx.doi.org/10.1080/03601231003799960] [PMID: 20512729]

[120] Falkowski PG, Fenchel T, Delong EF. The microbial engines that drive Earth's biogeochemical cycles. Science 2008; 320(5879): 1034-9.
[http://dx.doi.org/10.1126/science.1153213] [PMID: 18497287]

[121] Nicholson JK, Holmes E, Wilson ID. Gut microorganisms, mammalian metabolism and personalized health care. Nat Rev Microbiol 2005; 3(5): 431-8.
[http://dx.doi.org/10.1038/nrmicro1152] [PMID: 15821725]

[122] Van Tienderen PH. Evolution of generalists and specialists in spatially heterogeneous environments. Evolution 1991; 45(6): 1317-31.
[http://dx.doi.org/10.1111/j.1558-5646.1991.tb02638.x] [PMID: 28563821]

[123] Futuyma DJ, Moreno G. The evolution of ecological specialization. Annu Rev Ecol Syst 1988; 19(1): 207-33.
[http://dx.doi.org/10.1146/annurev.es.19.110188.001231]

[124] Székely AJ, Langenheder S. The importance of species sorting differs between habitat generalists and specialists in bacterial communities. FEMS Microbiol Ecol 2014; 87(1): 102-12.
[http://dx.doi.org/10.1111/1574-6941.12195] [PMID: 23991811]

[125] Zengler K, Zaramela LS. The social network of microorganisms — how auxotrophies shape complex communities. Nat Rev Microbiol 2018; 16(6): 383-90.
[http://dx.doi.org/10.1038/s41579-018-0004-5] [PMID: 29599459]

[126] Xu Q, Luo G, Guo J, Xiao Y, et al. Intraspecific variation and dormancy potential matter. Mol Ecol, Microbial generalist or specialist 2021; pp. 1-13.

[127] Lundquist JE, Camp AE, Tyrell ML, Seybold SJ, et al. Earth, wind, and fire: Abiotic factors and the impacts of global environmental change on forest health. In: Castello JD, Teale SA, Eds. Forest Health: An Integrated AQperspective. Cambridge, UK: Cambridge University Press 2011; pp. 195-243.
[http://dx.doi.org/10.1017/CBO9780511974977.008]

[128] McFall-Ngai M, Hadfield MG, Bosch TCG, et al. Animals in a bacterial world, a new imperative for the life sciences. Proc Natl Acad Sci USA 2013; 110(9): 3229-36.
[http://dx.doi.org/10.1073/pnas.1218525110] [PMID: 23391737]

[129] Vandenkoornhuyse P, Quaiser A, Duhamel M, Le Van A, Dufresne A. The importance of the microbiome of the plant holobiont. New Phytol 2015; 206(4): 1196-206.
[http://dx.doi.org/10.1111/nph.13312] [PMID: 25655016]

[130] Panke-Buisse K, Poole AC, Goodrich JK, Ley RE, Kao-Kniffin J. Selection on soil microbiomes reveals reproducible impacts on plant function. ISME J 2015; 9(4): 980-9.
[http://dx.doi.org/10.1038/ismej.2014.196] [PMID: 25350154]

[131] Karasov TL, Kniskern JM, Gao L, et al. The long-term maintenance of a resistance polymorphism through diffuse interactions. Nature 2014; 512(7515): 436-40.
[http://dx.doi.org/10.1038/nature13439] [PMID: 25043057]

[132] van der Heijden MGA, Klironomos JN, Ursic M, et al. Mycorrhizal fungal diversity determines plant biodiversity, ecosystem variability and productivity. Nature 1998; 396(6706): 69-72.

[http://dx.doi.org/10.1038/23932]

[133] Kembel SW, O'Connor TK, Arnold HK, Hubbell SP, Wright SJ, Green JL. Relationships between phyllosphere bacterial communities and plant functional traits in a neotropical forest. Proc Natl Acad Sci USA 2014; 111(38): 13715-20.
[http://dx.doi.org/10.1073/pnas.1216057111] [PMID: 25225376]

[134] Lindström ES, Langenheder S. Local and regional factors influencing bacterial community assembly. Environ Microbiol Rep 2012; 4(1): 1-9.
[http://dx.doi.org/10.1111/j.1758-2229.2011.00257.x] [PMID: 23757223]

[135] Fisher CK, Mehta P. The transition between the niche and neutral regimes in ecology. Proc Natl Acad Sci USA 2014; 111(36): 13111-6.
[http://dx.doi.org/10.1073/pnas.1405637111] [PMID: 25157131]

[136] Maignan L, Deforce EA, Chafee ME, Eren AM, Simmons SL. Ecological succession and stochastic variation in the assembly of Arabidopsis thaliana phyllosphere communities. MBio. 2014; 5: 01-23.

[137] Bodenhausen N, Horton MW, Bergelson J. Bacterial communities associated with the leaves and the roots of Arabidopsis thaliana. PLoS One 2013; 8(2): e56329.
[http://dx.doi.org/10.1371/journal.pone.0056329] [PMID: 23457551]

[138] Peiffer JA, Spor A, Koren O, *et al.* Diversity and heritability of the maize rhizosphere microbiome under field conditions. Proc Natl Acad Sci USA 2013; 110(16): 6548-53.
[http://dx.doi.org/10.1073/pnas.1302837110] [PMID: 23576752]

[139] Delmotte N, Knief C, Chaffron S, *et al.* Community proteogenomics reveals insights into the physiology of phyllosphere bacteria. Proc Natl Acad Sci USA 2009; 106(38): 16428-33.
[http://dx.doi.org/10.1073/pnas.0905240106] [PMID: 19805315]

[140] Franza T, Expert D. Role of iron homeostasis in the virulence of phytopathogenic bacteria: an 'à la carte' menu. Mol Plant Pathol 2013; 14(4): 429-38.
[http://dx.doi.org/10.1111/mpp.12007] [PMID: 23171271]

[141] Lebeis SL. Greater than the sum of their parts: characterizing plant microbiomes at the community-level. Curr Opin Plant Biol 2015; 24: 82-6.
[http://dx.doi.org/10.1016/j.pbi.2015.02.004] [PMID: 25710740]

[142] Aronson J. Restoration of Natural Capital: without reserves, there are no goods or services. Spanish Association of Terrestrial Ecology (AEET), Scienc tech J Ecol Environ. Volume 2007; 16: 15-24.

[143] LoSchiawo, A. Lessons learned from the first decade Adaptative Management in comprehensive Everglades Restauration. Ecol. Soc.(E&S), 2013; 18: 70.

Role of Microbes in Socio-Economic Development

CHAPTER 10

Microbiomes in Promoting a Sustainable Industrial Production System

Joseph Ezra John[1]**, Boopathi Gopalakrishnan**[2]**, Senthamizh Selvi**[3]**, Murugaiyan Sinduja**[1]**, Chidamparam Poornachandhra**[1,*]**, Ravi Raveena**[1] **and E. Akila**[4]

[1] *Department of Environmental Sciences, Tamil Nadu Agricultural University, Coimbatore, Tamil Nadu, India*

[2] *School of Atmospheric Stress Management, ICAR-National Institute of Abiotic Stress Management, Maharashtra, India*

[3] *Department of Agricultural Microbiology, Tamil Nadu Agricultural University, Coimbatore, Tamil Nadu, India*

[4] *Department of Agricultural Engineering, Tamil Nadu Agricultural University, Coimbatore, Tamil Nadu, India*

Abstract: The sustainable industrial revolution is the way forward to help humankind prolong its existence on Earth. The first step could be facilitating the natural process under a controlled environment to produce the desired products instead of chemicals. The industrial sectors, especially food and pharmaceuticals, depend on microbes for most of their production. Biocontrol, enzyme, and fuel production have been explored in recent years. Microbial production systems encompass the metabolites produced by bacteria, fungi, or viruses that facilitate industrial processes. These secondary metabolites have been noted to pose implications in many fields, including agriculture. After the advent of modern genetic engineering techniques, the utilization of microbiota in various activities is increasing due to their simplicity and cost-effectiveness. The gene mounting and biotechnological tolls have aided in manipulating these microbes' secondary metabolites, thereby improving productivity. Furthermore, multi-disciplinary and comprehensive approaches directed towards improving microbial production are described in this chapter.

Keywords: Bioproducts, Bioenzymes, Biofertilizers, Microbiomes, Pharmaceuticals, Sustainability.

* **Corresponding author Chidamparam Poornachandhra:** Department of Environmental Sciences, Tamil Nadu Agricultural University, Coimbatore, Tamil Nadu, India; E-mail: poorna155c@gmail.com

Govindaraj Kamalam Dinesh, Shiv Prasad, Ramesh Poornima, Sangilidurai Karthika, Murugaiyan Sinduja & Velusamy Sathya (Eds.)

INTRODUCTION

Global ecosystems are under much stress due to industrialization and population growth. Urbanization frequently occurs on previously farmed land, and to increa se food production, farmers frequently intensify their practices and use agrochemicals, which comprise a wide range of structurally different chemicals. Although the Green Revolution and the development of industries in the countries improved people's lives and boosted the world economy, environmental quality was generally degraded due to industrial waste's detrimental effects on the soil, water, and atmosphere. This affects the soil fertility in addition to the heavy metal contamination from industries. Toxic contaminants from polluted areas are now being reduced using techniques including recycling, disposal in a landfill, burning of wastes, and pyrolysis. Nevertheless, these techniques harm the ecosystem and produce intermediary chemical compounds that are much more poisonous and difficult to remediate. Microbes are widely used in food processing, food additives, alcoholic and non-alcoholic beverages, biofuels, metabolites, biofertilizers, chemicals, enzymes, bioactive molecules, vaccines, antibiotics, medicines, and other commercial products. A list of algal species involved in the production of essential products is presented in Table **1**.

Table 1. List of algal species involved in the production of important products.

S. No.	Algal Species	Products	Uses
1.	*Spirulina platensis*	Phycocyanins	Nutraceuticals, cosmetics
2.	*Chlorella vulgaris*	Ascorbic acid	Food supplement
3.	*Haematococcus pluvialis*	Carotenoids, astaxanthin	Nutraceuticals, pharmaceuticals, additives
4.	*Odontella aurita*	Fatty acids	Pharmaceuticals, cosmetics, baby food
5.	*Porphyridium cruentum*	Polysaccharides	Pharmaceuticals, cosmetics
6.	*Dunaliella salina*	Carotenoids	Nutraceuticals, food supplements, feed
7.	*Spirulina platensis, Dunaliella salina, Haematococcus pluvialis*	Phycobiliproteins	Pigments, cosmetics, vitamins
8.	*Chlorella minutissima Schizochytrium* sp.	PUFAs	Nutracceuticals, food supplement
9.	*Euglena gracilis Euglena gracilisa Prototheca moriformis*	Vitamins (Biotin, α-tocopherol and Vitamin C)	Nutrition

(Table 1) cont.....

S. No.	Algal Species	Products	Uses
10.	*Nostoc, Hapalosiphon*	Indole-3-acetic acid, or 3-methyl indole	Plant growth and development
11.	*Nostoc muscorum, Hapalosiphon fontinalis*	Vitamin B12	
12.	*Tolypothrix tenuis*		
13.	*Cylindrospermum* sp.		

Research on green technology is now being done to replace dangerous toxic pollutants. Microbes and microbial products in this field hold promise for ensuring food security in a dynamic environment and decomposing dangerous pollutants [1]. Using microbiological methods, agriculture can be developed sustainably. These microbes can enhance crop plants' access to nutrients through several mechanisms, enhancing plant growth. This chapter addresses the critical mechanisms that microbes could use to support environmental management and environmentally sound industrial production systems. In addition, the examples of effective applications of microorganisms for healthier, cleaner ecosystems without compromising crop yield or industrialization are shown in the sections below. These bioproducts and enzymes play essential roles in the food, bioenergy, cosmetics, and pharmaceutical industries. Traditionally, the glucose that is mainly hydrolyzed from food grain starch is used to make biofuels like bioethanol. Consequently, producing biofuels from molasses rather than glucose is a tactic to conserve food resources. However, some valuable products made by molasses-derived microorganisms have significant uses in food processing, preservation, safety, and nutrition.

NEED FOR MICROBES IN THE INDUSTRIAL PRODUCTION SYSTEM

Ecological connections bind nearly all living things and microorganisms together. There are many ways in which microorganisms are advantageous to people, and they also significantly impact both human welfare and the ecosystem. They also play a vital role in industrial production processes. They are employed in the energy sector to make fuels, chemicals, and energy, in the pharmaceutical sector to make insulin, antibiotics, and probiotics, in the food sector to make food and process it, and in the agricultural sector to make biofertilizers and biocontrol agents.

Energy Industry

In a significant part of energy conversion, microorganisms can convert waste biomass to bioenergy. It is used to create fuels that can be burned, including butanol, ethanol, hydrogen, methane gas, lipids, and others. Using certain bacteria to make hydrogen fuel from waste biomass is one of the alternative biomass conversion technologies that has recently been developed. It is considered a sustainable alternative to commercial hydrogen production from fossil energy sources.

Microbial Ethanol Fermentation: Fossil fuels and the associated environmental impacts have shifted the focus towards alternate energy sources like ethanol, isobutanol, alkanes, hydrogen gas, *etc.*, to achieve carbon neutrality. Ethanol is commercially produced from molasses in sugar industries by fermentation technique. Fermentation utilizes *Saccharomyces cerevisiae*, commonly called yeast, a fungal microorganism. Ethanol, used in various industries and as a blending fuel with petrol, was produced from petroleum industries and biomass. The importance of bioconversion using microorganisms gained attention, and commercial ethanol production was shifted from fossil fuel to biomass [2].

Microbial-derived alternate fuel is one of the research areas that has gained momentum recently. Corn-starch and waste *Pichia pastoris* were used as a semi-solid mixture to increase the butanol yield up to 10.5 g L^{-1}, increasing the total sugar utilization yield to >90% and digesting 53% of carbohydrates in the solids to achieve waste reduction [3]. *Cupriavidus basilensis* achieved complete conversion of carbohydrates from rice straw, producing 984.2 mg/g of reducing sugar and 482.7 mgL^{-1} of PHA as a result of the ligninolytic bacteria's conversion of the lignin [4]. *Trichoderma reesei* produced cellulase from wheat bran through solid-state fermentation in addition to CMCase production of 959,53 IU/gDS. Eighty percent of the sorghum stover's enzymatic hydrolysate was successfully converted to ethanol [5]. *Saccharomyces cerevisiae* (YSF2-19) had its genome shuffled, increasing glutathione output by 3.3 times (318.96 mgL^{-1}) above YS86. YSF2-19 had higher extracellular and intracellular GSH levels than YS86 (201.68 and 117.28 mg L^{-1}, respectively), and they exhibited faster growth and glucose uptake [6].

Microbial Fuel Cell: Microbial fuel cell (MFC) is a technology that uses microorganisms to convert chemical energy found as organic compounds in wastes into electrical energy. The entire process is considered eco-friendly as it provides a dual benefit of waste utilization and energy generation [7]. MFCs coupled with photocatalytic cells are increasingly used in energy production due to their high stability, improved efficiency, low toxicity, and lower cost. Septic tank wastewater was treated with co-cultures of *Serratia marcescens* and *Klebsiella pneumoniae,* which yielded an energy density of 869.11 ± 43 mA/m^2 and 84.5% methyl orange elimination efficiency and 0.119 Wm^{-2} power production were attained by MFC using a p-type Si nanowire photocathode and electrician-colonized bio-anode modified with lead nanoparticles in just 36 hours

[8]. Similarly, a double-chambered Photo-MFC with an AgBr/CuO hybrid photocathode and a bio-anode was able to attain 55.56% degradation efficiency for reactive black 5 and a power density of 61.11 mWm^{-2} under illuminated conditions [9]. A combination of photoelectric MFC and BiOCl/reduced graphene oxide photocathode was observed to achieve a power generation capacity of 696.51 ± 5.27 mV and oxytetracycline removal efficiency of 98.93 ± 0.15% after 4 days [10].

Microbial Electrosynthesis (MES): Methane-producing microbes catalyze the conversion of CO_2 into methane using electricity. However, the methane production rate remains very low for large-scale applications. Several studies have been carried out to improve performance through process modification, substrate modification, and bioengineering of microbes. The methane generation rate was increased to 12.5 Ld^{-1} using redox flow batteries with a large area-to-volume ratio [11]. CO_2 reduction and the subsequent conversion to acetic acid were improved by inoculum enriching and boosting the availability of CO_2 to biofilm, thereby increasing CO_2 conversion efficiency to >80% [12]. For the simultaneous generation of alcohols and carboxylic acids from CO_2, a dual-purpose MES cell with two biological cathode chambers and a common anode compartment was developed [13]. Optimized organics loading, operation period, and low-pH significantly increased the efficiency and conversion rate into alcohol by twenty times. This resulted in an increment in the system's efficiency [14]. A mutant strain of *Methylomonas* sp. was able to achieve higher cell development and 10 times more succinate production (195 mgL^{-1} succinate from 0.0789 g) compared to the wild strain [15]. An MES experiment with CO_2 rich impurity brewery gas showed that the co-culture of *Acetobacterium* generated 1.8±0.2 g L^{-1} CH_3COOH with 0.26±0.03 gL-catholyte/d, which outperformed the pure *Clostridium ljungdahlii* culture (1.1 ± 0.02 gL^{-1}; 0.138 ± 0.004 g Lcatholyte^{-1}d^{-1}) [16].

Microbial Hydrogen production: For the time being, the majority of the fossil fuels we use to produce hydrogen commercially are natural gas from coal beds or oil wells. In the recent development in energy technologies, using microorganisms to produce hydrogen from organic waste has become a sustainable alternative to fossil sources. Several conversion technologies are available for biohydrogen production, such as biophotolysis of water with microalgae, dark anaerobic fermentation and photo-fermentation, and microbial water–gas shift reaction of carbon monoxide. In the near future, it is evident that biohydrogen production from waste using microbes will be an inevitable part of the energy industry [17]. The process of producing hydrogen gas (H_2) as a metabolic by-product by certain microorganisms, such as bacteria and algae, from various substrates is known as microbial hydrogen generation. It is crucial to remember that for this process to produce hydrogen, an appropriate electron donor like organic matter or certain chemicals must be present. Given that hydrogen is a clean and adaptable energy carrier, this biological approach to produce hydrogen has drawn interest as a potential sustainable and renewable energy source.

The two primary biological processes involved in microbial hydrogen production are:

Fermentation: When organic matter is fermented, certain anaerobic bacteria, known as hydrogen-producing bacteria or hydrogenases, can make hydrogen. Hydrogen, carbon dioxide (CO_2), and other byproducts are produced when these bacteria break down complex organic substances, including sugars, starches, and other organic substrates. Some strains of *Escherichia coli* and *Clostridium* species are examples of bacteria that produce hydrogen [18]. The following is a representation of the reaction:

Organic substrate → Hydrogen + Carbon dioxide + Other byproducts

Photosynthesis: Green algae and cyanobacteria are two examples of photosynthetic microorganisms that have the ability to manufacture hydrogen through a process known as photobiological hydrogen generation. These bacteria absorb light energy during photosynthesis, transform it into chemical energy, and produce oxygen and hydrogen gas as by-products:

Light energy + Water → Hydrogen + Oxygen

Microbial hydrogen production has several advantages as an alternative energy source:

1. It provides a sustainable and renewable way to make hydrogen utilizing light energy or easily available organic substrates.

2. The CO_2 and oxygen that are produced as by-products of microbial hydrogen generation are safe for the environment.

3. The method can be connected with systems for the treatment of organic waste or other types of trash disposal, thus making it commercially viable.

4. Nevertheless, there are still a number of difficulties in expanding microbial hydrogen generation for industrial usage.

5. These include developing cost-effective technologies to gather and purify the hydrogen gas produced, as well as improving the rates and efficiency of hydrogen production.

6. As a potential renewable energy technology, microbial hydrogen generation is still undergoing research to increase its effectiveness and viability.

Biodiesel production: Biodiesel derived from microbes such as microalgae, yeast, and bacteria is a possible renewable substitute for petroleum-based diesel. Although biodiesel is now produced commercially from non-edible vegetable oils, recent knowledge of biodiesel synthesis from microalgae has attracted increasing interest due to its efficient conversion and waste reduction [19]. Over 95% of the world's biodiesel output is based on edible oils. Rapeseed oil used in Europe and Canada, sunflower oil used in Europe, palm oil used in Southeast Asia, and soybean oil used in the United States are the most often utilized edible vegetable oil feedstocks for global biodiesel manufacturing. Many non-edible oil-producing plants for biodiesel feedstock include *Jatropha curcas* (jatropha), *Azadirachta indica* (neem), *Pongamia pinnata* (karanja), *Hevea brasiliensis* (rubber), *Ricinus communis* (castor), *Nicotiana tabacum* (tobacco), and *Simmondsia chinensis* (jojoba).

Oleaginous microorganisms (OMs) can use low-cost feedstocks, such as waste substrates and lignocellulosic substrates (LCSs), to accumulate more lipids. These microorganisms take advantage of different renewable resources and turn them into microbial oil, which is then used to transesterify various renewable resources to create biodiesel. The single-cell oil is another name for the microbial oil. The first single-cell oil was made commercially in 1985 using the filamentous fungus *Mucor circinelloides*, although at the time, making biodiesel from it was not a consideration [20]. Incubation times for OM-based biodiesel production are shorter than those for plant and animal resources, lipid production is not affected by seasonal, climatic, or geographic variations, and OMs can be cultured all year long for microbial oil production [21].

Delignification, saccharification, and fermentation are the three main processes in the synthesis of bioethanol. After delignification, lignin that has been specifically broken down by enzymes can be utilized to make silver nanoparticles, antioxidants, glucose biosensors, reusable adsorbents, electric double-layer capacitors, *etc*. Pentoses are primarily left unused during fermentation. Therefore, adding these pentoses to OMs as a carbon source for the formation of lipids may be an effective strategy [22]. Additionally, these lipids can be transformed into FAMEs, such as glycerol and biodiesel, the latter of which can be used as a carbon source for the manufacturing of biodiesel (Fig. **1**). The biomass produced following lipid extraction can be used as biofertilizer and biomanure, for nutritional enrichment with cyanobacteria. The synthesis of industrial chemicals and biofuels with a zero-waste approach might, therefore, result from the integration of processes.

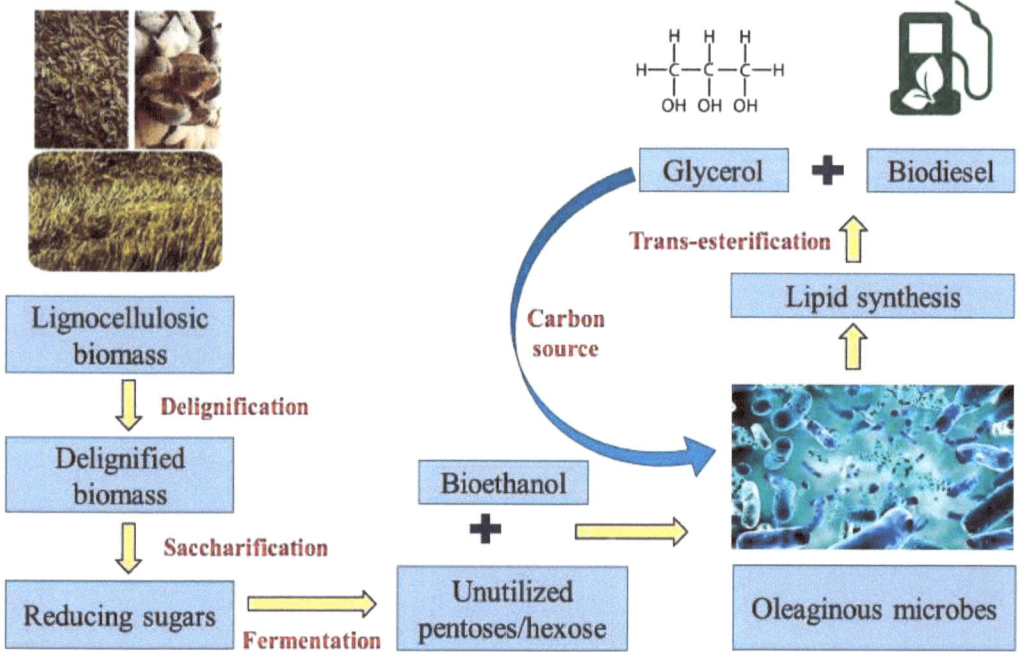

Fig. (1). Biodiesel production using Oleaginous microorganisms.

Biogas production: The digestion of organic waste, that is, manure, household or kitchen waste, and wet agricultural waste under anaerobic conditions produces biogas, a combination of methane (55-70%) and carbon dioxide (30-45%). The four stages of methane synthesis are namely Hydrolysis, Acidogenesis, Acetogenesis and Methanogenesis, where different anaerobic bacteria are involved (Fig. **2**). Using these bacteria, we can achieve two goals: effective waste management and sustainable energy production [23].

Among the four process, the first three are carried out by various *Eubacteria*, while the fourth is carried out by *Archaea*. In order to produce sugar intermediates, *Clostridia* and *Bacilli* participate in hydrolysis. To maximize the generation of biogas, it is necessary to study the traits and operation of these microbial communities. According to Ali *et al.* [24], adding a strain that produces hydrogen to the system may also increase the generation of biogas. The formation of hydrogen in the digester has been shown to benefit from *Caldicellulosiruptor saccharolyticus*. More hydrogen will result in methane-rich biogas and reduce the amount of carbon dioxide. Therefore, methane is a by-product of this anaerobic decomposition process, which attempts to digest organic acids and generate energy for the microbes [25]. Consequently, the primary pathways are: 1. Methane is produced through the decarboxylation of methyl alcohols, methyl amines, methyl sulfides, *etc.*, 2. Methane is produced *via* the reduction of H_2/CO_2 in a process known as hydrogenophilic or hydrogenotrophic methanogenesis, and 3. Acetotrophic or acetoclastic methanogenesis is the process of producing methane through acetate decarboxylation [26].

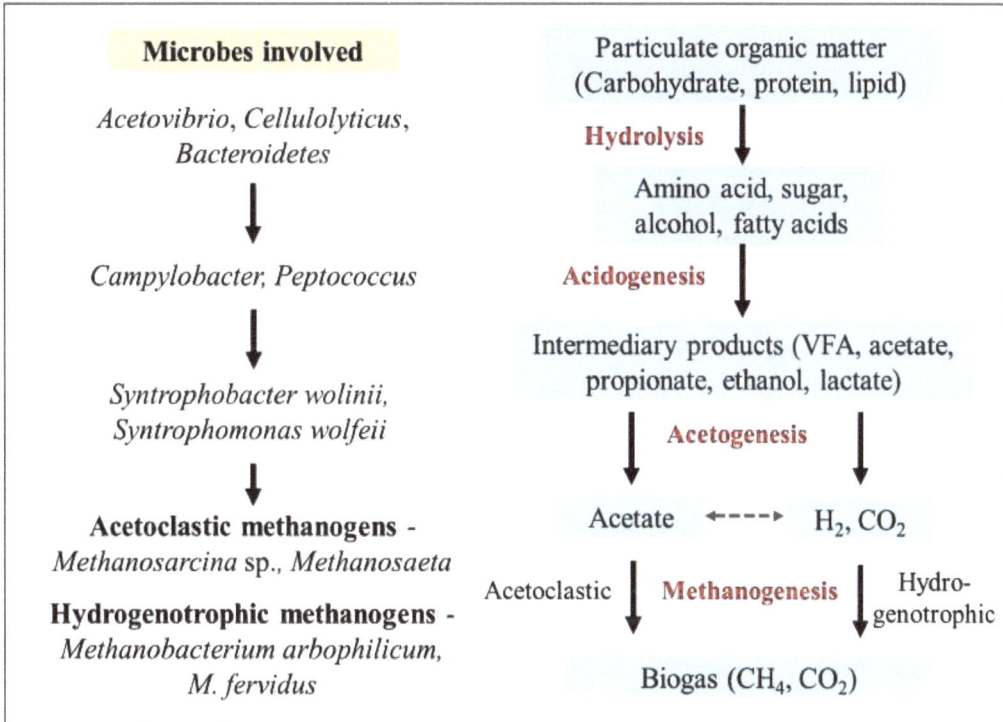

Fig. (2). Microbes involved in the anaerobic digestion process.

Food Industry

Although many microorganisms in the food industry are harmful, some are essential for manufacturing various fermented foods and drinks. Microorganisms, including bacteria, fungi, and yeasts, are employed to produce and provide food ingredients. In processing industries, they are used to increase the shelf life and to improve the flavor of vegetables and food materials. Also, natural pigments produced by microorganisms are used as food colorants, an alternative to synthetic colorants [27].

Diverse genera of microbes, including bacteria, fungi, algae, *etc.*, are a good source of microbial protein and can be used as food, feed, and value-added products [28]. They play a significant role in reducing the C footprint of industries and simultaneously generating feed, food, and environment-friendly chemicals in the process, thereby making the production system sustainable in the long run

[29]. For the development and synthesis of biomass, chemo-litho-autotrophs like hydrogen oxidizing bacteria employ hydrogen as an inorganic electron source and CO_2 as a carbon source. Methanotrophic bacteria utilize methane as both energy and carbon sources. Less than 70% of the protein in commercially available MP, like Uniprotein®, derived from *Methylococcus capsulatus*, comprises all the necessary amino acids. They are also environmentally friendly since they have a far less water and land footprint than plant-based or animal-based feed [30]. Among the yeast, *Saccharomyces cerevisiae* is primarily exploited by industries, followed by other species like *Candida utilis and Candida tropicalis,* which are raised for their protein and vitamin-rich biomass [31]. Several studies have explored the use of agrowaste, such as bagasse, orange, and potato peels, for yeast production so as to improve the sustainability of yeast-producing industries [32]. When exposed to artificial or natural light, photoautotrophic microalgae use CO_2 to produce energy through photosynthesis. The reason for using microalgae as food lies in the appreciable nutritive value, digestible protein, amino acids, vitamins, minerals, and PUFA content [33]. Microalgae like *Chlorella, Spirulina, Dunaliella, and Haematococcus* were the most popular commercially cultivated species for food, feed, and nutraceutical industries, with an annual production of 5000, 15000, 2000, and 1000 tons per annum, respectively [34]. *Chlorella* and *Spirulina* together contribute about 80% of the global biomass production of microalgae. Cyanobacteria like *Nostoc sphaeroides* and *Aphanizomenon flosaquae* are commercially produced for use in the food and cosmetic industry in China and USA, respectively. Microalgae have been increasingly used for fortifying food products like pasta, yogurt, bakery products, soft drinks, *etc* [35].

Bacteriocins are small peptides produced by certain types of bacteria to eliminate other bacteria that share their environment and compete for resources such as food and nutrients [36]. They are antimicrobial peptides produced by ribosomes and are typically used as food additives. During fermentation, the starter cultures either contribute these peptides to the food [37]. FDA-approved bacteriocins used in the food industry as preservatives include *Lactococcus lactis*, subsequently known as Nisin. Antimicrobial bacteriocins such as *Staphylococcus aureus* are used for treatment against human and animal microbial infections without showing side effects [38], followed by their use on pathogens of the respiratory tract, showing their effectiveness against the deadly infections of the respiratory column [39]. *Fermenticin, a* form of bacteriocin, is an effective sperm immobilizer in recent studies, paving the way for the formulation of bacteriocin in contraceptive products [40].

Pharmaceutical Industry

The chief input of microbes to the pharmaceutical industry is the development of antibiotics. In the past, all antibiotics were byproducts of microbial metabolism, but new genetic developments have made it feasible to create more potent and specific treatments. Contrarily, medicinal products like steroids and vaccines from microbial cultures are part of the medical industry. The use of microorganisms for medically significant investigations that aid in discovering cell processes is a further vital task of the pharmaceutical industry, antimicrobial drugs and other compounds that can be produced from microbes. Microbes are also employed in the pharmaceutical industry to create chemical medications and compounds (Table **2**) [41].

Table 2. Microbes and microbial products in the pharma industry.

Microbe	Purpose	Examples of Products Produced
Escherichia coli	Recombinant protein production	Insulin, Growth hormones, Enzymes, *etc.*
Saccharomyces cerevisiae	Protein expression, vaccine production	Hepatitis B vaccine, Recombinant proteins
Bacillus subtilis	Enzyme production	Alpha-amylase, Proteases, *etc.*
Pseudomonas sp.	Enzyme production	Lipases, Proteases, *etc.*
Streptomyces sp.	Antibiotic production	Streptomycin, Tetracycline, Erythromycin, *etc.*
Aspergillus niger	Organic acid production	Citric acid
Haemophilus influenzae	Vaccine production	Hib vaccine (*Haemophilus influenzae* type b)
Mycobacterium bovis	Vaccine production	BCG vaccine (Tuberculosis)

The convergence of cutting-edge technologies with advancements in genomics, metabolic engineering, and chemical synthesis is creating exciting new opportunities to harness the remarkable chemical diversity of small molecules from nature in the search for new drugs [42]. The discovery of physiologically active compounds from marine microorganisms has sparked interest in natural products derived from marine sources, which are currently being investigated in clinical studies. In most cases, microbial fermentation is more environmental friendly and financially feasible to produce the active pharmaceutical ingredient (API) than culturing slower-growing microorganisms [43].

Use of Microorganisms in Pharmaceuticals: Some evidence suggests that several other molecules in the drug development process may come from microorganisms. In other instances, the only supporting evidence is the chemical similarity between the marine-derived substance and recognized bacterial chemicals. For instance, symbiotic bacteria likely manufacture the cancer therapy drug Trabectedin, which the European Union has licensed. The semi-synthesis of Trabectedin is analogous to the *Pseudomonas fluorescens* production of cyanosafracin, and both Trabectedin and Zalypsis contain ring structures that are very similar to saframycin generated by bacteria. *Eribulin mesylate* is a kind of polyether halichondrin that is generated from sponges [38]. The likelihood of microbial manufacturing is increased by the structural similarities between bacterially generated polyethers and the halichondrin family of chemicals, such as tetronamycin and nigericin. Even though these compounds are important, only the biosynthesis routes for bryostatins and salinisporamides are known. There is optimism that the heterologous manufacture of bryostatin may result in significant yields for pharmacological treatment because the gene cluster responsible for its synthesis has been discovered [43].

Algal blooms in pharmaceuticals: Marine natural products exhibit a desirable attribute that can benefit human health. These compounds are promising prototypes for pharmacological development and conversion into functional medications because of their environmental and unique chemical structures, target specificness, and stability. Karlotoxins, for example, are both hazardous as a harmful algal bloom (HAB) toxin that kills big fish and fascinating prospects for developing cholesterol therapeutic leads. The *K. veneficum* produces a metabolite known as karlotoxins, which offers a special possibility for creating new medications. Both in laboratory cultures and the environment, these substances exertichthyotoxic, cytotoxic, and hemolytic effects. The Karlotoxins molecules provide an opportunity to investigate the development of medications to treat severe human health concerns involving cholesterol, such as heart disease and cancer [44]. The discovery of karlotoxins opens up the possibility of developing a safe cholesterol-based drug that might potentially move artery-bound cholesterol to the liver or kidneys for excretion.

Derivatives of Microbial Products as Commercial Pharma-products

There are various derivatives of microbial products as commercial pharma-products (i) Therapeutic products: Microorganisms, particularly bacteria of the Actinomycetales order and filamentous fungi, are a plentiful source of new antibiotics, drugs to lower cholesterol, anticancer treatments, and more. (ii) Antibiotics: Antibiotics have evolved over the past 80 years to become a crucial part of contemporary medicine [45]. Even with the considerable advancements in the development of antibiotics, infectious illnesses continue to be a leading cause of mortality worldwide. Bacterial infections primarily account for about 17 million fatalities yearly among children and the elderly. While the phrase was not utilized until after Alexander Fleming's discovery in 1928, which began the era of microbial drugs, Selman Waksman coined the term "antibiotic" in 1941 to describe any little molecule produced by bacteria that prevents the growth of other microbes. He found that a mould compound might kill *Staphylococcus aureus*-seeded bacteria in a Petri plate.

Penicillin, the mould's active ingredient that was subsequently named after it, isolated as a yellow powder, and utilized as a potent antibacterial drug during World War II, was discovered as *Penicillium notatum* [46]. *Streptomyces* is a group of bacteria that comprises well-known antibiotics such as kanamycin, cephalosporin, chloramphenicol, and streptomycin and accounts for over 80% of all antibiotics, demonstrating the importance of microbial systems to the antibiotic industry. It is estimated that screening around 10,000 actinomycetes could result in the discovery of nearly 2500 antibiotic-producing strains, with 2250 synthesizing streptothricin, 125 synthesizing streptomycins, and 40 synthesizing tetracyclines. Vancomycin, erythromycin, and the newest antibiotic, daptomycin, are among the antibiotics whose chances of discovery are believed to be one in 100,000, one in a million, and one in ten million, respectively. The production of antibiotics that are derived from natural sources is performed through the method of microbial fermentation in the industrial setting. Diarrhea is considered the primary cause of death among children worldwide, requiring antibiotics to prevent illness [47]. The need for antibiotics is further fueled by the introduction of novel illnesses like the Zika and Ebola viruses and other high-burden infectious diseases, including HIV/AIDS, pneumonia, malaria, and TB.

Biosurfactants: Although there are many sources of anticancer medications, microorganisms have drawn much attention because of their simple production and control and capacity to create a variety of bioactive metabolites, including biosurfactants [38]. Biosurfactants exhibit cytotoxic effects on cancer cells, possibly due to their amphiphilic nature, which allows them to interact effectively with fatty acids and phospholipids. This property is thought to be responsible for its anticancer activity. One example of a biosurfactant is surfactin, which has demonstrated potential cytotoxicity against human breast cancer cells by encouraging apoptosis by increasing the production of reactive oxygen species and decreasing the potential of mitochondrial membranes [48].

Nutraceuticals: In particular, age-related illnesses like osteoporosis, arthritis, depression, gastrointestinal diseases, diabetes, inflammation, cardiovascular diseases, and cancer are prevented by using nutraceuticals, which are frequently used for their health-promoting or disease-preventing benefits. The use of microbes is a crucial method for creating nutraceuticals. The rapid identification of various biosynthetic pathways for natural substances and the ability to genetically modify microbes have significantly expanded the use of microbes in producing numerous nutraceuticals [33]. Many substances produced by microbes, such as alkaloids, polyphenolic compounds, polysaccharides, terpenoids, and amino acids, can potentially benefit human health. Polyunsaturated fatty acids (PUFAs), which are necessary for human health but cannot be produced by the body and must be obtained through diet, are well known for being produced by several fungi, including *Aspergillus, Candida, Cryptococcus, Mortierella,* and *Mucor*. Probiotics and prebiotics are also essential nutraceuticals. Probiotics are living microbes that are given in sufficient amounts to provide a beneficial health effect to the host [38].

MICROBIAL PRODUCTION OF ORGANIC AND AMINO ACIDS

Humankind has benefited from the organic acids produced by microorganisms. Although these procedures appear plausible, it is not feasible to manufacture these organic acids on a large scale. The market share of microorganism-based biosynthetic chemical product synthesis is predicted to increase significantly. The chemical industry relies heavily on the functional groups of organic acids as a source of raw materials. These are a few examples of organic acids.

Lactic Acid: In the past, lactic acid was produced industrially using lactic acid bacteria. Lactic acid production has increased due to the manufacture of biodegradable plastic polylactide. In 2002, lactic acid production totaled over 150,000 tonnes, and lactic acid bacteria were responsible for 90% of that production [49]. The *Rhizopus oryzae,* lactic acid bacteria, could thrive on carbon sources, including starch, xylose, and mineral sources.

Citric Acid: The *Aspergillus niger* produces 1.6 tonnes of citric acid annually. Aspergillus can accumulate this acid in environments with high substrate levels and low nitrogen levels, as well as some essential elements in acidic environments while maintaining the DO. The improvised production of citric acid no longer generates satisfactory yields. Hence, a filamentous fungi, *Yarrowia lipolytica,* has been identified to synthesize citric acid in significant quantities by utilizing fatty acids and n-paraffins [50].

Succinic Acid: Succinic acid, which is generated from petroleum feedstocks, is also called amber acid. Succinic acid plays a crucial role as a raw material in producing various chemicals in food, medicine, and agriculture. Cost-effective and highly functional, succinic acid is synthesized biochemically from glucose using a genetically engineered bacterium [51].

L-Phenylalanine: In the culinary and pharmaceutical industries, it has numerous uses. Ten enzymatic processes are involved in biosynthetic pathways, five of which are feedback-regulated:aroG, aroK/L, CM-PDT, aroE, and tyrB. With strong promoter control, these genes can be overexpressed to circumvent feedback inhibition. Protein-directed evolution was used to improve production even further. This led to a significant reduction in the feedback inhibition caused by L-Phe and an increase in glucose production of 0.21 g g^{-1}. To produce an ultimate output of 47.0 g/L of L-Phe, the PTS system was deactivated, and the precursor supply was raised even further [52].

L-Tyrosine: An amino acid, L-tyrosine is a predecessor in producing phenylpropanoids, polymers, and melanin. In order to develop high-yielding L-tyrosine synthesizing microorganisms, many techniques were applied, including the abolition of response suppression and the overexpression of crucial enzymes ydiB, tkt, and aroK/L [53].

L-Threonine: It is a member of the nutritionally valuable amino acids, *i.e.*, the aspartate family, which is also used to produce chemical and medicinal reagents. Homoserine dehydrogenase I and II, Aspartate semialdehyde dehydrogenase, threonine synthase, and homoserine kinase are all examples of aspartokinases. They are also known as ThrA, met, LysC, and aspartokinase I, II, and III. Mutations at T342I in lysC and S345F in thrA have been added to L-threonin--producing strains to eliminate feedback inhibition. The *Escherichia coli* strain generated 52.0 g/L of L-threonine [54].

L-Valine: This is a necessary amino acid used in nutrition, medicine, and cosmetics. Isomeroreductase, dihydroxy acid dehydratase, acetohydroxy acid synthase, and transaminase B catalyze four processes necessary to convert pyruvate into L-valine (ilvE). By eliminating panB, ilvA, ilvBNC, and the ilvGMDA operon and removing feedback inhibition in ilvN, *Corynebacterium glutamicum* can produce valine [49].

ROLE OF MICROALGAE IN SUSTAINABLE AGRICULTURE AND ALLIED AGRICULTURAL INDUSTRY

Biofertilizers: In modern agriculture, we depend more on hybrid seeds, high-yielding varieties, and, most importantly, chemical fertilizers. But in recent days, the trend towards using chemical fertilizers has been decreasing as people are getting attracted to using natural or biofertilizers. Biofertilizers are primarily bacteria, fungi, or microalgae that improve vital nutrients available in the soil for the host plants to promote growth and development. A microorganism used as a biofertilizer should be a particular strain of one or more microorganisms, such as bacteria, fungus, algae, or mixtures [55]. For increased plant development and soil health for optimum yield and sustainable production, beneficial microorganisms offer a fair and sustainable substitute for mineral inputs. In agriculture worldwide, beneficial microbes, including *Rhizobia, Bacillus, Mycorrhizae, Azospirillum,*

*Trichoderma, Pseudomonas,*and many more, are frequently encouraged and exploited. In order to increase agricultural sustainability, soil fertility must be used and managed optimally, which depends on soil microbial activities and biodiversity.

Biopesticides: The role of biopesticides in plant growth is shown in Fig. **3**. United Nations Department of Economic and Social Affairs (UN DESA) reported that Earth's population will reach 10.12 billion people by the year 2100, and with such a large population comes a greater need for food, which necessitates more sophisticated and productive agricultural materials. The prescribed dose of fertilizers, suitable crop varieties, efficient disease and insect management, and increased agricultural productivity are all significant factors that play a vital role. Pest control is crucial for healthy, high-crop yields that can provide food for the world's expanding population. Food production has increased significantly worldwide because of the widespread use of chemical pesticides in recent decades. Comparatively, just 11% of pesticides come from natural sources. The excessive and disproportionate use of inorganic fertilizers and pesticides has increased grain yields. However, it has also led to the development of resistant organisms, eradicating natural enemies, deterioration of soil properties, biodiversity loss, and increased plant susceptibility to pests and pathogens. Overuse of pesticides has resulted in disease resurgence, increased pest resistance, loss of biodiversity, and environmental deterioration and pollution. In IPM (Integrated Pest Management), physical, chemical, mechanical, and biological technologies are used to eradicate pests and protect and produce crops. In order to promote the usage of biopesticides, a new, better approach known as Biointensive Integrated Pest Management (BIPM) has been proposed. In order to sustainably satisfy the needs for global food security, complementary approaches are required [56]. Enhancing the virtuous plant-associated microbiome is one strategy to create better and more sophisticated sustainable production methods.

Through the advancement of scientific understanding of plant and soil microbiomes, technologies drive hope for novel and efficient microbial pesticides. The great majority of microbial pesticides have yet to be found and have been promoted as a suitable replacement for chemically manufactured insecticides in recent years. According to projections, the market for microbial pesticides in India will increase from 69.62 million USD in 2022 to 130.37 million USD by 2029, growing at an Annual Growth Rate of 9.38%. ICAR Institutes have significantly impacted the sector by developing advanced microbial pesticides and biocontrol agents for various crop pests and diseases. For commercial manufacturing, the majority of microbial pesticide manufacturers rely on either a single strain or a small number of microorganisms. Several types of microbial pesticides may be developed according to their stability, longevity, and compatibility with microorganisms. These biopesticides can be either in dry or liquid formulations. The active component can be created by mixing stabilizers, synergists, stickers, surfactants, coloring agents, nutrients, dispersants, melting agents, and anti-freezing agents. It is claimed that *B. thuringiensis*, a single entomopathogenic bacterium, is the source of nearly 90% of biopesticides [57]. Around 200 biopesticides are sold in American markets, while 60 products are sold in the European Union. Due to their eco-friendly and host-specific character, these biopesticides are being produced and used at an ever-increasing rate on a global scale. Biopesticides have a limited shelf life, and it might be challenging to achieve it beyond 1-2 years in an ambient state, but in most circumstances, their mass manufacture and application are comparably extremely simple and inexpensive. It has also shown targeted action and broad-spectrum efficacy against certain insect pests at different phases of the disease organisms' life cycles. For the creation of better IPM strategies, interactions between insects and their related bacteria should be carefully investigated. Pathogens' tendency to spread themselves over space and time would unquestionably benefit sustainable production [58].

Fig. (3). Role of biopesticides in plant growth.

Because of many of their characteristics and the functions they perform in nature, microalgae are receiving increased attention from photosynthetic microorganisms. They are one of the most significant groups for capturing huge amounts of CO_2 and simultaneously producing large quantities of O_2 and oils [59]. Because microalgae are not conventional foods and feeds, their production takes less area and will not be able to compete with the arable crops, which can be grown in large open ponds or wastewater systems. Moreover, microalga biomass is an excellent source of biodiesel since it produces more oil than the greatest oilseed crops [60]. Microalgal biomass can be converted to energy sources for sustainable liquid fuels like bioethanol and biodiesel by several methods, *viz.*, chemical reaction, direct combustion, and thermochemical and biological conversion. Green microalgae with high lipid production, including *Chlorella vulgaris, Scenedesmus dimorphus, Nannochloris, Botryococcus braunii,* and *Chlorococcum littorale,* as well as diatoms like *Chaetoceros muelleri,* were thought to be suitable strains for producing various biofuels, including biodiesel, kerosene, and gasoline [61].

Algae have a significant part in the agricultural sector, where they may be employed as biofertilizers and soil stabilizers despite their capacity to generate larger amounts of lipids [62]. Moreover, many microalgae generate a variety of pharmaceuticals, vaccines, proteins, and nutrients that are otherwise unavailable or prohibitively expensive to get from sources, such as plants and animals. They are also recognized as a source of significant vitamins and precursors, including riboflavin, ascorbic acid, and tocopherol, which can be used as prospective candidates in the pharmaceutical and nutraceutical sectors. They are abundant and valuable in synthesizing secondary metabolites, including carotenoids, vitamins, and amino acids, which aid in developing plants. Interestingly, numerous unicellular algae like *Chlamydomonas pyrenoidosa* and *Chlorella vulgaris* have shown antimicrobial properties towards various pathogens in cell extracts and medium extracts [58].

Intensive use of chemical fertilizers heavily affects soil health, accelerating a significant reduction in agricultural production due to adverse effects on macro and microfauna in the soil system. Under these circumstances, microalgae-based biofertilizer would be an alternative solution to sustainable agriculture since it not only improves the fertility of soil and crop growth and development but also enhances the soil physico-chemical properties like soil particle agglomeration and water holding capacity [55]. Microalgal biomass could replace chemical fertilizers while enhancing crop productivity and soil quality to formulate slow-release bio-fertilizers. Microalgae may be produced in wastewater since they tolerate extreme environmental conditions and require few nutrients for growth, eventually decreasing the production cost and the benefit of wastewater treatment.

PIGMENTS AND COSMETICS INDUSTRIES

Microbial pigments are natural, biodegradable, and have excellent antioxidant and anti-aging potential, making them suitable candidates for cosmetic products. Carotenoids like β-carotene, astaxanthin, and lycopene are widely used in the preparation of photo protectants owing to their high antioxidant potential. Lycopene produced by the cyanobacterium *Anabaena vaginicola* is added to sunscreen products to neutralize the free radicals resulting from oxidative stress occurring in the skin [63]. Astaxanthin from *Haematococcus pluvialis* protects against oxidative stress and lipid peroxidation and has been used in cosmetics to treat wrinkles and skin spots [64]. Astaxanthin derived from marine yeast *Rhodotorula mucilaginosa* has the potential to reduce hyperpigmentation and prevent the occurrence of age spots in the skin. β-Cryptoxanthin produced by the microalgae *Dunaliella salina* induces hyaluronic acid production, which plays a major role in skin hydration [65]. A cosmetic formulation containing a combination of melanin from *Halomonas venusta* and extract from *Gelidium spinosum* (seaweed) exhibited high wound-healing efficiency with high Sun Protection Factor (SPF) values [66]. Phycobiliproteins from *Cyanobium* sp. have anti-collagenase and anti-hyaluronidase activities, making them an effective antiaging ingredient [67]. Fucoxanthin produced by *Laminaria japonica* inhibits tyrosinase activity and expression of skin mRNA linked to melanogenesis, making it ideal for skin-whitening products [68].

Pigments used in cosmetics, such as nail polish, gloss, and lipstick, are made from organic solvents that are considered harmful. In contrast, natural colors derived from microbes are safe and stable and have better acceptability among consumers. Ankaflavin from *Monascus* sp. has been used in skin formulations to create a durable pigment that resembles a natural skin tan [69]. Phycocyanin derived from *Spirulina* sp. has been commercially exploited as a coloring agent for eyeliners. Microalgae produce strong green pigments (chlorophyll), which can serve as both colorants and anti-inflammatory agents in cosmetics [70]. Spirulina (Cyanobacteria) also produces phycoerythrobilin, which has the potential to be used as a cosmetic pigment. *Porphyridium aerogineum*-derived phycobiliproteins are resistant to acidic pH and highly stable under light conditions, making them ideal for use in cosmetic colors [71]. Melanin pigment from *Streptomyces bellus* has been used in a variety of cosmetic products, including lip balms, to impart a dark color [72]. Red microalgae produce natural pigments that are used to derive various shades of purple and pink cosmetic colors. Chlorophyll from microalgae can mask foul smell, making it suitable for use in fragrances, dental creams, and other personal hygiene products [63].

Microbial pigments are versatile, with potential application in food, pharmaceutical, cosmetics, and textile industries as a promising natural source of colorant. Microbes are rich in fatty acids, enzymes, peptides, vitamins, lipopolysaccharides and pigments with beneficial properties for cosmetic applications. Bacterial pigments are a unique and sustainable source of bioactive color compounds used in cosmetics, food, textiles, printing, and pharmaceutical products. Microbial pigments are getting more attention due to their wide application in textiles dyeing, cosmetics, food colorants, painting, pharmaceuticals, and plastics, and it was assumed that bacterial pigments are to dominate the pigment industries and organic market in the near future [63].

CONCLUSION AND FUTURE PERSPECTIVES

Microbial techniques are quickly acquiring a key practical role in the energy-efficient synthesis of high-purity goods and numerous new molecules and chemicals. Novel strategies must be designed to increase the productivity and yield of essential compounds. The synthesis of natural goods and biologics by microorganisms faces numerous difficulties simultaneously. Low production of titters, challenging product isolation, or structural identification are some difficulties in natural product-based medication discovery. Similar improvements can be made to how recombinant proteins are expressed in microbial platforms. The synthesis of secondary metabolites can be enhanced through ribosome engineering and gene shuffles, two methods that can be combined to advance the field of microbial natural products.

Additionally, the exact quantification of biochemical alterations and metabolic pathways enabled by 'omics information integration in natural product drug development has significant potential. Metagenomics developments have made it possible to comprehend better a variety of complex microbiological sources, including sub-aerial sites, ice cores, lakes, rivers, and marine ecosystems. Overall, the diversity and importance of natural microbial products and biologics in human existence will continue to grow. The potential for recombinant pharmaceuticals is growing through the use of novel protein production platforms and initiatives to improve products. Microbial cells will continue to be successful as protein producers due to their adaptability and low cost. Despite the numerous obstacles, engineering techniques and recombinant DNA technologies will also expand the production of natural microbial products and recombinant proteins. The further development of natural product analogs will open opportunities to discover substances with superior biological activity to their natural equivalents. Microbial natural products, which are a reliable source for novel sustainable products, can be developed further using modern technology.

CONSENT FOR PUBLICATON

The authors hereby give consent to Bentham Science Publishers Pt. Ltd. to publish this book chapter, and there are no other consents to declare.

ACKNOWLEDGEMENTS

The authors thank the management of Tamil Nadu Agricultural University, Coimbatore & ICAR- National Institute of Abiotic Stress Management, Maharashtra, for their continual encouragement and unflinching support.

REFERENCES

[1] Timmusk S, Behers L, Muthoni J, Muraya A, Aronsson AC. Perspectives and challenges of microbial application for crop improvement. Front Plant Sci 2017; 8: 49.
[http://dx.doi.org/10.3389/fpls.2017.00049] [PMID: 28232839]

[2] Jiang Y, Chu N, Qian DK, Jianxiong Zeng R. Microbial electrochemical stimulation of caproate production from ethanol and carbon dioxide. Bioresour Technol 2020; 295: 122266.
[http://dx.doi.org/10.1016/j.biortech.2019.122266] [PMID: 31669871]

[3] Ding J, Xu M, Xie F, Chen C, Shi Z. Efficient butanol production using corn-starch and waste Pichia pastoris semi-solid mixture as the substrate. Biochem Eng J 2019; 143: 41-7.
[http://dx.doi.org/10.1016/j.bej.2018.12.017]

[4] Si M, Yan X, Liu M, *et al.* *In situ* lignin bioconversion promotes complete carbohydrate conversion of rice straw by Cupriavidus basilensis B-8. ACS Sustain Chem& Eng 2018; 6(6): 7969-78.
[http://dx.doi.org/10.1021/acssuschemeng.8b01336]

[5] Idris ASO, Pandey A, Rao SS, Sukumaran RK. RETRACTED: Cellulase production through solid-state tray fermentation, and its use for bioethanol from sorghum stover. Bioresour Technol 2017; 242: 265-71.
[http://dx.doi.org/10.1016/j.biortech.2017.03.092] [PMID: 28366693]

[6] Yin H, Ma Y, Deng Y, *et al.* Genome shuffling of Saccharomyces cerevisiae for enhanced glutathione yield and relative gene expression analysis using fluorescent quantitation reverse transcription polymerase chain reaction. J Microbiol Methods 2016; 127: 188-92.
[http://dx.doi.org/10.1016/j.mimet.2016.06.012] [PMID: 27302037]

[7] Xu X, Zhao Q, Wu M, Ding J, Zhang W. Biodegradation of organic matter and anodic microbial communities analysis in sediment microbial fuel cells with/without Fe(III) oxide addition. Bioresour Technol 2017; 225: 402-8.
[http://dx.doi.org/10.1016/j.biortech.2016.11.126] [PMID: 27956331]

[8] Thulasinathan B, Ebenezer JO, Bora A, *et al.* Bioelectricity generation and analysis of anode biofilm metabolites from septic tank wastewater in microbial fuel cells. Int J Energy Res 2021; 45(12): 17244-58.
[http://dx.doi.org/10.1002/er.5734]

[9] Ahmadpour T, Aber S, Hosseini MG. Improved dye degradation and simultaneous electricity generation in a photoelectrocatalytic microbial fuel cell equipped with AgBr/CuO hybrid photocathode. J Power Sources 2020; 474: 228589.
[http://dx.doi.org/10.1016/j.jpowsour.2020.228589]

[10] Zhang J, Wang Z, Chu L, *et al.* Unified photoelectrocatalytic microbial fuel cell harnessing 3D binder-free photocathode for simultaneous power generation and dual pollutant removal. J Power Sources 2021; 481: 229133.
[http://dx.doi.org/10.1016/j.jpowsour.2020.229133]

[11] Geppert F, Liu D, Weidner E, Heijne A. Redox-flow battery design for a methane-producing bioelectrochemical system. Int J Hydrogen Energy 2019; 44(39): 21464-9.
[http://dx.doi.org/10.1016/j.ijhydene.2019.06.189]

[12] Mateos R, Sotres A, Alonso RM, Morán A, Escapa A. Enhanced CO2 conversion to acetate through microbial electrosynthesis (MES) by continuous headspace gas recirculation. Energies 2019; 12(17): 3297.
[http://dx.doi.org/10.3390/en12173297]

[13] Vassilev I, Kracke F, Freguia S, *et al.* Microbial electrosynthesis system with dual biocathode arrangement for simultaneous acetogenesis, solventogenesis and carbon chain elongation. Chem Commun (Camb) 2019; 55(30): 4351-4.
[http://dx.doi.org/10.1039/C9CC00208A] [PMID: 30911739]

[14]   Gavilanes J, Reddy CN, Min B. Microbial electrosynthesis of bioalcohols through reduction of high concentrations of volatile fatty acids. Energy Fuels 2019; 33(5): 4264-71.
[http://dx.doi.org/10.1021/acs.energyfuels.8b04215]

[15]   Nguyen DTN, Lee OK, Hadiyati S, Affifah AN, Kim MS, Lee EY. Metabolic engineering of the type I methanotroph Methylomonas sp. DH-1 for production of succinate from methane. Metab Eng 2019; 54: 170-9.
[http://dx.doi.org/10.1016/j.ymben.2019.03.013] [PMID: 30986511]

[16]   Roy M, Yadav R, Chiranjeevi P, Patil SA. Direct utilization of industrial carbon dioxide with low impurities for acetate production *via* microbial electrosynthesis. Bioresour Technol 2021; 320(Pt A): 124289.
[http://dx.doi.org/10.1016/j.biortech.2020.124289] [PMID: 33129088]

[17]   Srivastava RK, Shetti NP, Reddy KR, Aminabhavi TM. Sustainable energy from waste organic matters *via* efficient microbial processes. Sci Total Environ 2020; 722: 137927.
[http://dx.doi.org/10.1016/j.scitotenv.2020.137927] [PMID: 32208271]

[18]   Kucera J, Lochman J, Bouchal P, *et al.* A model of aerobic and anaerobic metabolism of hydrogen in the extremophile Acidithiobacillus ferrooxidans. Front Microbiol 2020; 11: 610836.
[http://dx.doi.org/10.3389/fmicb.2020.610836] [PMID: 33329503]

[19]   Heeres AS, Picone CSF, van der Wielen LAM, Cunha RL, Cuellar MC. Microbial advanced biofuels production: overcoming emulsification challenges for large-scale operation. Trends Biotechnol 2014; 32(4): 221-9.
[http://dx.doi.org/10.1016/j.tibtech.2014.02.002] [PMID: 24630476]

[20]   Chintagunta AD, Zuccaro G, Kumar M, *et al.* Biodiesel production from lignocellulosic biomass using oleaginous microbes: Prospects for integrated biofuel production. Front Microbiol 2021; 12: 658284.
[http://dx.doi.org/10.3389/fmicb.2021.658284] [PMID: 34475852]

[21]   Kumar B, Bhardwaj N, Agrawal K, Chaturvedi V, Verma P. Current perspective on pretreatment technologies using lignocellulosic biomass: An emerging biorefinery concept. Fuel Process Technol 2020; 199: 106244.
[http://dx.doi.org/10.1016/j.fuproc.2019.106244]

[22]   Kumar D, Singh B, Korstad J. Utilization of lignocellulosic biomass by oleaginous yeast and bacteria for production of biodiesel and renewable diesel. Renew Sustain Energy Rev 2017; 73: 654-71.
[http://dx.doi.org/10.1016/j.rser.2017.01.022]

[23]   Saratale GD, Saratale RG, Shahid MK, *et al.* A comprehensive overview on electro-active biofilms, role of exo-electrogens and their microbial niches in microbial fuel cells (MFCs). Chemosphere 2017; 178: 534-47.
[http://dx.doi.org/10.1016/j.chemosphere.2017.03.066] [PMID: 28351012]

[24]   Ali SN, Anwar MN, Nizami AS, Baqar M. Microbial and technological advancements in biogas production. Curr Dev Biotechnol Bioeng 2020; pp. 137-61.
[http://dx.doi.org/10.1016/B978-0-444-64309-4.00006-4]

[25]   Goswami R, Chattopadhyay P, Shome A, Banerjee SN, Chakraborty AK, Mathew AK, *et al.* An overview of physico-chemical mechanisms of biogas production by microbial communities: a step towards sustainable waste management. 3 Biotech 2016;6:1–12.

[26]   Schnürer A. Biogas production: microbiology and technology. Anaerobes Biotechnol 2016; pp. 195-234.

[27]   Aman Mohammadi M, Ahangari H, Mousazadeh S, Hosseini SM, Dufossé L. Microbial pigments as an alternative to synthetic dyes and food additives: a brief review of recent studies. Bioprocess Biosyst Eng 2022; 45(1): 1-12.
[http://dx.doi.org/10.1007/s00449-021-02621-8] [PMID: 34373951]

[28]   Matassa S, Boon N, Pikaar I, Verstraete W. Microbial protein: future sustainable food supply route

with low environmental footprint. Microb Biotechnol 2016; 9(5): 568-75.
[http://dx.doi.org/10.1111/1751-7915.12369] [PMID: 27389856]

[29]   Lippolis A, Bussotti L, Ciani M, Fava F, Niccolai A, Rodolfi L, *et al.* Microbes: Food for the Future. 2019.

[30]   Cumberlege T, Blenkinsopp T, Clark J. Assessment of Environmental Impact of FeedKindTM Protein. 2016.

[31]   Kieliszek M, Kot AM, Bzducha-Wróbel A, BŁażejak S, Gientka I, Kurcz A. Biotechnological use of Candida yeasts in the food industry: A review. Fungal Biol Rev 2017; 31(4): 185-98.
[http://dx.doi.org/10.1016/j.fbr.2017.06.001]

[32]   COFALEC. Carbon Footprint of Yeast produced in the European Union. Sustain - Yeast carbon Footprint. 2012.

[33]   Bishop WM, Zubeck HM. Evaluation of microalgae for use as nutraceuticals and nutritional supplements. J Nutr Food Sci 2012; 2: 1-6.
[http://dx.doi.org/10.4172/2155-9600.1000147]

[34]   Hu Q. Current status, emerging technologies, and future perspectives of the world microalgal industry. Florence, Italy: B. Abstr. AlgaEurope Conf. Eur. Algae Biomass Assoc. 2019; p. 139.

[35]   Niccolai A, Shannon E, Abu-Ghannam N, Biondi N, Rodolfi L, Tredici MR. Lactic acid fermentation of Arthrospira platensis (spirulina) biomass for probiotic-based products. J Appl Phycol 2019; 31(2): 1077-83.
[http://dx.doi.org/10.1007/s10811-018-1602-3]

[36]   Juturu V, Wu JC. Microbial production of bacteriocins: Latest research development and applications. Biotechnol Adv 2018; 36(8): 2187-200.
[http://dx.doi.org/10.1016/j.biotechadv.2018.10.007] [PMID: 30385277]

[37]   Chikindas ML, Weeks R, Drider D, Chistyakov VA, Dicks LMT. Functions and emerging applications of bacteriocins. Curr Opin Biotechnol 2018; 49: 23-8.
[http://dx.doi.org/10.1016/j.copbio.2017.07.011] [PMID: 28787641]

[38]   Kapoor D, Sharma P, Sharma MMM, Kumari A, Kumar R. Microbes in pharmaceutical industry. Microb Divers Interv Scope 2020; pp. 259-99.

[39]   Ahmad V, Khan MS, Jamal QMS, Alzohairy MA, Al Karaawi MA, Siddiqui MU. Antimicrobial potential of bacteriocins: in therapy, agriculture and food preservation. Int J Antimicrob Agents 2017; 49(1): 1-11.
[http://dx.doi.org/10.1016/j.ijantimicag.2016.08.016] [PMID: 27773497]

[40]   Kaewnopparat S, Dangmanee N, Kaewnopparat N, Srichana T, Chulasiri M, Settharaksa S. *In vitro* probiotic properties of Lactobacillus fermentum SK5 isolated from vagina of a healthy woman. Anaerobe 2013; 22: 6-13.
[http://dx.doi.org/10.1016/j.anaerobe.2013.04.009] [PMID: 23624069]

[41]   Acevedo-Rocha CG, Gronenberg LS, Mack M, Commichau FM, Genee HJ. Microbial cell factories for the sustainable manufacturing of B vitamins. Curr Opin Biotechnol 2019; 56: 18-29.
[http://dx.doi.org/10.1016/j.copbio.2018.07.006] [PMID: 30138794]

[42]   Koehn FE, Carter GT. The evolving role of natural products in drug discovery. Nat Rev Drug Discov 2005; 4(3): 206-20.
[http://dx.doi.org/10.1038/nrd1657] [PMID: 15729362]

[43]   Waters AL, Hill RT, Place AR, Hamann MT. The expanding role of marine microbes in pharmaceutical development. Curr Opin Biotechnol 2010; 21(6): 780-6.
[http://dx.doi.org/10.1016/j.copbio.2010.09.013] [PMID: 20956080]

[44]   Kalsoom M, Ur Rehman F, Shafique T, *et al.* Biological importance of microbes in agriculture, food and pharmaceutical industry: A review. Innovare Journal of Life Sciences 2020; 8: 1-4.

[http://dx.doi.org/10.22159/ijls.2020.v8i6.39845]

[45]   Bush K, Courvalin P, Dantas G, *et al.* Tackling antibiotic resistance. Nat Rev Microbiol 2011; 9(12): 894-6.
[http://dx.doi.org/10.1038/nrmicro2693] [PMID: 22048738]

[46]   Demain AL, Vaishnav P. Production of recombinant proteins by microbes and higher organisms. Biotechnol Adv 2009; 27(3): 297-306.
[http://dx.doi.org/10.1016/j.biotechadv.2009.01.008] [PMID: 19500547]

[47]   Zhang C, Kim SK. Application of marine microbial enzymes in the food and pharmaceutical industries. Adv Food Nutr Res 2012; 65: 423-35.
[http://dx.doi.org/10.1016/B978-0-12-416003-3.00028-7] [PMID: 22361204]

[48]   Chiewpattanakul P, Phonnok S, Durand A, Marie E, Thanomsub BW. Bioproduction and anticancer activity of biosurfactant produced by the dematiaceous fungus Exophiala dermatitidis SK80. J Microbiol Biotechnol 2010; 20(12): 1664-71.
[PMID: 21193821]

[49]   Paul PEV, Sangeetha V, Deepika RG. Emerging trends in the industrial production of chemical products by microorganisms. Recent Dev Appl Microbiol Biochem 2019; pp. 107-25.
[http://dx.doi.org/10.1016/B978-0-12-816328-3.00009-X]

[50]   Chen S, Yang Y, Liu C, Dong F, Liu B. Column bioleaching copper and its kinetics of waste printed circuit boards (WPCBs) by Acidithiobacillus ferrooxidans. Chemosphere 2015; 141: 162-8.
[http://dx.doi.org/10.1016/j.chemosphere.2015.06.082] [PMID: 26196406]

[51]   Santos CNS, Xiao W, Stephanopoulos G. Rational, combinatorial, and genomic approaches for engineering L-tyrosine production in *Escherichia coli*. Proc Natl Acad Sci USA 2012; 109(34): 13538-43.
[http://dx.doi.org/10.1073/pnas.1206346109] [PMID: 22869698]

[52]   Chu HS, Kim YS, Lee CM, *et al.* Metabolic engineering of 3□hydroxypropionic acid biosynthesis in *Escherichia coli*. Biotechnol Bioeng 2015; 112(2): 356-64.
[http://dx.doi.org/10.1002/bit.25444] [PMID: 25163985]

[53]   Choi JM, Han SS, Kim HS. Industrial applications of enzyme biocatalysis: Current status and future aspects. Biotechnol Adv 2015; 33(7): 1443-54.
[http://dx.doi.org/10.1016/j.biotechadv.2015.02.014] [PMID: 25747291]

[54]   Yuan P, Cao W, Wang Z, Chen K, Li Y, Ouyang P. Enhancement of l-phenylalanine production by engineered Escherichia coli using phased exponential l-tyrosine feeding combined with nitrogen source optimization. J Biosci Bioeng 2015; 120(1): 36-40.
[http://dx.doi.org/10.1016/j.jbiosc.2014.12.002] [PMID: 25553973]

[55]   Cao TND, Mukhtar H, Le LT, *et al.* Roles of microalgae-based biofertilizer in sustainability of green agriculture and food-water-energy security nexus. Sci Total Environ 2023; 870: 161927.
[http://dx.doi.org/10.1016/j.scitotenv.2023.161927] [PMID: 36736400]

[56]   Jaiswal S, Singh DK, Shukla P. Gene editing and systems biology tools for pesticide bioremediation: a review. Front Microbiol 2019; 10: 87.
[http://dx.doi.org/10.3389/fmicb.2019.00087] [PMID: 30853940]

[57]   Vadiveloo A, Shayesteh H, Bahri PA, Moheimani NR. Comparison between continuous and daytime mixing for the treatment of raw anaerobically digested abattoir effluent (ADAE) and microalgae production in open raceway ponds. Bioresour Technol Rep 2022; 17: 100981.
[http://dx.doi.org/10.1016/j.biteb.2022.100981]

[58]   Sharma P, Sharma N. Industrial and biotechnological applications of algae: a review. Journal of Advances in Plant Biology 2017; 1(1): 530.
[http://dx.doi.org/10.14302/issn.2638-4469.japb-17-1534]

[59]   Hayes CJ, Burgess DR Jr, Manion JA. Combustion pathways of biofuel model compounds: A review

of recent research and current challenges pertaining to first-, second-, and third-generation biofuels. Adv Phys Org Chem 2015; 49: 103-87.
[http://dx.doi.org/10.1016/bs.apoc.2015.09.001]

[60]   Cheng P, Zhou C, Feng Y, Chu R, Wang H, Bo Y, *et al.* Algal cultivation and algal residue conversion to bioenergy and valuable chemicals Biomass, Biofuels, Biochem. Elsevier 2022; pp. 115-30.

[61]   Baek J, Choi J, Park H, Lim S, Park SJ. Isolation and proteomic analysis of a Chlamydomonas reinhardtii mutant with enhanced lipid production by the gamma irradiation method. J Microbiol Biotechnol 2016; 26(12): 2066-75.
[http://dx.doi.org/10.4014/jmb.1605.05057] [PMID: 27586532]

[62]   Kumar A, Singh JS. Microalgal bio-fertilizers Handb Microalgae-Based Process Prod. Elsevier 2020; pp. 445-63.
[http://dx.doi.org/10.1016/B978-0-12-818536-0.00017-8]

[63]   Ding J, Wu B, Chen L. Application of marine microbial natural products in cosmetics. Front Microbiol 2022; 13: 892505.
[http://dx.doi.org/10.3389/fmicb.2022.892505] [PMID: 35711762]

[64]   Mendes-Silva TCD, Andrade RFS, Ootani MA, *et al.* Biotechnological potential of carotenoids produced by extremophilic microorganisms and application prospects for the cosmetics industry. Adv Microbiol 2020; 10(8): 397-410.
[http://dx.doi.org/10.4236/aim.2020.108029]

[65]   Mourelle M, Gómez C, Legido J. The potential use of marine microalgae and cyanobacteria in cosmetics and thalassotherapy. Cosmetics 2017; 4(4): 46.
[http://dx.doi.org/10.3390/cosmetics4040046]

[66]   Poulose, N., Sajayan, A., Ravindran, A., Sreechithra, T. V., Vardhan, V., Selvin, J., & Kiran, G. S. Photoprotective effect of nanomelanin-seaweed concentrate in formulated cosmetic cream: With improved antioxidant and wound healing properties. Journal of Photochemistry and Photobiology B: Biology 2020; 205: 111816.

[67]   Pagels F, Almeida C, Vasconcelos V, Guedes AC. Cosmetic Potential of Pigments Extracts from the Marine Cyanobacterium *Cyanobium* sp. Mar Drugs 2022; 20(8): 481.
[http://dx.doi.org/10.3390/md20080481] [PMID: 36005483]

[68]   Wang HMD, Chen CC, Huynh P, Chang JS. Exploring the potential of using algae in cosmetics. Bioresour Technol 2015; 184: 355-62.
[http://dx.doi.org/10.1016/j.biortech.2014.12.001] [PMID: 25537136]

[69]   Aruldass CA, Dufossé L, Ahmad WA. Current perspective of yellowish-orange pigments from microorganisms- a review. J Clean Prod 2018; 180: 168-82.
[http://dx.doi.org/10.1016/j.jclepro.2018.01.093]

[70]   Yarkent Ç, Gürlek C, Oncel SS. Potential of microalgal compounds in trending natural cosmetics: A review. Sustain Chem Pharm 2020; 17: 100304.
[http://dx.doi.org/10.1016/j.scp.2020.100304]

[71]   Chandra R, Parra R, Iqbal HM. MN Iqbal H. Phycobiliproteins: a novel green tool from marine origin blue-green algae and red algae. Protein Pept Lett 2017; 24(2): 118-25.
[http://dx.doi.org/10.2174/0929866523666160802160222] [PMID: 27491380]

[72]   Srinivasan M, Keziah SM, Hemalatha M, Devi CS. Pigment from Streptomyces bellus MSA1 isolated from marine sediments. IOP Conference Series: Materials Science and Engineering, 2017, 263(2), pp. 022049.

CHAPTER 11

Role of Microbes and Microbiomes in Human and Animal Health Security

A. Ch. Pradyutha[1,*] and S. Chaitanya Kumari[2]

[1] *Department of Microbiology, Raja Bahadur Venkata Rama Reddy Women's College, Narayanguda, Hyderabad, Telangana, India*

[2] *Department of Microbiology, Bhavan's Vivekananda College of Science, Humanities & Commerce, Sainikpuri, Secunderabad, Telangana, India*

Abstract: Most of the various categories of bacteria and fungi that comprise the human microbiota are primarily incapable of causing diseases. Human beings and animal microbiomes can influence their health and homeostasis through the synthesis of necessary nutrients and vitamins, metabolism of drugs, guarding against pathogenic microbes, additional production of bile acids from the host, immune response, vulnerability to illness, and consistent behavior change. Animal species harbor distinctive microbiomes and possess greater complexity compared to the human microbiome. Living organisms are somewhat exposed to microbes in the newborn stage, at the time of delivery from the birth passage or vagina, and through breastfeeding. The kind of microbes the infant carries relies exclusively on the species seen in the mother. Further, changes in the microbiota of animals and humans depend on exposure to the environment and type of diet. This change can help benefit the health of the host or put one at a more significant chance for disease. This transformation of the microbiome in earlier life holds possible health importance to developing the immune system, influencing health effects including gastroenteritis, asthma, hay fever (allergic rhinitis), and chronic illnesses like diabetes. In addition to the genes of the family, surroundings, medication use, and diet greatly determine what microbiota is present in animals and humans. All of these aspects construct a particular microbiome from individual to individual. An adult living being is colonized by multiple species of bacteria. The total biomass of these microorganisms is typically estimated at around 0.2 kg in adults. The microbiomes present in human and animal bodies serve several functions. They contribute to the breakdown of food, allowing for the digestion of complex carbohydrates, fiber, and other substances that our bodies cannot process alone. Additionally, these microbiomes produce essential nutrients that are made available to us. They also play a vital role in neutralizing toxins or harmful compounds, promoting detoxification, and safeguarding our well-being. Using microorganisms in therapies is one of the clinical revolutions in the 21st century. Numerous research studies have revealed the crucial functions of microbes and microbiomes in human and animal health security.

****** Corresponding author A. Ch. Pradyutha:** Department of Microbiology, Raja Bahadur Venkata Rama Reddy Women's College, Narayanguda, Hyderabad, Telangana, India; E-mail: pradyutha.g@gmail.com

Keywords: Allergic rhinitis, Diseases, Gastroenteritis, Immune system, Microbiome, Toxins.

INTRODUCTION

Within the depths of the human and animal body resides a formidable and enigmatic kingdom of microorganisms [1]. Bacteria, fungi, viruses, and their intricate interactions, metabolites, and genetic components coexist on or within the human and animal body, collectively referred to as microbiomes [2]. These countless microbial communities thrive in various regions, forging a complex and diverse microbiome crucial for maintaining individual well-being. Disruptions in this delicate balance can lead to a host of ailments, including autoimmune conditions, cardiovascular disorders, and even cancers [3]. Consequently, analyzing the human and animal microbiome is paramount in understanding health. Microbiome research has witnessed an unprecedented surge in recent decades, captivating the attention of both the scientific community and the general public [4]. This burgeoning field of study has kindled profound curiosity and fascination among scientists and individuals alike. An all-encompassing investigation is currently underway to categorize and unravel the functional roles of the human microbiome, cementing its status as an indispensable powerhouse in the realm of health and disease [5].

The human microbiome plays a vital role in human development, immunity, and nutrition, where beneficial bacteria establish themselves as colonizers rather than destructive invaders. For example, vaginal Lactobacilli produce lactic acid, maintaining a low pH to protect against pathogen growth. Disorders in the microbiome have been linked to autoimmune conditions such as diabetes, rheumatoid arthritis, muscular dystrophy, multiple sclerosis, and fibromyalgia. Intriguingly, research suggests that these autoimmune diseases may be transmitted not through DNA inheritance but by inheriting the family's microbiome. The human microbiome also exhibits significant effects such as preventing pathogen invasion by inducing competition for resources and space, boosting the host's immune response, producing pathogen-harming antibiotics (Colicins), synthesizing essential vitamins like K and B, and occasionally becoming pathogenic when host defenses weaken. Additionally, the ubiquitous presence of the human microbiome and its resemblance to certain pathogens can sometimes lead to diagnostic confusion.

Recognizing the immense importance of the human microbiome, the National Institutes of Health (NIH) launched the Human Microbiome Project (HMP) in 2008. This ambitious five-year initiative, with a budget of around $115-$150 million, aimed to map the microbial composition of healthy individuals using genome sequencing techniques. On June 13, 2012, the HMP investigators established a reference database and identified the range of normal microbiome variation in humans. The NIH continues to support the Human Microbiome Project, funding research to catalog the diverse bacteria that inhabit humans and investigate correlations between microbiome changes and human health. The objectives of the HMP encompass the comprehensive characterization of the human microbiome, including its composition, diversity, function, and gene sequencing, as well as exploring the association between infections and microbiome modifications. The project also focuses on developing new computational methods and devices, establishing a resource repository, and addressing the ethical, legal, and social implications of human microbiome research.

This chapter aims to provide an overview of the significant role of microbes and microbiomes in ensuring the security of human and animal health, shedding light on their profound impact on various aspects of well-being.

OUTLINE OF HUMAN AND ANIMAL MICROBIOME (NORMAL FLORA)

The human body harbors a diverse array of microorganisms known as normal flora, which peacefully coexist without causing disease [6]. In a fascinating revelation, it has been discovered that our bodies are inhabited by over 10 times more microbial cells, totaling nearly 100 trillion cells, than human cells [7]. Astonishingly, despite their abundance, the collective weight of the microbiome is approximately 200 grams. These microbiomes constitute intricate ecological communities consisting of symbiotic organisms, commensals, and potentially pathogenic microorganisms that reside in various regions of our bodies. Among these regions, the gut is home to the most prominent microbial residents, while the skin and genitals also host significant populations. What is truly remarkable is that these microorganisms, along with their genetic material, establish their presence within us from the moment of birth and persist throughout our entire lives [8].

Microbes begin to populate the human body from the moment of birth. Newborns acquire microorganisms through various means, primarily from their mothers, during the delivery process. This transmission occurs either through a passage through the birth canal or by coming into contact with the mother's skin during a cesarean procedure. Among the crucial bacteria involved in this microbial transfer is Lactobacilli, a resident flora naturally inhabiting the mother's vagina. These beneficial bacteria colonize the baby's gut, playing a vital role in enhancing the digestion of lactose sugar found in milk [9]. What makes each person's microbiome truly unique is the fact that no two individuals share the exact same composition. Each person carries more than 100 different types of microbes, and the microbiome of one person can significantly differ from that of another [9] (Fig. **1**) Various factors contribute to the divergence in the structure of the microbiome. These factors include the age of the host, environmental influences, health status, genetic makeup, socioeconomic factors, geographical location, and dietary habits [10].

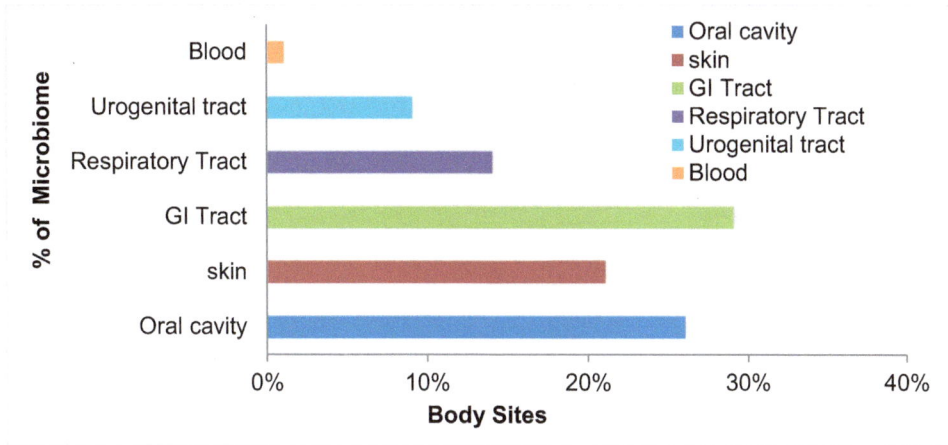

Fig. (1). Percent of microbiomes at various body sites.

The human microbiome can be categorized into two groups: (i) Resident Microbiome and (ii) Transient Microbiome. A resident microbiome is a collection of microorganisms permanently inhabiting different body regions in a healthy individual. The residential flora consists of a constant group of microbes that can be found in specific body sites at different stages of a person's life (Fig. **2**). The microbes in the human microbiome are predominantly stable and establish a mutually beneficial relationship with their host. The human body offers a suitable environment and nourishment that supports the growth and persistence of these microbiomes, while microbes inhabit a place of a potential pathogen that may cause disease. They help in digestion and metabolism and control pathogens from causing diseases by competing for colonization sites and nutrients. The microbiome also can synthesize and produce essential vitamins, such as vitamin K, which the host utilizes. In addition, the human microbiome was responsible for stimulating the production of antibodies [11]. In some circumstances, the body's normal flora can become an opportunistic pathogen (which causes disease in severely ill patients) or can be invaded by other pathogens to cause disease [12].

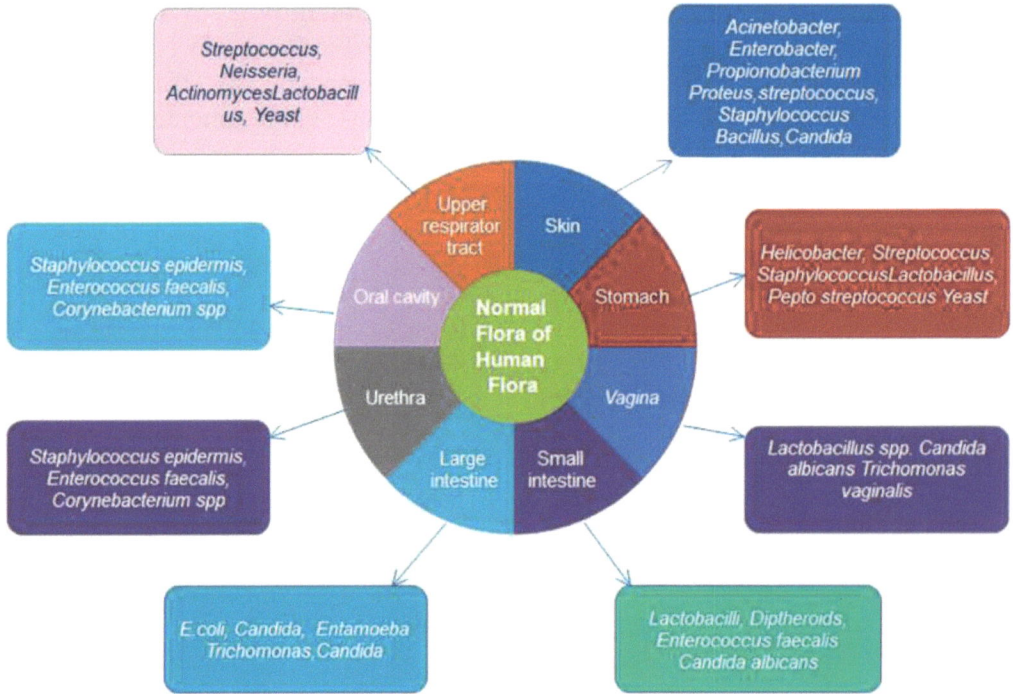

Fig. (2). Human microbiome at different sites.

A transient microbiome means microorganisms that are commonly not seen in the host. For example, the transient microflora of the skin can be specified as microbes that are temporarily transferred from other parts of the body rather than being common inhabitants. The distribution of transient bacteria predominantly appears through direct skin contact or indirect contact by touching and sharing various objects. They commonly struggle to compete when it comes to colonizing specific areas of the body due to the existence of a resident microbiome. However, given the possibility, they will compromise our bodies and cause infections [13].

The microbial composition of an animal microbiome is closely linked to its lifestyle and environment. However, exploring these microbial communities may furnish helpful insights, such as how the changes in behavior, food, and environment and disruptions in habitat influence animal health [14]. Like in humans, microbiomes living within and on animals play essential functions in their hosts' health. Microbiomes control functions associated with disease resistance, digestion and reproduction. Unfortunately, they can negatively influence animals' health by rendering illnesses and infectious diseases. Recent investigations and research propose that the microbiome community colonizing various areas of the body, from animals' guts to their skin, may help inform and support preservation approaches [15].

Through close contact, microbial communities can exchange between human beings and animals. A research study provided evidence of sharing microbiomes, specifically skin microbial flora, between humans and dogs living in the same home [16]. Another study revealed that individuals residing in homes with pets exhibited notable similarities in their nose and skin microbiomes, resembling those of people who did not have pets. This finding suggests that pets play a significant role in facilitating the exchange of microorganisms [17]. Analogous practices were noticed in the distribution of microbiomes between humans and surrounding anima stock. Animal husbandry has also been identified to affect the microbiome of farmers significantly [18].

FUNCTIONS OF THE MICROBIOME IN TERMS OF HUMAN AND ANIMAL HEALTH

Role of Human Microbiome in Health Security

Numerous research investigations have disclosed the crucial functions of human gut microorganisms in increasing the capacity to extract energy from food and improving the absorbance of nutrients from food. It also confirmed that the human microbiome could alter the hunger sign and also help in synthesizing vitamins. Recent studies also proved that microbiomes could metabolize multiple materials, including xenobiotics, as the human microbiome produces different types of specified enzymes and has adapted to various biochemical paths [19]. In addition, the gut microbiome is concerned with multiple essential biological functions, including regulating epithelial growth, modulation of the host metabolism, and stimulating natural immunity. The microbes shield the body from pathogens through competitive colonization or by secretion of antimicrobial substances such as bacteriocins that kill disease-causing microorganisms [20, 21]. All these functions of the microbiome help humans to maintain good health.

Role of Animal Microbiome in Health Security

The microbial composition of invertebrates and vertebrates plays a significant role in animal health. The associations between the host and its animal microbiome can take numerous states, including mutualistic or pathogenic relationships. Mutualistic microbes can be helpful to the hosts in multiple ways, such as nutrient supply and activating the host's defense mechanism [22]. Food is an excellent changing factor in gut microbial composition in animals. Providing animals with unique dietary compounds demonstrates assurance for modulating the microbiome and their influence on the host by preventing pathogen growth and decreasing the risk of diseases. The gut microorganisms of animals play a vital function in the digestion of food material. The microbiome has been correlated with animal physiology, metabolism, and immunity. Early life microbe habitat in the newborn progeny is vital for gut development and maturation of immunity and also plays a substantial role in animal strength to protect from pathogens in further life [23].

Microbiome to Treat Animal Diseases

It is a tremendous challenge for wild animals to protect from deadly pathogenic organisms, habitat loss, environmental change, and pollution. The animal microbiome also protects the animals from pathogenic fungi. For example, *Batrachochytrium dendrobatidis* (Bd), a pathogenic fungus, kills amphibians mercilessly. This fungus inhibits the capacity to breathe *via* its skin. This fungus spreads like wildfire—responsible for the global population downfall of over 500 amphibian species [24]. Yet, many bacterial members of the amphibian microbiome of the skin, such as the *Janthino bacterium lividum* and *Pseudomonas fluorescens,* protect the amphibian against Bd fungus. Scientists are doing meticulous research on this microbiome, proving that these microbiomes can be used to control Bd infections of amphibians [25].

MICROBIOME-HUMAN AND ANIMAL DISEASES

The human microbiome comprises diverse communities of non-pathogenic microorganisms, such as bacteria, archaea, and microbial eukaryotes. These microorganisms can impact human health and homeostasis through multiple mechanisms. They contribute to nutrient and drug metabolism, synthesize vitamins, defend against pathogens, aid in processing host bile acids, modulate the immune system, influence susceptibility to infection, and even modify behavior. The intricate interactions between the human body and its microbiome underscore the importance of these microorganisms in maintaining overall health and functioning [26].

Humans have established symbiotic relationships with various microbial species in the environment during evolution. However, it is only recently that scientists have fully grasped the extent of the diversity of microorganisms associated with humans, their capacity to express functional genes, and the wide range of biochemicals they generate within and on the human body. The growing understanding of these microbial communities highlights their profound influence on human health and well-being [27]. Advanced techniques now enable researchers to directly analyze the genetic material in a given sample, offering insights into the identity and functional capabilities of microorganisms. The advent of high-throughput sequencing methods has revolutionized the sequencing and analysis of large volumes of genetic data, facilitating the discovery and characterization of previously unknown microbial species. These techniques have greatly enhanced our ability to explore and understand the diverse microbial world.

In addition to the Human Microbiome Project (HMP), which focused on characterizing the human body's microbiome, similar sequencing efforts have been conducted for other non-human vertebrate and invertebrate species microbiomes. These studies have revealed the importance of microbial communities in maintaining the health and functioning of various organisms across the animal kingdom. Advancements in molecular biology techniques have played a crucial role in advancing our understanding of the complex and diverse microbial communities that make up the microbiome. Traditional cultivation methods were limited in capturing the full diversity of microorganisms, as many species are difficult or impossible to culture in the laboratory. However, modern molecular techniques, such as DNA sequencing and metagenomic analysis, have provided new avenues for studying non-cultivable microbial communities.

High-throughput experiments have revealed that the healthy human body consists of diverse niche-specific microbiomes distributed across various body habitats. In healthy individuals, each habitat, such as the skin or gastrointestinal tract, harbors a distinct microbiota composition closely associated with specific environmental conditions, such as substrate availability, water content, and pH levels. For example, the skin microbiome exhibits variations in composition based on factors such as the availability of nutrients and water and the acidity or alkalinity of different skin regions. Similarly, the gastrointestinal tract comprises multiple compartments, each with its unique microbial community.

Recent studies focusing on the development of the infant microbiome have shown that niche-specific microbial assemblies can be observed as early as six weeks of age [28]. This suggests that early-life factors significantly influence the formation of niche-specific microbial communities during infancy. One mechanism that influences the early-life assembly of microbiomes is pioneer microbial colonization. During this process, certain microorganisms first colonize the body's various niches, potentially affecting the subsequent formation of niche-specific microbial communities. These pioneer microorganisms may utilize local nutrients and produce chemicals that control compatibility between different species, influencing the competitive colonization of other microorganisms.

The identification of niche-specific microbial communities in various body habitats and the understanding of early-life microbial colonization processes offer valuable insights into the factors that influence the composition and functioning of the human microbiome. Microbiota dysbiosis can lead to one of the most common disorders, which is infection. The human microbiota is significantly impacted by infectious diseases and their treatment, subsequently influencing the interaction between the infectious disease and the human host. Invasive pathogens colonize the intestinal mucosa, triggering a strong inflammatory response and potentially resulting in the translocation of intestinal bacteria. This highlights the intricate interplay between the microbiota, infectious diseases, and host responses (Fig. **3**).

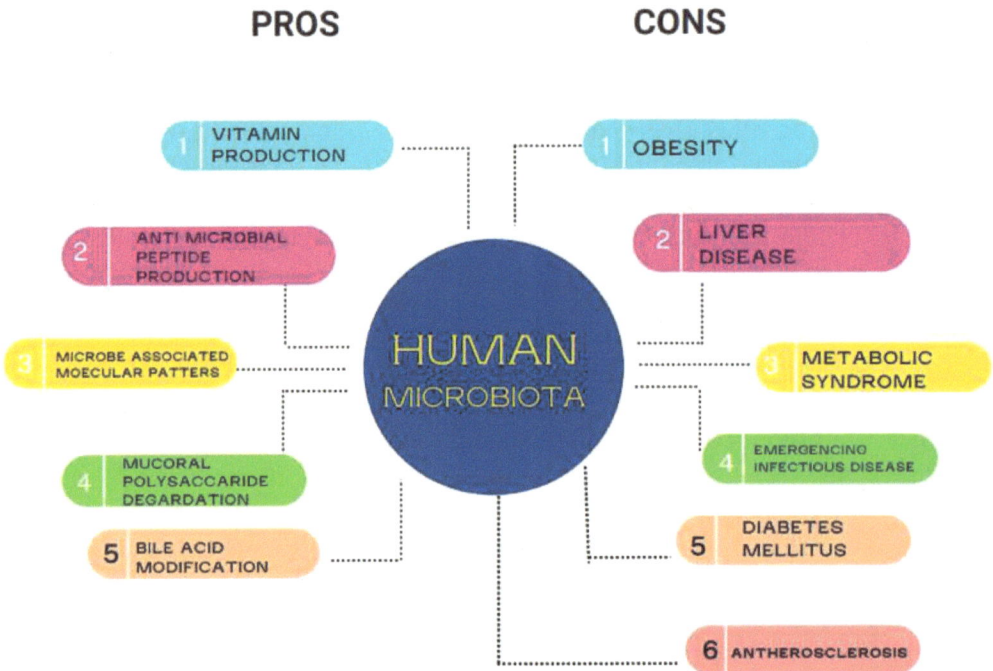

PROS   **CONS**

1 VITAMIN PRODUCTION   1 OBESITY

2 ANTI MICROBIAL PEPTIDE PRODUCTION   2 LIVER DISEASE

3 MICROBE ASSOCIATED MOECULAR PATTERS   HUMAN MICROBIOTA   3 METABOLIC SYNDROME

4 MUCORAL POLYSACCARIDE DEGARDATION   4 EMERGENCING INFECTIOUS DISEASE

5 BILE ACID MODIFICATION   5 DIABETES MELLITUS

6 ANTHEROSCLEROSIS

Fig. (3). Pros and Cons of Human Microbiota.

Numerous scientific investigations have provided substantial evidence supporting the strong correlation between infection and dysbiosis of the microbiota, underscoring the interplay between infections, viruses, and the microbiome. Notably, individuals suffering from *Clostridium difficile* infection (CDI) exhibit significant variations in their gut microbiotas compared to those unaffected. Disturbances in the normal microbial equilibrium within the gut can create a favorable environment for the growth and proliferation of *C. difficile*, leading to the occurrence of CDI [29] (Fig. **4**).

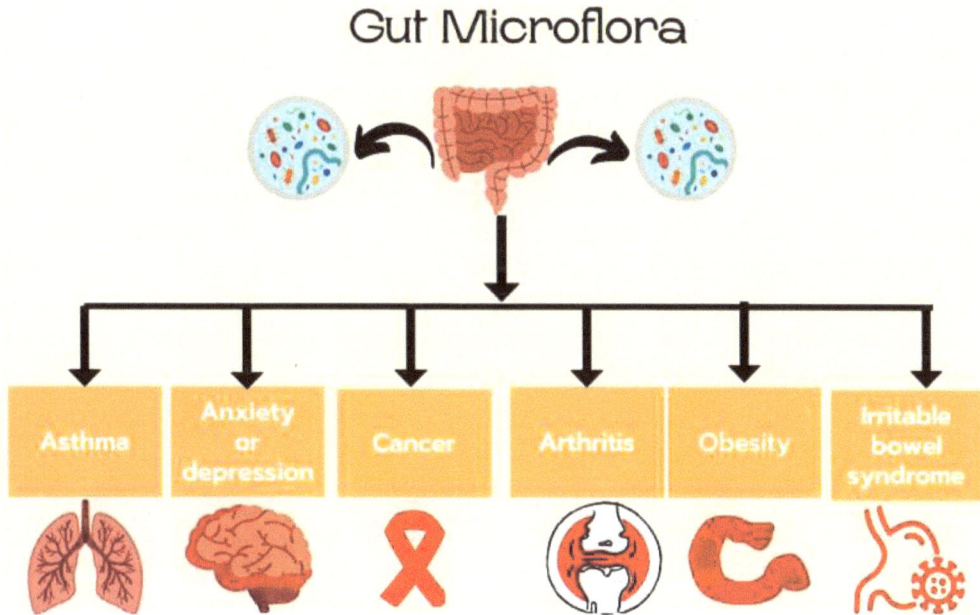

Fig. (4). Microbiome in human diseases.

Impact of Vaginal Microbiome on Childbirth

Preterm birth is a medical condition characterized by the delivery of a baby before completing 37 weeks of pregnancy. It is the second leading cause of neonatal mortality, referring to the death of newborns within the first 28 days of life. HMP's first model approach proved it-"Vaginal Microbiome Consortium Multi-Omic Microbiome Studies" [30, 31]. According to their report, a pregnant female's beneficial vaginal microbiota influences the successful birth of a healthy child and proves that change in the vaginal microbiome is responsible for preterm birth.

Obesity

According to a mouse study, depending on the nutritional conditions, laboratory mice can pick up obesity or leanness from the gut microbes of obese or lean humans. Although there is strong evidence that genetics and other factors play a vital role in the development of obesity, the community of bacteria that dwell in the gut and their collective bacterial genomes, or gut microbiome, may impact and reflect a person's health and nutritional state [32].

Cardiovascular Diseases

Cardiovascular disease, including conditions like coronary heart disease, cerebrovascular disease, and peripheral artery disease, is a major global health concern with high morbidity and mortality rates. While traditional risk factors like atherosclerosis, hypertension, obesity, diabetes, dyslipidemia, and mental illness are well-established, emerging research indicates a potential contribution of the microbiota to cardiovascular health. Studies exploring microbiota transplantation, microbiota-dependent pathways, and metabolites have revealed the microbiota's influence on host metabolism and its potential involvement in cardiovascular disease. This highlights the importance of investigating the microbiota's role in cardiovascular health and disease progression [33].

Inflammatory Bowel Disease

Crohn's disease and ulcerative colitis are examples of inflammatory bowel diseases caused by the imbalance between the immune system and gut microbes. Although there are environmental and genetic risk factors for inflammatory bowel disease, they are insufficient to account for the substantial rise in Inflammatory bowel disease over the past 50 years. Inflammatory bowel disease and the gut microbiota are related, yet neither condition seems to be brought on by a specific pathogen. Instead, inflammatory bowel illness has been linked to a general disruption of the gut [34].

Cancer and Cirrhosis

Cancer is a condition that can affect practically every part of the body and is defined by the uncontrollable and rapid multiplication of abnormal cells. Over 10 million people will die from cancer worldwide, making it the most significant cause of death at the moment. Despite the fact that carcinogenesis is a complex process, it is well-known that the leading causes of cancer are tobacco, microorganisms, alcohol, and radiation. According to recent studies, the microbiota substantially influences how cancer develops, mainly by altering immune system function, altering host metabolism, and regulating host cell proliferation and death. According to studies, individuals with cancer have a different gut microbiota than healthy individuals, including greater quantities of pathogens [35].

The final stage of all chronic liver illnesses, cirrhosis, is characterized by tissue fibrosis and the development of aberrant nodules instead of the normal liver's natural architecture. Recent studies suggest a potential role of the oral and intestinal microbiota in chronic liver disorders. Imbalances in these microbiotas have been associated with conditions such as NAFLD, ALD, and cirrhosis, impacting liver inflammation and metabolism. Further research is needed to understand the mechanisms and develop targeted interventions.

Anxiety and Depression

Anxiety and depression disorders currently present a significant medical challenge, prompting the exploration of new personalized strategies for antidepressant medication. Recently, the gut-brain axis has emerged as a topic of great importance in understanding various diseases. Both pre-clinical and clinical research have shed light on the influence of intestinal bacteria on this axis. Multiple systems allow gut bacteria to interact with the brain and influence behavior, including amino acid metabolism, short-chain fatty acids, the vagus nerve, endocrine signaling, and immunological responses. Experimental studies have further confirmed the causal role of intestinal microorganisms in mood disorders and anxiety. For instance, the gut microbiota transfer from individuals with significant depression to animal models has demonstrated their impact on mood and anxiety-related behaviors. These findings emphasize the significance of the gut microbiota in these disorders and provide insights into potential therapeutic avenues for their management [36].

Ruminants Animals

Ruminants have made significant contributions to human development. It is crucial to comprehend the rumen microbiome because it is closely tied to animal nutrition and health, the products and by-products they produce, and the emission of greenhouse gases. Investigated the effectiveness of employing non-edible plant by-products, like maize stover, and evaluated how this feed impacts milk output and quality using applied metabolomics, metagenomics, and transcriptomics. The impact of bacteria on the stomach, hepatic, mammary glands, and biofluids—all vital components of milk production—is evaluated using more modern methods. To see if there were any potential correlative relationships between the rumen microbiome and increased weight, researchers also examined the bacterial and fungal makeup of the gut microbiome in beef cattle. A study that collected samples of the perinatal gut microbiome from calves at delivery and one week after delivery showed how the microbiome's makeup changed over time. At birth, Firmicutes, Proteobacteria, Actinobacteria, and Bacteroidetes comprise most of the microbiome. The calf gut microbes had changed to resemble the fecal microbiome more closely one week after delivery [37].

Poultry

Humans primarily get their protein from poultry. Although little research has been done on it, we understand that the poultry gut enables chemical and mechanical food digestion. One of the early investigations on an animal's microbiome focused on a chicken's ceca. Except for the ceca, where Bacteroidetes was the predominant phylum, all locations had more than 60% of their sequences made up of firmicutes. Due to poultry's short meal retention time, there is less microbial diversity than in other animal species. The species found in the poultry intestine can change depending on factors such as age, diet, and antibiotic use [38].

Aquaculture

Today, fish, crustaceans, and mollusks are raised in aquaculture and fisheries farms worldwide. Studies on market quality, growth, and microbiome research to catalog the symbiotic organisms and pests that cause sickness in fish stocks are frequent. The study examined the diversity of the fish microbiota, the diversity of microbes in various tissues, and alters in the microbes during development. The influence of the host's genotype on the microbiome, the environment, feeding methods, the consequences of the microflora on fish health, the direct and indirect impacts of the microflora on disease-causing organisms, and nutrient absorption were also studied [39]. Oysters are popular food sources and pearls; hence, much microbiota research has been done on these species. A recent review investigated probiotics in edible oysters to increase shelf life, decrease bacterial pollutants that would harm humans, increase yields, and decrease disease in oysters to increase marketability. A recent study used the 16S rRNA to assess the bacterial communities' assemblages of the black-lipped pearl oyster. The study aimed to evaluate the diversity and function of the microbiomes present in various tissues. Furthermore, various functional characteristics were found among the tissues. One intriguing finding was the existence of the symbiotic Endozoicomonas, which is absent in all other oyster species despite being prevalent in coral reef animals [40].

MICROBIOTA AND DISEASE TREATMENT

With the expanding knowledge of microbiota, there has been a growing interest in the potential of modifying the microbiota to treat various diseases. The human gut microbiota, in particular, has been identified as a key player in numerous physiological processes. This has prompted considerable attention towards manipulating the human gut microbiota as a potential therapeutic approach for preventing or treating associated disorders. As a result, more clinical trials are being conducted to explore the feasibility and efficacy of microbiota manipulation in the context of disease treatment and prevention. The aim is to harness the potential of microbiota-based interventions to improve human health outcomes (Fig. **5**). The use of probiotics as an adjuvant treatment in people with cancer also showed excellent results. Although the success rate is positive, it should be remembered that the trials primarily consist of preliminary studies with limited sample sizes. In order to improve the trial design and customize the treatment, further research into the underlying cause for a complete response was also necessary [41].

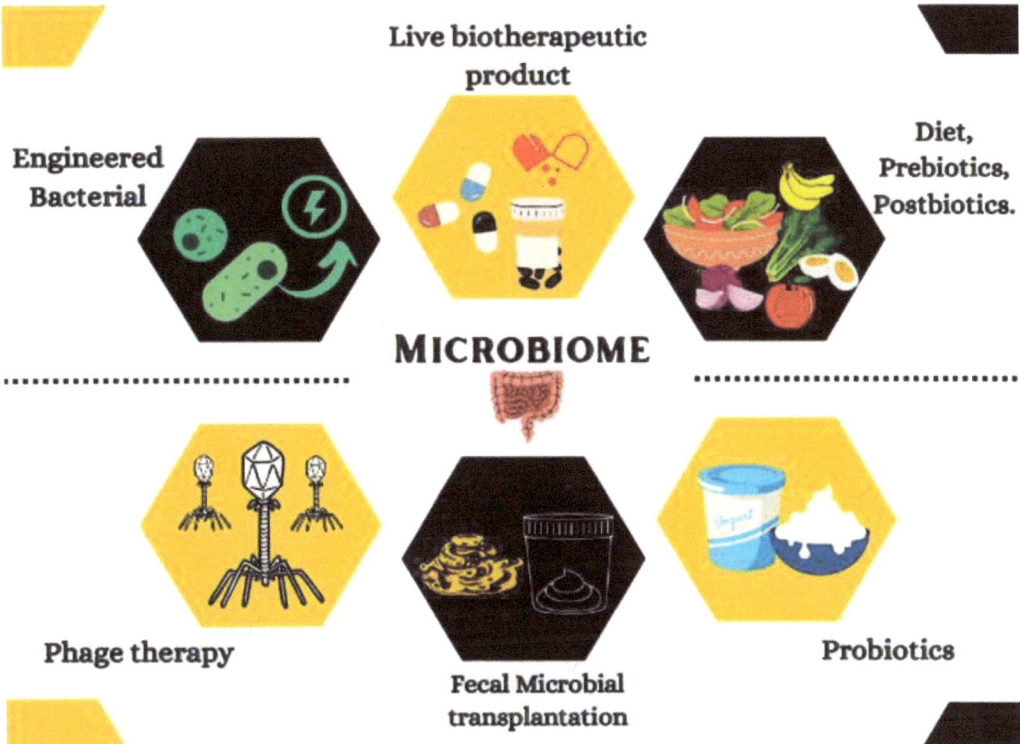

Fig. (5). Microbiota in the treatment of diseases.

The utilization of gene-editing technology such as CRISPR has expanded the application of engineered bacteria in microbiota modulation. These advancements have paved the way for the development of innovative antibacterial strategies. One study conducted by Hwang *et al*. demonstrated the effectiveness of using the exonuclease CRISPR-associated protein 3 in creating a probiotic that can selectively eliminate pathogenic bacteria. This technique enables modified bacteria to deliver transcription factors, peptide vaccines, or proteins into host cells, offering a novel approach to restoring the balance of gut flora. However, it is crucial to recognize that the impact of a specific supplement or medication on the host-microbiota interface is just one piece of the complex puzzle in the human environment. To fully understand the implications of interventions, it is crucial to consider their interactions with the host's nutrition, DNA, immune system, and resident symbionts. While further research is necessary before technologies like

CRISPR can be effectively employed in therapeutic settings, they hold tremendous potential and open up new avenues for microbiome-based therapies [42].

Probiotics as Therapeutics

Using microorganisms in therapies is one of the clinical revolutions in the 21st century. Our gut is a metabolic hub for trillions of microorganisms. These cells occupy our gastrointestinal tract and boost physiological function and immunity through a close symbiotic association with the host. Many research findings establish that modifying the host microbiome is a potential treatment tool or prophylaxis for numerous health ailments [43]. In probiotic therapies, generally used bacterial strains are *Lactobacillus*, *Bifidobacterium*, and other probiotic bacteria. Numerous Research studies have analyzed the possible efficacy of probiotics in preventing and treating various diseases. (Table **1**, Fig. **6**) [44 - 46].

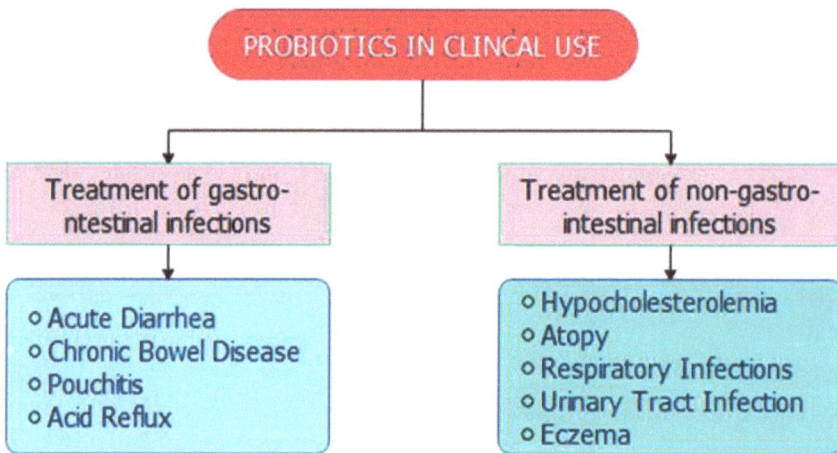

Fig. (6). Probiotics in Clinical Use.

Table 1. List of probiotics used to treat various diseases.

Diseases	Bacterial microbial strains used to treat diseases
Diarrhea	*Lactobacillus rhamnosus, Saccharomyces boulardii, Bifidobacterium lactis, and Lactobacillus casei*
IBS(Irritable bowel syndrome)	*Escherichia coli and Lactobacillus*
Pouchitis (Inflammation that occurs in the lining of a pouch created during surgery)	*L. rhamnosus*
Acid reflux	*L.reuteri, Enterococcus faecium and Bacillus subtilis*
Hypocholesteraeamia	*L. plantarum, L. Acidophilus*
Atopic dermatitis	*L.ehamnosus, L.reuteri*

(Table 1) cont.....

Diseases	Bacterial microbial strains used to treat diseases
Respiratory Infections	*L.rhamnosus, Bifidobacterium breve*
UTI	*L.rhamnosus and L.reuteri*
Eczema	*L. rhamnosus, Bifidobacterium lactis, Bifidobacterium breve and L.salivarius*

Fecal Microbiota Transplantation (FMT)

Fecal microbiota transplantation (FMT) is a successful technique for treating recurrent Clostridium difficile infections. It involves the transfer of beneficial fecal bacteria from a healthy donor to a patient, aiming to restore a diverse and healthy gut microbiome. FMT is employed when conventional treatments like antibiotics fail and aim to alleviate specific gut disorders' severity. This clinical practice replenishes beneficial bacteria in the colon by administering stool from a healthy donor through colonoscopy or enema. FMT has also shown promise in treating conditions such as constipation, colitis, and irritable bowel syndrome [47].

BIOLOGICAL RELEVANCE OF HUMAN AND ANIMAL MICROBIOME AND ITS INFLUENCING FACTORS

While much of the research on microbiomes has predominantly focused on the human microbiome, a relatively limited amount of research has been dedicated to studying the microbiomes of animals. However, in recent times, there has been a growing interest in comprehending the animal microbiome. This increased focus on animal microbiomes can enhance our understanding of the variations between species, populations, and ecosystems. By studying the animal microbiome, we can gain valuable insights into the complex interactions between hosts and their microbial communities and how they contribute to animals' overall health and functioning in various environments. The animal microbiome can be categorized into symbiotic and pathogenic microbiomes based on their effects on their host. Several factors, such as diet, stage of life, the genotype of the host, and physiological and ecological functions, determine the composition of the bacterial communities on the host. For example, some studies have shown that microbiomes are essential in supplementing nutrients, developing the immune system, and tolerance to environmental stresses [48] (Fig. **7**).

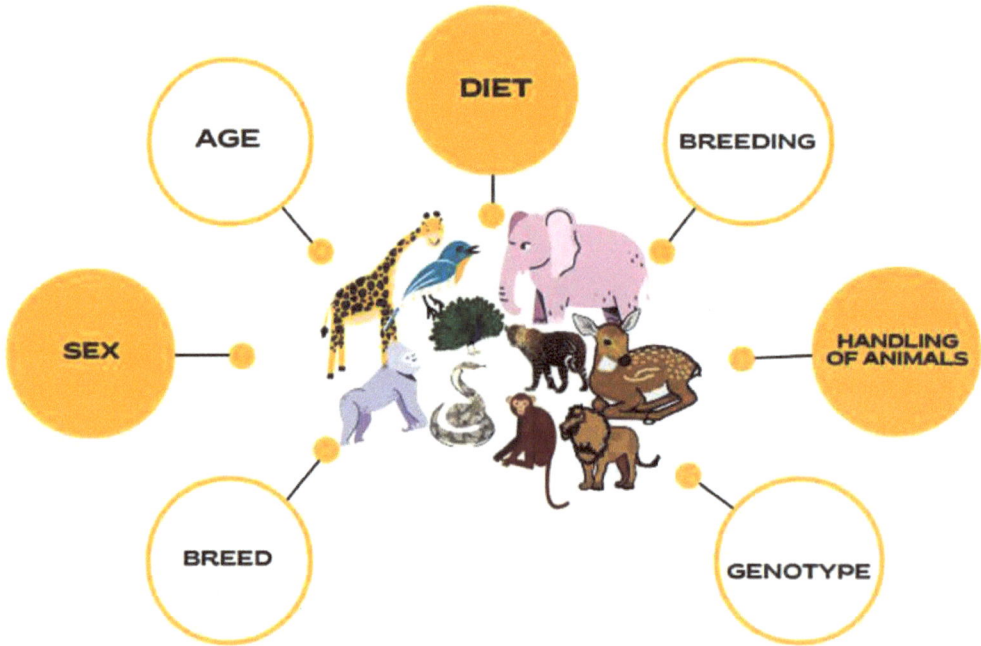

Fig. (7). Factors affecting the human and animal microbiome.

Importance of Conservation and its Effects

Microorganisms play a crucial role in various aspects of human, animal, and plant life, with their impact extending to sustainable farming, greenhouse gas regulation, and the functioning of biogeochemical and ecological processes. It has been observed that any alterations in the composition of microbiomes can significantly impact the fitness of their hosts. Research has demonstrated that species kept in captivity, in laboratory settings, or subjected to habitat fragmentation tend to have less diverse microbiomes compared to their wild counterparts. Therefore, a species needs to study the impact of habitat fragmentation and less diverse microbiomes. Previously, microbiological studies in the natural environment were limited, but with the help of advanced molecular techniques such as metagenomics and next-generation sequencing, our understanding of microbial communities has dramatically increased [49].

Compared to species found in the wild, species across taxa raised in laboratories or that suffer from habitat fragmentation exhibit fewer microbiomes. Therefore, species are in danger due to direct habitat degradation, decreased resource availability, and indirect microbiota decline. Future research must therefore, focus on the microbiome, how habitat fragmentation affects the microbiomes of various species, and how species with less diversified microbiomes fare in these circumstances. In a similar vein, breeding populations and animal captivity are expected to have an impact on animal microbiomes. This is frequently done to reintroduce species into the wild by protecting or increasing the abundance of uncommon species [50].

It has been proposed that manipulating microbiomes can benefit the health of both plants and animals. However, it is unknown how this can be applied to conservation. Basic genetic concepts are frequently incorporated into breeding plans for threatened species in zoos or other captive environments, but conservation biology has overlooked the evolutionary potential of microbiomes.

CONCLUSION

The limited ability to grow the majority of bacteria found in or on other animals has stimulated the progress of molecular biology techniques, opening up new avenues for studying and constructing complex microbial communities. One commonly utilized method for investigating diverse microbiomes is amplicon sequencing, which focuses on the 16S ribosomal RNA (16S rRNA) genes as a bacterial marker. While extensive research has highlighted the significant influence of the microbiome on human health, there has been a growing emphasis on studying the microbiomes of non-human species in recent years. However, compared to research on the human microbiome, exploring the microbiome's impact on animal health remains relatively limited. It is evident that studying host-microbiota interactions should go beyond characterizing community composition and strive for a comprehensive understanding of the relationships between community members. This is important considering the wide-ranging influence of microbes throughout the human body.

Advancements in the field will significantly enhance our comprehension of the interactions between human microbiota and human development. They will also provide insights into the potential contributions of specific microbiota in various diseases, such as liver diseases, bacterial infections, cancer, psychiatric diseases, and metabolic diseases. These future developments will be facilitated by innovative techniques that predict microbiota function, new models for studying microbiota interactions, and novel analytical and simulation approaches.

Furthermore, the field of personalized medicine will benefit from microbiome-based diagnostic and therapeutic strategies, allowing for more targeted and effective treatments. Additionally, a deeper exploration of the pivotal roles played by the human microbiota will be undertaken. The evolution of Next Generation Sequencing-Metagenomics has proven valuable in addressing several clinically relevant infectious diseases, offering significant advantages in early diagnosis. As these advancements continue to unfold, they hold great potential for unraveling the intricate relationship between the microbiota and human health, providing insights into disease mechanisms, and paving the way for innovative approaches to diagnosis and treatment.

CONSENT FOR PUBLICATON

The authors hereby give consent to Bentham Science Publishers Pt. Ltd. to publish this book chapter, and there are no other consents to declare.

ACKNOWLEDGEMENT

The authors thank the management of Raja Bahadur Venkata Rama Reddy Women's College, Narayanguda, Hyderabad-Telangana, and Bhavan's Vivekananda College of Science, Humanities & Commerce, Telangana, for their continual encouragement and unflinching support.

REFERENCES

[1]   Young, V. B. The role of the microbiome in human health and disease: an introduction for clinicians. BMJ, 2017; 15:356: j831.
[http://dx.doi.org/10.1136/bmj.j831]

[2]   Hair M, Sharpe J. Fast facts about the human microbiome. Available from: http://depts.washington.edu/ceeh/downloads/FF_Microbiome.pdf

[3]   Zhang YJ, Li S, Gan RY, Zhou T, Xu DP, Li HB. Impacts of gut bacteria on human health and diseases. Int J Mol Sci 2015; 16(4): 7493-519.
[http://dx.doi.org/10.3390/ijms16047493] [PMID: 25849657]

[4]   Berg G, Rybakova D, Fischer D, *et al.* Microbiome definition re-visited: old concepts and new challenges. Microbiome 2020; 8(1): 1-22.
[PMID: 31901242]

[5]   Bulsiewicz, W. Fiberfueled: The plant-based gut health program for losing weight, restoring your health, and optimizing your microbiome. 2020; pp.201-230.

[6]   Tannock GW. Normal microflora: an introduction to microbes inhabiting the human body. Springer Science & Business Media 1995; pp. 61-73.

[7]   Zhu B, Wang X, Li L. Human gut microbiome: the second genome of human body. Protein Cell 2010; 1(8): 718-25.
[http://dx.doi.org/10.1007/s13238-010-0093-z] [PMID: 21203913]

[8]   Haque SZ, Haque M. The ecological community of commensal, symbiotic, and pathogenic gastrointestinal microorganisms – an appraisal. Clin Exp Gastroenterol 2017; 10: 91-103.
[http://dx.doi.org/10.2147/CEG.S126243] [PMID: 28503071]

[9]   Nuriel-Ohayon M, Neuman H, Koren O. Microbial changes during pregnancy, birth, and infancy. Front Microbiol 2016; 7: 1031.
[http://dx.doi.org/10.3389/fmicb.2016.01031] [PMID: 27471494]

[10]   Moran-Ramos S, Lopez-Contreras BE, Villarruel-Vazquez R, *et al.* Environmental and intrinsic factors shaping gut microbiota composition and diversity and its relation to metabolic health in children and early adolescents: A population-based study. Gut Microbes 2020; 11(4): 900-17.
[http://dx.doi.org/10.1080/19490976.2020.1712985] [PMID: 31973685]

[11]   Kong HH, Segre JA. Skin microbiome: looking back to move forward. J Invest Dermatol 2012;

132(3): 933-9.
[http://dx.doi.org/10.1038/jid.2011.417] [PMID: 22189793]

[12]   Casadevall A, Pirofski L. Host-pathogen interactions: redefining the basic concepts of virulence and pathogenicity. Infect Immun 1999; 67(8): 3703-13.
[http://dx.doi.org/10.1128/IAI.67.8.3703-3713.1999] [PMID: 10417127]

[13]   Weese JS. The canine and feline skin microbiome in health and disease. Veterinary Dermatology. 2013 Feb; 24(1): 137 e31.
[http://dx.doi.org/10.1002/9781118644317.ch19]

[14]   Proctor LM. The national institutes of health human microbiome project. Semin Fetal Neonatal Med 2016; 21(6): 368-72.
[http://dx.doi.org/10.1016/j.siny.2016.05.002] [PMID: 27350143]

[15]   Peterson J, Garges S, Giovanni M, *et al.* The NIH human microbiome project. Genome Res 2009; 19(12): 2317-23.
[http://dx.doi.org/10.1101/gr.096651.109] [PMID: 19819907]

[16]   Rivera-Amill V. The human microbiome and the immune system: An ever-evolving understanding. J Clin Cell Immunol 2014; 5(6): e114.
[PMID: 27088046]

[17]   Clayton JB, Al-Ghalith GA, Long HT, *et al.* Associations Between Nutrition, Gut Microbiome, and Health in A Novel Nonhuman Primate Model. Sci Rep 2018; 8(1): 11159.
[http://dx.doi.org/10.1038/s41598-018-29277-x] [PMID: 29311619]

[18]   Trevelline, B. K., Fontaine, S. S., Hartup, B. K., & Kohl, K. D. Conservation biology needs a microbial renaissance: a call for considering host-associated microbiota in wildlife management practices—proceedings of the Royal Society B, 2019; 286(1895), 20182448.

[19]   Song SJ, Lauber C, Costello EK, *et al.* Cohabiting family members share microbiota with one another and with their dogs. eLife 2013; 2: e00458.
[http://dx.doi.org/10.7554/eLife.00458] [PMID: 23599893]

[20]   Misic AM, Davis MF, Tyldsley AS, *et al.* The shared microbiota of humans and companion animals as evaluated from Staphylococcus carriage sites. Microbiome 2015; 3(1): 2.
[http://dx.doi.org/10.1186/s40168-014-0052-7] [PMID: 25705378]

[21]   Kraemer JG, Ramette A, Aebi S, Oppliger A, Hilty M. Influence of pig farming on the human nasal microbiota: The crucial role of airborne microbial communities. Appl Environ Microbiol 2018; 84(6): e02470-17.
[http://dx.doi.org/10.1128/AEM.02470-17] [PMID: 29330190]

[22]   Raoult D. Obesity pandemics and the modification of bacterial digestive flora. Eur J Clin Microbiol Infect Dis 2008; 27(8): 631-4.
[http://dx.doi.org/10.1007/s10096-008-0490-x] [PMID: 18322715]

[23]   Garcia-Gutierrez E, Mayer MJ, Cotter PD, Narbad A. Gut microbiota as a source of novel antimicrobials. Gut Microbes 2019; 10(1): 1-21.
[http://dx.doi.org/10.1080/19490976.2018.1455790] [PMID: 29584555]

[24]   Flandroy L, Poutahidis T, Berg G, *et al.* The impact of human activities and lifestyles on the interlinked microbiota and health of humans and of ecosystems. Sci Total Environ 2018; 627: 1018-38.
[http://dx.doi.org/10.1016/j.scitotenv.2018.01.288] [PMID: 29426121]

[25]   Thimm T, Hoffmann A, Borkott H, Charles Munch J, Tebbe CC. The gut of the soil microarthropod Folsomia candida (Collembola) is a frequently changeable but selective habitat and a vector for microorganisms. Appl Environ Microbiol 1998; 64(7): 2660-9.
[http://dx.doi.org/10.1128/AEM.64.7.2660-2669.1998] [PMID: 9647845]

[26]   McKenney EA, Koelle K, Dunn RR, Yoder AD. The ecosystem services of animal microbiomes. Mol

Ecol 2018; 27(8): 2164-72.
[http://dx.doi.org/10.1111/mec.14532] [PMID: 29427300]

[27]   Green M, Arora K, Prakash S. Microbial medicine: prebiotic and probiotic functional foods to target obesity and metabolic syndrome. Int J Mol Sci 2020; 21(8): 2890.
[http://dx.doi.org/10.3390/ijms21082890] [PMID: 32326175]

[28]   Zhu L, Wang J, Bahrndorff S. The wildlife gut microbiome and its implication for conservation biology. Front Microbiol 2021; 12: 697499.
[http://dx.doi.org/10.3389/fmicb.2021.697499] [PMID: 34234768]

[29]   Lloyd-Price J, Abu-Ali G, Huttenhower C. The healthy human microbiome. Genome Med 2016; 8(1): 51.
[http://dx.doi.org/10.1186/s13073-016-0307-y] [PMID: 27122046]

[30]   Hsiao EY, McBride SW, Hsien S, *et al.* Microbiota modulate behavioral and physiological abnormalities associated with neurodevelopmental disorders. Cell 2013; 155(7): 1451-63.
[http://dx.doi.org/10.1016/j.cell.2013.11.024] [PMID: 24315484]

[31]   Ganguly P. Microbes in us and their role in human health and disease. NIH Natl Hum Genome Res Inst. 2019.

[32]   Jean S, Huang B, Parikh HI, *et al.* Multi-omic microbiome profiles in the female reproductive tract in early pregnancy. Infectious Microbes and Diseases 2019; 1(2): 49-60.
[http://dx.doi.org/10.1097/IM9.0000000000000007]

[33]   Hyman RW, Fukushima M, Jiang H, *et al.* Diversity of the vaginal microbiome correlates with preterm birth. Reprod Sci 2014; 21(1): 32-40.
[http://dx.doi.org/10.1177/1933719113488838] [PMID: 23715799]

[34]   Emanuelsson F, Claesson BEB, Ljungström L, Tvede M, Ung KA. Faecal microbiota transplantation and bacteriotherapy for recurrent Clostridium difficile infection: A retrospective evaluation of 31 patients. Scand J Infect Dis 2014; 46(2): 89-97.
[http://dx.doi.org/10.3109/00365548.2013.858181] [PMID: 24354958]

[35]   Ghosh PN, Brookes LM, Edwards HM, *et al.* Cross-disciplinary genomics approaches to studying emerging fungal infections. Life (Basel) 2020; 10(12): 315.
[http://dx.doi.org/10.3390/life10120315] [PMID: 33260763]

[36]   Pickard JM, Zeng MY, Caruso R, Núñez G. Gut microbiota: Role in pathogen colonization, immune responses, and inflammatory disease. Immunol Rev 2017; 279(1): 70-89.
[http://dx.doi.org/10.1111/imr.12567] [PMID: 28856738]

[37]   Dethlefsen L, McFall-Ngai M, Relman DA. An ecological and evolutionary perspective on human–microbe mutualism and disease. Nature 2007; 449(7164): 811-8.
[http://dx.doi.org/10.1038/nature06245] [PMID: 17943117]

[38]   Yao Y, Cai X, Ye Y, Wang F, Chen F, Zheng C. The role of microbiota in infant health: from early life to adulthood. Front Immunol 2021; 12: 708472.
[http://dx.doi.org/10.3389/fimmu.2021.708472] [PMID: 34691021]

[39]   Wang B, Yao M, Lv L, Ling Z, Li L. The human microbiota in health and disease. Engineering (Beijing) 2017; 3(1): 71-82.
[http://dx.doi.org/10.1016/J.ENG.2017.01.008]

[40]   Li Z, Zhang B, Wang N, Zuo Z, Wei H, Zhao F. A novel peptide protects against diet-induced obesity by suppressing appetite and modulating the gut microbiota. Gut 2023; 72(4): 686-98.
[PMID: 35803703]

[41]   Makover ME, Shapiro MD, Toth PP. There Is Urgent Need to Treat Atherosclerotic Cardiovascular Disease Risk Earlier, More Intensively, and with Greater Precision. A Review of Current Practice and Recommendations for Improved Effectiveness. American Journal of Preventive Cardiology. 2022 Aug 6:100371.

[42]   Danese S, Fiocchi C. Etiopathogenesis of inflammatory bowel diseases. World J Gastroenterol 2006; 12(30): 4807-12.
[http://dx.doi.org/10.3748/wjg.v12.i30.4807] [PMID: 16937461]

[43]   Jain T, Sharma P, Are AC, Vickers SM, Dudeja V. New insights into cancer–microbiome–immune axis: decrypting a decade of discoveries. Front Immunol 2021; 12: 622064.
[http://dx.doi.org/10.3389/fimmu.2021.622064] [PMID: 33708214]

[44]   Clapp M, Aurora N, Herrera L, Bhatia M, Wilen E, Wakefield S. Gut microbiota's effect on mental health: The gut-brain axis. Clin Pract 2017; 7(4): 987.
[http://dx.doi.org/10.4081/cp.2017.987] [PMID: 29071061]

[45]   Matthews C, Crispie F, Lewis E, Reid M, O'Toole PW, Cotter PD. The rumen microbiome: a crucial consideration when optimising milk and meat production and nitrogen utilisation efficiency. Gut Microbes 2019; 10(2): 115-32.
[http://dx.doi.org/10.1080/19490976.2018.1505176] [PMID: 30207838]

[46]   Wickramasuriya SS, Park I, Lee K, *et al.* Role of physiology, immunity, microbiota, and infectious diseases in the gut health of poultry. Vaccines (Basel) 2022; 10(2): 172.
[http://dx.doi.org/10.3390/vaccines10020172] [PMID: 35214631]

[47]   Diwan AD, Harke SN, Gopalkrishna , Panche AN. Aquaculture industry prospective from gut microbiome of fish and shellfish: An overview. J Anim Physiol Anim Nutr (Berl) 2022; 106(2): 441-69.
[http://dx.doi.org/10.1111/jpn.13619] [PMID: 34355428]

[48]   Prado S, Romalde JL, Barja JL. Review of probiotics for use in bivalve hatcheries. Vet Microbiol 2010; 145(3-4): 187-97.
[http://dx.doi.org/10.1016/j.vetmic.2010.08.021] [PMID: 20851536]

[49]   Rowland I, Gibson G, Heinken A, *et al.* Gut microbiota functions: metabolism of nutrients and other food components. Eur J Nutr 2018; 57(1): 1-24.
[http://dx.doi.org/10.1007/s00394-017-1445-8] [PMID: 28393285]

[50]   Mu Y, Zhang C, Li T, *et al.* Development and Applications of CRISPR/Cas9-Based Genome Editing in *Lactobacillus.* Int J Mol Sci 2022; 23(21): 12852.
[http://dx.doi.org/10.3390/ijms232112852] [PMID: 36361647]

SUBJECT INDEX

A

Acid(s) 16, 29, 39, 40, 241
  fulvic 16
  hydrocyanic 241
  sulfuric 29, 39, 40
Activities 17, 18, 19, 29, 44, 51, 70, 84, 131, 147, 150, 171, 188, 190, 194, 249, 286, 292
  anticancer 286
  antiviral 194
  electrochemical 84
  industrial 44
  metabolic 29, 51, 70, 84, 147, 150, 171, 188, 190
  nitrite reductase 249
  oxygenase 131
  soil enzyme 17, 18, 19
  tyrosinase 292
Agents, stress-causing 55
Agricultural soils 5, 12, 156, 157, 171, 173
  organic 5
Agricultural waste 49, 119
Airborne transport 187
Ammonia 51, 75, 80, 149, 151, 152, 173, 190, 211, 249
  oxidation 249
Anaerobic 38, 51, 67, 69, 204, 209, 210, 215, 221, 222, 281, 282
  conditions 38, 51, 67, 204, 209, 210, 215, 221, 222, 281
  digestion process 282
  sludge 69
Anthropogenic pressure 220
Anti-cytotoxic action 194
Antibiotics biosensors 88
Automotive gas oil 133

B

Biodiesel pathways 126

Biofilm(s) 45, 73, 75, 77, 78, 89
  electroactive 45, 73, 75, 77, 89
  electrogenic 77
  microbial diversity 78
Biofuel(s) 105, 107, 111, 112, 119, 123, 124, 127, 128, 129, 130, 134, 135, 276
  industry 123
  processes 124
  producing 107, 112, 276
  production 105, 111, 119, 123, 124, 127, 128, 129, 130, 134, 135
Biointensive integrated pest management (BIPM) 289
Bioleaching techniques 56
Biological 67, 211, 212, 223, 225
  fuel cells 67
  nitrogen fixation (BNF) 211, 212, 223, 225
Biomass 154, 276, 277, 291
  microalgal 154, 291
  waste 276, 277
Bioremediation 48, 57, 195
  methods 195
  process, microbial-mediated 57
  techniques 48
Biosensors 88
  heavy metal 88
  organic toxicants 88
Biosorption, industrial 45
Biosparging process 48

C

Carcinogenesis 309
Cellulase production 18
Charcoal, bamboo 80
Chemicals 86, 106, 108, 154, 156, 274, 275, 276, 279, 280, 282, 285, 287, 293
  industrial 280
  toxic 86
Chemoautotrophic microbes 169
Chemolithoautotrophic microbes 38

**Govindaraj Kamalam Dinesh, Shiv Prasad, Ramesh Poornima, Sangilidurai Karthika, Murugaiyan Sinduja &
Velusamy Sathya (Eds.)**
All rights reserved-© 2024 Bentham Science Publishers

www.ingramcontent.com/pod-product-compliance
Lightning Source LLC
Chambersburg PA
CBHW050807220326
41598CB00006B/138